ALTERNATIVE
AGRICULTURE

Committee on the Role of
Alternative Farming Methods
in Modern Production Agriculture

Board on Agriculture

National Research Council

NATIONAL ACADEMY PRESS
Washington, D.C.   1989

National Academy Press   2101 Constitution Avenue, NW   Washington, DC 20418

NOTICE: The project that is the subject of this report was approved by the Governing Board of the National Research Council, whose members are drawn from the councils of the National Academy of Sciences, the National Academy of Engineering, and the Institute of Medicine. The members of the committee responsible for the report were chosen for their special competences and with regard for appropriate balance.

This report has been reviewed by a group other than the authors according to procedures approved by a Report Review Committee consisting of members of the National Academy of Sciences, the National Academy of Engineering, and the Institute of Medicine.

This project was supported by the W. K. Kellogg Foundation; the Rockefeller Brothers Fund; the Cooperative State Research Service of the U.S. Department of Agriculture, under Agreement Number 59-3159-5-37; and the Wallace Genetic Foundation, Inc. Dissemination of the report was assisted by the Joyce Foundation and the Jessie Smith Noyes Foundation. The project was additionally supported by the National Research Council (NRC) Fund, a pool of private, discretionary, nonfederal funds that is used to support a program of Academy-initiated studies of national issues in which science and technology figure significantly. The NRC Fund consists of contributions from a consortium of private foundations including the Carnegie Corporation of New York, the Charles E. Culpeper Foundation, the William and Flora Hewlett Foundation, the John D. and Catherine T. MacArthur Foundation, the Andrew W. Mellon Foundation, the Rockefeller Foundation, and the Alfred P. Sloan Foundation; the Academy Industry Program, which seeks annual contributions from companies that are concerned with the health of U.S. science and technology and with public policy issues with technological content; and the National Academy of Sciences and the National Academy of Engineering Endowments.

Library of Congress Cataloging-in-Publication Data

Alternative agriculture / Committee on the Role of Alternative Farming
   Methods in Modern Production Agriculture, Board on Agriculture,
   National Research Council.
       p.   cm.
   Bibliography: p.
   Includes index.
   ISBN 0-309-03987-8.—ISBN 0-309-03985-1 (pbk.)
     1. Agricultural systems—United States.   2. Agricultural ecology—
   United States.   3. Agriculture—Economic aspects—United States.
   4. Agriculture and state—United States.   5. Agricultural systems—
   United States—Case studies.   I. National Research Council (U.S.).
   Committee on the Role of Alternative Farming Methods in Modern
   Production Agriculture.
   S441.A46   1989   88-26997
   630'.973—dc19   CIP

Cover photograph by Larry Lefever from Grant Heilman

Printed in the United States of America

Committee on the Role of
Alternative Farming Methods
in Modern Production Agriculture

JOHN PESEK, *Chairman*, Iowa State University
SANDRA BROWN, University of Illinois
KATHERINE L. CLANCY, Syracuse University
DAVID C. COLEMAN, University of Georgia
RICHARD C. FLUCK, University of Florida
ROBERT M. GOODMAN, Calgene, Inc.
RICHARD HARWOOD, Winrock International
WILLIAM D. HEFFERNAN, University of Missouri
GLENN A. HELMERS, University of Nebraska
PETER E. HILDEBRAND, University of Florida
WILLIAM LOCKERETZ, Tufts University
ROBERT H. MILLER, North Carolina State University
DAVID PIMENTEL, Cornell University
CALVIN O. QUALSET, University of California, Davis
NED S. RAUN, Winrock International
HAROLD T. REYNOLDS, University of California, Riverside (retired)
MILTON N. SCHROTH, University of California, Berkeley

Staff

RICHARD WILES, *Project Director*
SUSANNE E. MASON, *Senior Project Assistant*

Board on Agriculture

iv

Preface

The 1980s have been a time of change in U.S. agriculture. The financial viability of many farms and rural communities declined during the mid-1980s as crop prices and land values fell. More than 200,000 farms went bankrupt. Since 1986, increasing market prices and exports of major farm commodities have improved the farm economy, but this recovery would not have been possible without record levels of government support.

The environmental consequences of farming have also become increasingly important to policymakers, farmers, and the public. The Environmental Protection Agency has identified agriculture as the largest nonpoint source of water pollution. Pesticides and nitrates from fertilizers and manures have been found in the groundwater of most states. The issue of pesticide and antibiotic residues in food remains unresolved. Soil erosion, salinization, and depletion of aquifers for irrigation are significant problems in some regions.

In 1984, the Board on Agriculture appointed a committee to study the science and policies that have influenced the adoption of alternative production systems designed to control these problems. The committee found that many farmers have taken steps to reduce the costs and adverse environmental effects of their operations. Some have improved conventional techniques, and others have adopted alternatives.

Farmers who have adopted alternatives try to take greater advantage of natural processes and beneficial on-farm biological interactions, reduce off-farm input use, and improve the efficiency of their operations. Many farmers have tried alternative systems. Some have succeeded; others have failed. It appears, however, that a growing number of farmers and agricultural researchers are seeking innovative ways to reduce costs and protect human health and the environment.

For the rest of this century, agricultural producers and policymakers will focus on three goals: (1) keeping U.S. farm exports competitive; (2) cutting production costs; and (3) reducing the environmental consequences of farming. The committee's report examines the scientific and economic viability of alternative systems that can help farmers and policymakers achieve these goals.

Chapter 1 describes the dimensions of U.S. agriculture in the domestic and world economies and its evolution since World War II. The committee discusses changes in input use, including fertilizers, pesticides, antibiotics, and irrigation water. Trade policy, federal commodity price and income support programs, and regulatory and tax policy are discussed, as is their influence on farm practices.

Chapter 2 outlines some of the economic and environmental consequences of agricultural practices and federal government policies. The committee describes problems in the farm economy, agricultural pollution of surface water and groundwater, pest resistance to pesticides, aquifer depletion, soil erosion and salinization, and pesticide and antibiotic residues in food.

Chapter 3 examines the basic science supporting farming practices widely used in alternative agriculture: crop rotations, alternative crop nutrient sources and management strategies, integrated pest management, biological pest control, and alternative animal management systems. Much of the evidence presented comes from the agricultural research system. The results of most scientific research, however, have not been sufficiently integrated into systems designed to solve on-farm problems. This chapter discusses the need for an interdisciplinary problem-solving research system.

Chapter 4 analyzes the economic potential of alternative systems. The committee discusses methods of economic analysis, regional differences in production costs, and the relationship between federal commodity programs and production inefficiencies. Using midwestern corn and northwestern wheat production as examples, the committee examines commodity program biases and their influence on the profitability of conventional and alternative systems. Additionally, the economic benefits of integrated pest management, biological pest control, and alternative livestock systems are discussed.

The report concludes with 11 case studies describing 14 farms managed with an efficient combination of alternative and conventional practices. Detailed descriptions of the practices and financial performance of five crop and livestock operations, seven fruit and vegetable farms, one western beef operation, and one rice farm are presented. The case studies provide insights into the operation of alternative farms in different regions producing different crops by the use of different methods. Each farm is tailored to the limitations and potential of its soil, water, and climate and the local economy.

Farmers have a history of adopting new systems. While much work remains to be done, the committee believes that farmers, researchers, and policymakers will perceive the benefits of the alternative systems described in this report and will work to make them tomorrow's conventions.

JOHN PESEK
Chairman

Acknowledgments

Farms across the country face major decisions each growing season: What to plant? How to control pests and weeds? How to maintain or improve soil fertility, deal with erosion, and manage water sources? What production goals to set? What level of expenditure to absorb and debt to incur? In this report, the committee provides technical and policy information that can aid farmers in making these decisions. The committee also makes recommendations that will aid policymakers in shaping a strong and efficient agricultural sector.

The collection and analysis of the quantities of data in this report could not have been accomplished without the assistance of committee consultants Patrick Madden, Edward Schaefer, and Kevin Finneran. The work of Patrick Madden is particularly appreciated.

The committee also acknowledges the cooperation of the owners and managers of the farms described in the 11 case studies contained in this report. These individuals were exceptionally generous with their time and with detailed information about their farms.

The committee members thank Executive Director Charles M. Benbrook for his continued support and interest in our deliberations. We are particularly grateful to past chairman of the Board on Agriculture William L. Brown. His vision of the future of U.S. agriculture and encouragement through all stages of the study were vital to the successful completion of this project.

Finally, we thank Project Director Richard Wiles. His thoughtful analyses and diligent effort were critical to the successful completion of this project.

The National Academy of Sciences is a private, nonprofit, self-perpetuating society of distinguished scholars engaged in scientific and engineering research, dedicated to the furtherance of science and technology and to their use for the general welfare. Upon the authority of the charter granted to it by the Congress in 1863, the Academy has a mandate that requires it to advise the federal government on scientific and technical matters. Dr. Frank Press is president of the National Academy of Sciences.

The National Academy of Engineering was established in 1964, under the charter of the National Academy of Sciences, as a parallel organization of outstanding engineers. It is autonomous in its administration and in the selection of its members, sharing with the National Academy of Sciences the responsibility for advising the federal government. The National Academy of Engineering also sponsors engineering programs aimed at meeting national needs, encourages education and research, and recognizes the superior achievements of engineers. Dr. Robert M. White is president of the National Academy of Engineering.

The Institute of Medicine was established in 1970 by the National Academy of Sciences to secure the services of eminent members of appropriate professions in the examination of policy matters pertaining to the health of the public. The Institute acts under the responsibility given to the National Academy of Sciences by its congressional charter to be an adviser to the federal government and, upon its own initiative, to identify issues of medical care, research, and education. Dr. Samuel O. Thier is president of the Institute of Medicine.

The National Research Council was organized by the National Academy of Sciences in 1916 to associate the broad community of science and technology with the Academy's purposes of furthering knowledge and advising the federal government. Functioning in accordance with general policies determined by the Academy, the Council has become the principal operating agency of both the National Academy of Sciences and the National Academy of Engineering in providing services to the government, the public, and the scientific and engineering communities. The Council is administered jointly by both Academies and the Institute of Medicine. Dr. Frank Press and Dr. Robert M. White are chairman and vice chairman, respectively, of the National Research Council.

Contents

PART TWO

TABLES

FIGURES

ALTERNATIVE
AGRICULTURE

PART ONE

Executive Summary

I N THE 1930S, CROP YIELDS in the United States, England, India, and Argentina were essentially the same. Since that time, researchers, scientists, and a host of federal policies have helped U.S. farmers dramatically increase yields of corn, wheat, soybeans, cotton, and most other major commodities. Today, fewer farmers feed more people than ever before. This success, however, has not come without costs.

The U.S. Environmental Protection Agency (EPA) has identified agriculture as the largest nonpoint source of surface water pollution. Pesticides and nitrate from fertilizers are detected in the groundwater in many agricultural regions. Soil erosion remains a concern in many states. Pest resistance to pesticides continues to grow, and the problem of pesticide residues in food has yet to be resolved. Purchased inputs have become a significant part of total operating costs. Other nations have closed the productivity gap and are more competitive in international markets. Federal farm program costs have risen dramatically in recent years.

Because of these concerns, many farmers have begun to adopt alternative practices with the goals of reducing input costs, preserving the resource base, and protecting human health. The committee has reviewed the dimensions and structure of U.S. agriculture, its problems, and some of the alternatives available to farmers to resolve them.

Many components of alternative agriculture are derived from conventional agronomic practices and livestock husbandry. The hallmark of an alternative farming approach is not the conventional practices it rejects but the innovative practices it includes. In contrast to conventional farming, however, alternative systems more deliberately integrate and take advantage of naturally occurring beneficial interactions. Alternative systems emphasize management; biological relationships, such as those between the pest

3

and predator; and natural processes, such as nitrogen fixation instead of chemically intensive methods. The objective is to sustain and enhance rather than reduce and simplify the biological interactions on which production agriculture depends, thereby reducing the harmful off-farm effects of production practices.

Alternative agriculture is any system of food or fiber production that systematically pursues the following goals:

- More thorough incorporation of natural processes such as nutrient cycles, nitrogen fixation, and pest-predator relationships into the agricultural production process;
- Reduction in the use of off-farm inputs with the greatest potential to harm the environment or the health of farmers and consumers;
- Greater productive use of the biological and genetic potential of plant and animal species;
- Improvement of the match between cropping patterns and the productive potential and physical limitations of agricultural lands to ensure long-term sustainability of current production levels; and
- Profitable and efficient production with emphasis on improved farm management and conservation of soil, water, energy, and biological resources.

Alternative agriculture is *not* a single system of farming practices. It includes a spectrum of farming systems, ranging from organic systems that attempt to use no purchased synthetic chemical inputs, to those involving the prudent use of pesticides or antibiotics to control specific pests or diseases. Alternative farming encompasses, but is not limited to, farming systems known as biological, low-input, organic, regenerative, or sustainable. It includes a range of practices such as integrated pest management (IPM); low-intensity animal production systems; crop rotations designed to reduce pest damage, improve crop health, decrease soil erosion, and, in the case of legumes, fix nitrogen in the soil; and tillage and planting practices that reduce soil erosion and help control weeds. Alternative farmers incorporate these and other practices into their farming operations. Successful alternative farmers do what all good managers do—they apply management skills and information to reduce costs, improve efficiency, and maintain production levels.

Some examples of practices and principles emphasized in alternative systems include

- Crop rotations that mitigate weed, disease, insect, and other pest problems; increase available soil nitrogen and reduce the need for purchased fertilizers; and, in conjunction with conservation tillage practices, reduce soil erosion.
- IPM, which reduces the need for pesticides by crop rotations, scouting, weather monitoring, use of resistant cultivars, timing of planting, and biological pest controls.

- Management systems to control weeds and improve plant health and the abilities of crops to resist insect pests and diseases.
- Soil- and water-conserving tillage.
- Animal production systems that emphasize disease prevention through health maintenance, thereby reducing the need for antibiotics.
- Genetic improvement of crops to resist insect pests and diseases and to use nutrients more effectively.

Alternative systems are often diversified. Diversified systems, which tend to be more stable and resilient, reduce financial risk and provide a hedge against drought, pest infestation, or other natural factors limiting production. Diversification can also reduce economic pressures from price increases for pesticides, fertilizers, and other inputs; drops in commodity prices; regulatory actions affecting the availability of certain products; and pest resistance to pesticides.

Alternative farming practices can be compatible with small or large farms and many different types of machinery. Differences in climate and soil types, however, affect the costs and viability of alternative systems. Alternative practices must be carefully adapted to the biological and physical conditions of the farm and region. For example, it is relatively easy for corn and soybean farmers in the Midwest to reduce or eliminate routine insecticide use, a goal much harder for fruit and vegetable growers in regions with long production seasons, such as the hot and humid Southeast. Crop rotation and mechanical tillage can control weeds in certain crops, climates, and soils, but herbicides may be the only economical way to control weeds in others. Substituting manure or legume forages for chemical fertilizers can significantly reduce fertilizer costs. However, a local livestock industry is often necessary to make these practices economical.

FINDINGS

In assessing current conventional and alternative farming practices in U.S. agriculture the committee

- Studied the potential influence of alternative farming practices on national economic, environmental, and public health goals;
- Identified and evaluated the factors, including government programs and policies, that influence adoption of alternative farming practices; and
- Reviewed the state of scientific and economic knowledge of alternative farming practices to determine what further research is needed.

Based on its study, the committee arrived at four major findings.

1. A small number of farmers in most sectors of U.S. agriculture currently use alternative farming systems, although components of alternative systems are used more widely. Farmers successfully adopting these systems generally derive significant sustained economic and environmental benefits.

Wider adoption of proven alternative systems would result in even greater economic benefits to farmers and environmental gains for the nation.

2. A wide range of federal policies, including commodity programs, trade policy, research and extension programs, food grading and cosmetic standards, pesticide regulation, water quality and supply policies, and tax policy, significantly influence farmers' choices of agricultural practices. As a whole, federal policies work against environmentally benign practices and the adoption of alternative agricultural systems, particularly those involving crop rotations, certain soil conservation practices, reductions in pesticide use, and increased use of biological and cultural means of pest control. These policies have generally made a plentiful food supply a higher priority than protection of the resource base.

3. A systems approach to research is essential to the progress of alternative agriculture. Agricultural researchers have made important contributions to many components of alternative as well as conventional agricultural systems. These contributions include the development of high-yielding pest-resistant cultivars, soil testing methods, conservation tillage, other soil and water conservation practices, and IPM programs. Little recent research, however, has been directed toward many on-farm interactions integral to alternative agriculture, such as the relationship among crop rotations, tillage methods, pest control, and nutrient cycling. Farmers must understand these interactions as they move toward alternative systems. As a result, the scientific knowledge, technology, and management skills necessary for widespread adoption of alternative agriculture are not widely available or well defined. Because of differences among regions and crops, research needs vary.

4. Innovative farmers have developed many alternative farming methods and systems. These systems consist of a wide variety of integrated practices and methods suited to the specific needs, limitations, resource bases, and economic conditions of different farms. To make wider adoption possible, however, farmers need to receive information and technical assistance in developing new management skills.

Incentives for the Adoption of Alternatives

Major segments of U.S. agriculture entered a period of economic hardship and stress in the early and mid–1980s. This period followed more than 30 years of growth in farm size and production following World War II. Export sales after 1981 slumped well below the record levels of the late 1970s. This was caused by the rising value of the dollar, a period of worldwide recession, high and rigid federal commodity program loan rates, and increases in agricultural production and exports from developed and certain developing countries. As food surpluses grew in some regions of the world, the industrialized nations promoted agricultural exports with a variety of subsidies. Many U.S. farmers, particularly specialized producers of major export crops such as corn, soybeans, cotton, and wheat, suffered financial hardship.

Some farmers, caught by the abrupt downward turn in commodity prices and land values, were unable to pay debts. Many were forced to leave farming. A substantial increase in federal price and income support payments beginning in 1983, coupled with stronger export demand, has helped insulate row-crop and small-grain producers from further economic losses. Nonetheless, tens of thousands of farms are still struggling, particularly middle-sized family farms with little or no off-farm income.

Apart from economic hardship, other adverse effects of conventional agriculture are being felt in some regions. Specialization and related production practices, such as extensive synthetic chemical fertilizer and pesticide use, are contributing to environmental and occupational health problems as well as potential public health problems. Insects, weeds, and pathogens continue to develop resistance to some commonly used insecticides, herbicides, and fungicides. Insects and pathogens also continue to overcome inbred genetic resistance of plants. Nitrate, predominantly from fertilizers and animal manures, and several widely used pesticides have been found in surface water and groundwater, making agriculture the leading nonpoint source of water pollution in many states. The decreasing genetic diversity of many major U.S. crops and livestock species (most notably dairy cattle and poultry) increases the potential for sudden widespread economic losses from disease.

Evaluating Alternative Farming Methods and Systems

A review of the literature led the committee to conduct a set of case studies to further explore and illustrate the principles and practices of alternative agriculture. Some farmers who have adopted alternative practices have been very successful, while others have tried and failed. Some who have successfully adopted alternatives experienced setbacks during the transition. Experience and research have led to a detailed understanding of some alternative methods. But many others are not well understood. Consequently, it is hard to predict where and how specific alternative practices might be useful. Although science has accumulated a great base of knowledge of potential benefit to alternative agriculture, research and extension have not focused on integrating this knowledge into practical solutions to farmers' problems.

It is difficult to estimate the economic impact of many alternative farming practices, particularly those that influence several facets of the farm, such as soil fertility and pest populations. The task of isolating the impact of a new practice requires detailed knowledge of a farm's biological and agronomic characteristics. Even more difficult is the task of predicting and measuring the economic effects of the transition to alternative methods.

During the transition period, it is often unclear how well and how quickly alternative practices will become effective.

The aggregate effects of alternative agriculture need to be evaluated in the context of market forces and government policies that determine farm profitability. In spite of obstacles, however, innovative farmers will continue to broaden and refine alternative farming practices, with increasingly significant benefits for agriculture, the economy, and the environment. With appropriate changes in farm policy and expanded and redirected research and extension efforts, the rate of progress in developing and adopting alternative systems could be markedly accelerated.

CONCLUSIONS

Alternative Farming Practices and Their Effectiveness

Farmers who adopt alternative farming systems often have productive and profitable operations, even though these farms usually function with relatively little help from commodity income and price support programs or extension.

The committee's review of available literature and commissioned case studies illustrates that alternative systems can be successful in regions with different climatic, ecological, and economic conditions and on farms producing a variety of crops and livestock. Further, a small number of farms using alternative systems profitably produce most major commodities, usually at competitive prices, and often without participating in federal commodity price and income support programs. Some of these farms, however, depend on higher prices for their products. Successful alternative farmers often produce high per acre yields with significant reductions in costs per unit of crop harvested. A wide range of alternative systems and techniques deserves further support and investigation by agricultural and economic researchers. With modest adjustments in a number of federal agricultural policies many of these systems could become more widely adopted and successful.

> *Alternative farming practices are not a well-defined set of practices or management techniques. Rather, they are a range of technological and management options used on farms striving to reduce costs, protect health and environmental quality, and enhance beneficial biological interactions and natural processes.*

Farmers adopting alternative practices strive for profitable and ecologically sound ways to use the particular physical, chemical, and biological potentials of their farms' resources. To these ends, they make choices to diversify their operations, make the fullest use of available on-farm resources, protect themselves and their communities from the potential hazards of agricultural chemicals, and reduce off-farm input expenses. Instead of rejecting modern agricultural science, farmers adopting alternative systems rely on increased knowledge of pest management and plant nutrition, improved genetic and biological potential of cultivars and livestock, and better management techniques.

A fuller understanding of biological and ecological interactions, nutrient cycles, and management systems geared toward sustaining and maximizing on-farm resources is often prerequisite for a successful transition to an alternative system. The transition can occur rapidly in some cases; however, most farmers adopt alternative practices gradually as they learn to integrate these practices into more profitable farm management systems.

> *Well-managed alternative farming systems nearly always use less synthetic chemical pesticides, fertilizers, and antibiotics per unit of production than comparable conventional farms. Reduced use of these inputs lowers production costs and lessens agriculture's potential for adverse environmental and health effects without necessarily decreasing—and in some cases increasing—per acre crop yields and the productivity of livestock management systems.*

Farmers can reduce pesticide use on cash grains through rotations that disrupt the reproductive cycle, habitat, and food supply of many crop insect pests and diseases. By altering the timing and placement of nitrogen fertilizers, farmers can often reduce per acre application rates with little or no sacrifice in crop yields. Further reductions are possible in regions where leguminous forages and cover crops can be profitably grown in rotation with corn, soybeans, and small grains. This usually requires the presence of a local hay market. Fruit and vegetable growers can often dramatically decrease pesticide use with an IPM program, particularly in dry or northern regions. Subtherapeutic use of antibiotics can be reduced or eliminated without sacrificing profit in most beef and swine production systems not reliant on extreme confinement rearing. Significant reduction of antibiotic use in poultry production is possible, but will be more difficult without major changes in the management and housing systems commonly used in intensive production.

> *Alternative farming practices typically require more information, trained labor, time, and management skills per unit of production than conventional farming.*

Alternative farming is not easy. Grain farmers who add livestock to their farms may find it more difficult to balance demands on their time during certain peak work seasons. Labor needs, particularly for trained personnel, typically increase on farms using alternative systems. Marketing plans take more time to develop and implement. Alternative farming practices also require more attention to unique farm conditions. Scouting for pests and beneficial insects, using biological controls, adopting rotations, and spot spraying insecticides or herbicides require more knowledge and management than simply treating entire fields on a programmed schedule.

The development of optimum rotations or planting schedules for specific climatic and soil conditions demands careful observation of crop response and precise management. Preventive health care for livestock requires greater knowledge of animal health and accurate diagnoses of health problems. Monitoring soil nutrient levels through soil and crop tissue testing is a reliable way to estimate more precisely fertility needs and calibrate fertilizer applications. Such testing and analysis, however, require time, knowledge, money, and, in many cases, specialized skills.

The Effect of Government Policy

Many federal policies discourage adoption of alternative practices and systems by economically penalizing those who adopt rotations, apply certain soil conservation systems, or attempt to reduce pesticide applications. Federal programs often tolerate and sometimes encourage unrealistically high yield goals, inefficient fertilizer and pesticide use, and unsustainable use of land and water. Many farmers in these programs manage their farms to maximize present and future program benefits, sometimes at the expense of environmental quality.

Commodity program rules have an enormous influence on agriculture. Through provisions governing allowable uses of base acres, these programs promote specialization in one or two crops, rather than more varied rotations. Between 80 and 95 percent of all acreage producing corn, other feed grains, wheat, cotton, and rice (or about 70 percent of the nation's cropland) are currently enrolled in federal commodity programs.

All acres enrolled in the federal commodity income and price support programs are subject to specific crop program rules that determine eligibility. The most crucial and basic rule determines eligible base acres. A farm's base acres are those eligible for program participation and benefits. They are calculated as an average of acreage enrolled in a particular crop program each year during the past 5 years. Thus, any practice that reduces acreage counted as planted to a program crop will reduce the acreage eligible for federal subsidies for the next 5 years. For example, if a farmer rotates all of his or her base acreage one year to a legume that will fix and supply nitrogen and conserve soil, fewer acres will be eligible for program payments in subsequent years. In general, under this scenario, benefits would be reduced 20 percent per year for the next 5 years. Payment reductions could be even greater in subsequent years.

Another rule, cross-compliance, passed in the Food Security Act of 1985, has had a great influence on a farmer's choice of crops. Cross-compliance stipulates that to receive any benefits from an established crop acreage base, a farmer must not exceed his or her acreage base for any other program crop. In general, cross-compliance discourages diversification into rotations involving other program crops. For example, if a farm is enrolled in the corn program and has no other program crop base acreage, the farm would lose all corn program benefits that year if *any* other program crop were planted on the farm. Farmers wishing to diversify into rotations with other program crops must generally forfeit program payments from crops currently in the program. If a farm had base acreage for two or more crops when cross-compliance went into effect in 1986, it must meet two criteria to retain eligibility for maximum program benefits: (1) the farm may not be planted with any other program crops and (2) the farm must stay enrolled in both programs each year. Oats are currently exempt from cross-compliance to encourage production. And in 1989, farmers have the option of planting 10 to 25 percent of feedgrain base acres to soybeans with no reduction in feedgrain base acres in subsequent years.

The government also sets per bushel target prices for program crops. Farmers enrolled in the programs are paid the difference between the target price and the crop-specific loan rate or market price, whichever is less, in the form of a per bushel (per hundredweight for rice, per pound for cotton) deficiency payment. This is paid in addition to what a farmer receives on the market or for placing the crop under loan with the U.S. Department of Agriculture's (USDA) Commodity Credit Corporation. Often these deficiency payments are a substantial portion of gross farm income. For example, in 1986 and 1987, corn deficiency payments were $1.11 and $1.21 per bushel, while market prices averaged $1.92 and $1.82, respectively. Wheat deficiency payments in 1986 and 1987 were $1.98 and $1.78 per bushel, while market prices averaged $2.40 and $2.60, respectively.

Farmers in these programs manage their land to maximize future eligibility for farm program benefits. They are often far more responsive to subtle economic effects of the farm programs than to the biological and physical constraints of their land. Two principal objectives of farmers participating in the commodity programs are to sustain or expand eligible base acres and to maximize yields on those acres, thus maximizing per acre payments. These goals are usually achieved by growing the same crop or crops year after year and striving for the highest possible yield on the greatest possible acreage.

Shifts in international market demand driven by economic policy changes in the United States, including devaluation of the dollar and changes in the tax code and deficits, can also have significant, unintended effects on the land. During the export boom of the 1970s and early 1980s, land previously considered unsuitable for cultivation, primarily because of erosion, was brought into cultivation. About 25 million acres of this land has been recently idled under the Conservation Reserve Program (CRP), but much remains in production.

Fertilizers and pesticides are often applied at rates that cannot be justified economically without consideration of present or future farm program payments.

The committee identified two major forms of input inefficiency encouraged by federal commodity programs: (1) excess input use to achieve higher yields and maximize government program payments and (2) use of inputs to expand crop production onto marginal lands or to support the production of crops in regions poorly suited to a particular crop.

Efficiency of input use, total variable costs, and per unit production costs differ widely among growers and regions. The committee's review of selected cost of production studies resulted in the following conclusions that warrant further study to help improve farm profitability and reform farm policies:

- Within a given region for a specific crop, average production costs per unit of output on the most efficient farms are typically 25 percent less, and often more than 50 percent less, than average costs on less efficient farms. There is a great range in the economic performance of seemingly similar or neighboring farms.
- Average production costs per unit of output also vary markedly among regions, although not as dramatically as among individual farms.
- High-income and low-cost farms are often larger. The causes and effects of this, however, deserve study.
- Certain variable production expenses—machinery, pesticides, fertilizers, and interest (excluding land)—account disproportionately for differences in per unit production costs.

Federal grading standards, or standards adopted under federal marketing orders, often discourage alternative pest control practices for fruits and vegetables by imposing cosmetic and insect-part criteria that have little if any relation to nutritional quality. Meat and dairy grading standards continue to provide economic incentives for high-fat content, even though considerable evidence supports the relationship between high consumption of fats and chronic diseases, particularly heart disease.

Most fruits and vegetables are marketed under orders that set specific criteria for cosmetic damage and other quality criteria that rarely affect the safety or nutritional value of the food. Commodity producer organizations generally support these standards as a way of reducing market supply and increasing price; food processors favor them as a quality control mechanism and because they can offer a lower price for food that does not meet the highest cosmetic standards. In many cases, pesticides are applied solely to meet grading criteria. Although IPM methods permit successful maintenance or even enhancement of crop yields, in many cases they are less effective than routine spraying for controlling cosmetic damage. Pesticides applied solely to meet cosmetic or insect fragment standards increase pest control costs to producers and may increase residues of pesticides in food and hazards to agricultural workers. Repercussions from pesticide use may become more serious as new pests encroach on major fruit- and vegetable-producing regions, and as insects and plant diseases become resistant to currently effective products.

Many animal feeding and management systems and technologies currently exist to reduce the fat content of meat and dairy products. These practices also often help cut costs. Producers are unlikely to adopt them, however, without changes in grading standards and higher prices for lower fat products. Some progress is under way in this area, particularly in the beef and pork industries, but further reform of the rules is needed.

> *Current federal pesticide regulatory policy applies a stricter standard to new pesticides and pest control technologies than to currently used older pesticides approved before 1972. This policy exists in spite of the fact that a small number of currently used pesticides appears to present the vast majority of health and environmental risks associated with pesticides. This policy inhibits the marketing of biologically based or genetically engineered products and safer pesticides that may enhance opportunities for alternative agricultural production systems.*

Federal pesticide regulatory procedures and standards are increasingly expensive and time-consuming. Many scientific issues remain unresolved, complicating decisions to allow new pesticides onto the market and remove older pesticides from the market. Pesticide benefits assessments, for example, are an extremely challenging area for research. Neither the EPA nor the USDA has developed formal procedures to calculate the economic benefits of pesticides under regulatory review. This often leads to uncertainty, controversy, and delay in regulatory decisions on older pesticides. The benefits assessments that are typically developed tend to overestimate the actual value of pesticides under review for health and environmental effects, by not fully accounting for IPM and nonchemical alternatives. This policy helps to preserve market share for older compounds known to pose health and environmental hazards. This in turn discourages the development and adoption of biological, cultural, or other alternative pest control practices.

Current and pending regulations need to be improved to provide greater opportunity for the development of naturally occurring pest control agents and those that rely in some way on genetic engineering. Uncertainty over the definition of a genetically altered organism has resulted in some confusion in registration of nonpathogenic microflora that can help control pests biologically. One possible outcome of this confusion is delay in efforts to select and produce strains of naturally occurring bacteria for many purposes, including more efficient fixation of atmospheric nitrogen by legumes and control of plant pests.

The State of Research and Extension

The results and design of basic, discipline-oriented research programs often are not sufficiently integrated into practical interdisciplinary efforts to understand agricultural systems and solve some major agricultural problems.

Many would agree that the United States has been slow to marshal certain new scientific capabilities, such as biotechnology, to develop agricultural products and technologies. This is largely due to declining support for applied research and extension and difficulty in maintaining facilities and incentives for multidisciplinary research. While the decline of the heavy industry and manufacturing sectors is perhaps the most dramatic example of the erosion of U.S. technological leadership, many fear that agriculture will be added to the list in the early 1990s.

U.S. agriculture has always taken pride in its ability to apply science and technology in overcoming the everyday problems of farmers. Many states, however, are losing by retirement and attrition the multidisciplinary agricultural research and education experts capable of bridging the gap between laboratory advances and practical progress on the farm. These individuals, frequently cooperative extension system employees, have traditionally played an important role in informing research scientists of the problems faced by farmers and in integrating research advances into production programs on the farm.

Insufficient numbers of young scientists are pursuing careers in interdisciplinary or systems research. This is in part because higher education, peer review, the agricultural research systems, and their funding sources tend to encourage narrow intradisciplinary research over interdisciplinary work. As a result, agricultural scientists often lack the skills and insights to understand fully on-farm problems or how farmers can most readily overcome them. The lack of support for on-farm systems research is creating a serious problem for the cooperative extension system. The cooperative extension system's ability to carry out its traditional role has eroded substantially in the last decade. This trend is likely to continue unless there are changes in research and development, educational policies, and increased financial support.

The committee is nonetheless encouraged by the growing interest in alternative farming practices among research and extension personnel. Without additions to existing programs and new research and educational initiatives, however, the current system will not be able to provide farmers the kind of information, managerial assistance, and new technologies needed to support widespread adoption of alternative agricultural practices. An effective alternative agricultural research program will require the participation of and improved communication among problem-solving and systems-oriented researchers, innovative farmers, farm advisers, and a larger cadre of extension specialists.

Research and extension program funds to study, develop, and promote alternative farming practices are inadequate. It is unrealistic to expect more rapid progress in developing and transferring alternative practices to farmers without increased funding.

A shortage of public funds in support of agricultural research has discouraged work on alternative agriculture. With shrinking funds, publicly supported research and extension services have not been able to provide adequate regional or farm-specific information about alternative farming practices. Increasing production efficiency through the use of off-farm inputs to achieve higher yields has been a dominant objective, in part because private funds were available to support these efforts.

During the last two decades, research support has increased for biological research, especially in molecular biology. This work has made possible advances in the understanding of plants and animals at the subcellular level. During the same period, however, government support for field and applied research and extension in farming systems has not kept pace with the need, or even with inflation. This applied research and extension is vital to improving agricultural practices and dealing with agriculture's adverse environmental effects.

State support for research, which tends to emphasize applied research adapted to local crops and field conditions, is stable, at best, in many states. Land-grant colleges, which receive much of their support from the states, have had to find other sources of funds (including commodity organizations and agribusiness firms) to support adaptive field research. Despite some success in securing private industry funding in support of some applied research on specific products, private funds are rarely provided to support the multidisciplinary research needed to advance alternative agriculture.

The committee believes that farming systems research promises significant short- and long-term returns. Inadequate funding, however, has postponed work in several areas, including the development of monitoring processes and analytical tools, biological control methods, cover crops, alternative animal care systems, rotations, plant health and nutrition, and many others. Without increased funding and a change in the intradisciplinary orientation in the tenure and promotion systems of major research universities, farming systems research and extension will remain limited, and progress toward alternatives will be much slower than otherwise possible.

There is inadequate scientific knowledge of economic, environmental, and social costs and thresholds for pest damage, soil erosion, water contamination, and other environmental consequences of agricultural practices. Such knowledge is needed to inform farm managers of the tradeoffs between on-farm practices and off-farm consequences.

Farmers are told too little about the ecological, biological, and economic relationships associated with the use of agricultural chemicals. Farmers generally follow the guidelines offered by the input manufacturers, but these typically do not explain alternatives or the many conditions that may reduce the need for a pesticide or a fertilizer. Farmers receive little guidance in evaluating the economics of input use with respect to shifts in the market price for a commodity or those inputs. Eradicating as many pests as possible, for example, is rarely the most economical option and often ignores the long-range impact of pesticides on the environment. When fertilizer costs are low, higher per acre nitrogen fertilizer applications may seem like a prudent investment. Applications in excess of need, however, are not completely used by crops and can aggravate water quality problems.

Many agricultural practices have an off-farm impact on society and the environment. Common agricultural practices have degraded surface water quality, and, to a lesser degree, groundwater quality in most major farming regions. In recent years, state and federal agencies have recognized that off-farm costs of certain agricultural practices must be reduced, especially the costs associated with some pesticides, tillage methods, and excessively high rates of manure and nitrogen fertilizer application. But methods and models for measuring the costs and benefits of conventional and alternative farming practices are simplistic. Moreover, many policy goals, such as conserving soil and increasing exports, are often at odds. Farmers need guidance and management tools to balance stewardship and production objectives. To help farmers make these choices, reliable cost-benefit comparisons between conventional and alternative systems are needed. Developing improved information and techniques for calculating on- and off-farm costs, benefits, and tradeoffs inherent in different farming systems and technologies must be a priority.

> *Research at private and public institutions should give higher priority to development and use of biological and genetic resources to reduce the use of chemicals, particularly those that threaten human health and the environment.*

Genetic research has greatly increased the productivity of plants and animals in agriculture. Conventional plant breeding research such as hybridization has produced many crop cultivars that are naturally resistant to various diseases and insects. Genetic engineering techniques such as gene transfer mediated by bacteria and viruses and direct transfer methods promise further improvements.

Financial incentives exist for the development of crop cultivars that produce higher yields. But there is less incentive and more risk for private industry to produce cultivars designed to reduce input use and make various alternative farming practices more feasible and profitable. Thus, the federal government must increase its support for this type of research.

Examples of genetically engineered products that could reduce the need for purchased inputs include legumes and bacteria that more effectively fix nitrogen, diagnostic tools and preventative measures for major infectious animal diseases, crop cultivars with genetic resistance to insects and other pests, and enhancement of the allelopathic capability of crops to suppress weeds. In these areas, genetic research could greatly reduce pesticide use, increase the profitability of legumes and cover crops in crop rotations, and lessen chemical levels in the food supply and the environment. While it is too early to tell how biotechnology will influence agriculture, the committee believes that biotechnology has much to offer farmers looking to adopt alternative production practices.

Greater support for research on biological controls and improved plant nutrition is also needed. Research on and implementation of biological control lags far behind total support for other pest control methods, even though several important pests remain difficult or costly to control by current methods. Better understanding of the role of plant nutrition and health in resisting pests, utilizing available soil nutrients, and improving yields could be of great benefit to farmers. Greater public support is needed, however, to support research designed specifically to achieve these goals and reduce input costs and the environmental consequences of current practices.

RECOMMENDATIONS

Farm and Environmental Policy

A variety of farm programs and policies have had a profound, continuing influence on U.S. agriculture. Over the years, policies have had intended and unintended effects. One important unintended effect is the variety of financial penalties that farmers must overcome when adopting alternative and resource-conserving production practices. These include the potential loss of farm program subsidies, the inability of publicly supported research institutions to provide information on alternative farming systems, and the way current policies tolerate external environmental and public health costs associated with contemporary production practices. Many changes in commodity and regulatory policies will be required to neutralize their bias against the adoption of alternative farming systems.

> *Federal commodity programs must be restructured to help farmers realize the full benefits of the productivity gains possible through many alternative practices. These practices include wider adoption of rotations with legumes and nonleguminous crops, the continued use of improved cultivars, IPM and biological pest control, disease-resistant livestock, improved farm machinery, lower-cost management strategies that use fewer off-farm and synthetic chemical inputs, and a host of alternative technologies and management systems.*

A number of government policies and programs have strongly encouraged farmers to specialize and deterred them from adopting diversified farming practices. This is particularly true for farmers growing major commodities covered by price support programs. In many regions, the need to retain eligibility for future government program payments has become more important than the inherent efficiency or immediate profitability of a production system in the absence of government program payments.

The committee recommends that a primary goal of commodity program reform be the removal of the existing disincentives to alternative farming practices. This step would ensure that farmers who employ crop rotations and recommended resource conservation practices are not deprived of farm income support. For the Congress, this means that

- Existing commodity programs, if retained, should be revised to eliminate penalties for farmers adopting rotations. These revisions should allow more flexibility in substituting or adjusting base acreage allotments to accommodate crop rotations, acceptance of forage crops in rotations as satisfying set-aside requirements, and harvesting or grazing of forage crops grown during such rotations;
- Mandatory production controls, if enacted, should not require land retirement for participation because this discourages crop rotations. Farmers should be free to decide how to produce the allotted level of output over a 2- to 5-year period; and
- Decoupling of income support from crop production, if enacted, should ensure that all farming systems and rotations are treated equitably.

Natural Resource Management

Despite five decades of federally supported soil conservation programs, soil erosion and water quality deterioration continue. Agricultural and conservation policies have not consistently supported the stewardship of natural resources. This inconsistency among policies should be changed. The committee recommends that

Provisions in the Food Security Act of 1985 designed to protect erodible lands and wetlands must be fully and fairly implemented.

Future farm programs should offer no new incentives to manage these and other fragile lands in a way that impairs environmental quality.

Surface water and groundwater quality monitoring must be more systematic and coupled with educational and regulatory policies that prevent future water contamination.

Cost-effective water quality protection provisions must be incorporated into existing conservation and commodity programs.

Regulations that require farmers to maintain soil and water conserva-

tion practices and structures installed with government technical or financial assistance must be enforced.

Adjustments in regional cropping patterns must be facilitated when such changes are necessary in order to make progress toward profitable and environmentally sustainable production systems.

Regulatory Change

Procedures for review and approval of the safety of existing and new agricultural chemicals and other agents used in production agriculture must be implemented to achieve more rapid progress toward safer working conditions, improved environmental quality, and reduced chemical residues in foods and water.

Existing policies permit pesticides with known risks to human health but approved years ago under less stringent criteria to remain in use, while new effective and safer substitutes are sometimes kept off the market by the regulatory approval process. *Regulating Pesticides in Food: The Delaney Paradox*, a report of the National Research Council published in 1987, presents detailed recommendations for a consistent policy for regulating dietary exposure to pesticides.

> *A set of guidelines for assessing the benefits of pesticides under regulatory review should be developed. This procedure must include a definition of beneficiaries as well as an assessment of the costs and benefits of other available pest control alternatives. Benefits of control methods must be assessed as they accrue to growers, consumers, taxpayers, the public health, and the environment. As a basic rule, the benefits of any pest control method should be characterized as the difference between its benefits and those of the next best alternative, which may involve an alternative cropping system that requires little or no pesticide use. The dollar costs of the health and environmental consequences of each pest control method should be weighed against its benefits.*

> *Public information efforts should explain to consumers the relationship of appearance to food quality and safety. Alternate means of controlling the supply and price of fruits and vegetables should be developed. Cosmetic and grading standards should be revised to emphasize the safety of food and deemphasize appearance and other secondary criteria.*

Federally approved grading standards and marketing orders for fruits and vegetables usually allow few surface blemishes on fresh produce or extremely low levels of insect parts in processed food. Consequently, farmers use more pesticides to meet these standards and guarantee receipt of a top price. This increases worker exposure to pesticides and may result in increased food residues. Cosmetic standards, however, often have no relation to nutritional quality, flavor, or food safety. Furthermore, these standards discourage alternative pest control practices that may not be as effective in meeting their rigid criteria.

Research and Development

Exploring the interactions and integration of agricultural practices is vital to the understanding and development of alternative farming systems. Investigation must begin with on-farm studies that address relationships among practices that supply nutrients, conserve soil and water, control pests, and sustain livestock health and productivity.

Long-term monitoring of commercial farms using alternative methods must be added to farm management record studies to evaluate the environmental, agronomic, and economic effects and viability of specific alternative farming systems. Farming systems research must also take into account the effects of policies and management decisions on resource conservation, environmental integrity, farm worker health, food safety, and economic sustainability.

The committee recommends the following strategy to encourage research and development in support of alternative farming practices:

> *Develop a regional, multidisciplinary, long-term research, demonstration, and extension program such as that initiated by the USDA's low-input sustainable agriculture (LISA) initiative. This program should focus on alternative farming practices and systems tailored for each region's major types of crop and livestock operations.*

The research program must include on-farm studies of farming systems, with participating farmers cooperating with researchers and extension personnel in conducting field tests and demonstrations. The program should establish at least six research and demonstration farm sites in each of the four Cooperative State Research Service (CSRS) administrative regions. Within each region, grants from between $100,000 and $1 million would support research at each site. State agricultural experiment stations would manage or coordinate farm site research.

In addition, centers for sustainable or alternative agriculture should be instituted in these four CSRS regions. These centers would establish a network of physical, chemical, biological, and social scientists from government, academia, and foundations. In cooperation with participating farms, these centers would determine and oversee the research agenda of the research and demonstration farms.

*Substantial annual funding—at least $40 million—should be allocated
for alternative farming research. The USDA should distribute the
money through its competitive grants program to scientists from universities, private research institutions, foundations, and industry.*

A new competitive grants program is essential to accelerate work in support of alternative agriculture. New funding should give priority to basic
and applied multidisciplinary research involving scientists at public and
private universities and private research institutions and foundations. The
specific research areas for an expanded competitive grants program should
include biological, genetic, and ecological research priorities and social science research objectives focusing on the economic performance and consequences of alternative systems. Priorities for the competitive grants program are:

- Nutrient cycling research to assess plant nutrient availability and increase the efficiency of nutrient use; establishment of economically and
environmentally optimum levels and methods of fertilization with emphasis on leguminous crops; identification of points in the nutrient
cycle where nutrients are lost; exploration of how the efficiency of
nutrient uptake is affected by the source of nutrients, plant health, and
plant cultivars; and evaluation of the role of soil structure, tilth, and
soil biota in plant nutrient use and availability.
- Analysis of the effect of alternative tillage systems on weed and erosion
control, nutrient availability, fertilizer and pest control needs, cultivation costs, and compatibility with leguminous and nonleguminous cover
crops and specific soils.
- Development of new pest management strategies that take advantage of
cultural practices; rotations; allelopathy; beneficial insect, parasite, and
pathogen species; and other biological and genetic pest control mechanisms.
- Analysis of the effect of crop rotations, including leguminous forages,
on plant vigor; disease, insect, and weed damage; allelopathy; soil
microorganisms; nutrient levels; and the effectiveness of strip intercropping, overseeding, and relay cropping.
- Development of improved crop and livestock species' resistance to diseases and pests through genetic engineering or classical breeding techniques.
- Development and modification of farm equipment to meet the needs of
alternative farming practices and development of better processing and
handling systems for plant residues, animal wastes, and other biomass
to recycle plant nutrients into the soil.
- Research on the economics of alternative agricultural systems to determine their effect on net return to the farm family; per unit production
costs; the profitability of conventional versus alternative systems with
reduction or elimination of government support; the effect of alternative

agriculture on labor demand, supply, and rural development; and the influence of widespread adoption of alternative systems on U.S. agriculture's competitiveness in international markets.

- Development of computer software and systems to aid farmers in the management and decision making needed to adopt alternative systems.

Economics and Markets

Data bases and economic research on the profitability of alternative farming systems are minimal. Meaningful research on the effect of these systems on the international competitiveness of U.S. agriculture is not available. The results of most studies to date are not relevant. They often compare the performance of conventional production systems that differ primarily in the level of inputs applied per acre. They do not compare conventional systems with successful alternative systems. An objective assessment of the macroeconomic impacts of widespread adoption of highly productive alternative farming practices has not been undertaken.

Recent economic studies of IPM demonstrate its profitability. However, studies also highlight the fact that IPM requires continuous refinement as new crop production methods are adopted or when new pests become established. IPM systems can also change as old pests develop resistance to pesticides, regulations are imposed, and prices paid and received by farmers fluctuate. Studies of the economics of whole-farm systems, once common in farm management research and extension, are now rare, and the necessary data bases are seriously neglected in all but a handful of states, crops, and enterprise types.

Compared with conventional systems, alternative farming systems usually require new management skills along with greater reliance on skilled and unskilled labor. How these demands will affect net income and rural economies, however, is not known and is difficult to predict. The committee's case studies and review of available data illustrate that alternative farming is often profitable, but the sample is too small and unrepresentative to justify conclusions about the precise economic effects of widespread adoption of specific practices or systems. The goal of sustaining a viable operation during transition from conventional to alternative farming also deserves more study.

The aggregate, health-related, and environmental costs and benefits to society of alternative farming practices must be documented more fully. More reliable estimates are needed of the long-term costs of soil erosion, water pollution, human exposure to pesticides, certain animal health care practices, and other off-farm consequences.

The committee recommends that

> *More resources should be allocated to collect and disseminate data on yields, profits, labor requirements, human health risks, threats to*

water quality, and other environmental hazards of conventional and alternative farming practices within a given region. These data will help policymakers and farmers make more informed choices.

Research should be undertaken to predict the long-term impacts of various levels of adoption of alternative farming practices on the total production and prices of various agricultural commodities; use and prices of various farm inputs; international trade; employment, economic development, and incomes of various categories of farmers; and the overall structure of agriculture and viability of rural communities.

Research should be expanded on consumer attitudes toward paying slightly higher prices for foods with lower or no pesticide residues, even though such foods may not meet contemporary standards for appearance.

THE FUTURE OF ALTERNATIVE FARMING

Current scientific, technological, economic, social, and environmental trends are causing farmers to reconsider their practices and look for alternatives. Many farmers are turning to farming practices that reduce purchased off-farm input costs and the potential for environmental damage through more intensive management and efficient use of natural and biological resources.

The success of some of these farmers indicates that these alternative farming practices hold promise for many other farmers and potentially significant benefits for the nation. How fast and how far this transformation of U.S. agriculture will go depends on economic opportunities and incentives, which are shaped by farm policies, market forces, research priorities, and the importance society places on achieving environmental goals.

Government policies that discourage the adoption of alternative practices must be reformed. Information about alternative practices and new policies to encourage their wider adoption must be disseminated effectively to farmers. Experimentation must provide the basic physical, biological, and economic understanding of agroecosystems on which alternative practices and systems are built.

Ultimately, farmers will be the ones to decide. However, significant adoption of alternative practices will not occur until economic incentives change. This change will require fundamental reforms in agricultural programs and policies. Regulatory policy may play a role, particularly in raising the cost of conventional practices to reflect more closely their full social and environmental costs. On-farm research will have to be increased and directed toward systems that achieve the multiple goals of profitability, continued productivity, and environmental safety. Farmers will also have to acquire the new knowledge and management skills necessary to implement successful alternative practices. If these conditions are met, today's alternative farming practices could become tomorrow's conventional practices, with significant benefits for farmers, the economy, and the environment.

1

Agriculture and the Economy

ONE OF THE STRENGTHS OF U.S. AGRICULTURE is the willingness of farmers to adopt proven alternatives. This constant evolution and adoption of new practices has helped the United States become a global leader in agricultural research, technology, and production. Many of today's common practices were the alternative practices of the postwar era. One example is monocultural production, which synthetic chemical fertilizers and pesticides made possible. The widespread adoption of these alternatives, referred to internationally as the "Green Revolution," led to dramatic increases in per acre yield and overall agricultural production in the United States and many other countries.

The historical pattern is clear: today's alternatives are tomorrow's conventions. The committee believes that this is true for many of the agricultural alternatives described in this report. For example, some farming systems—such as corn and soybean production using ridge tillage, rotations, and mechanical cultivation—include new and old practices and satisfy this committee's definition of alternative agriculture (see the boxed article, "Definition of Alternative Agriculture"). Nonetheless, much can be done to improve most production systems and to accelerate the widespread adoption of farming methods specifically designed to achieve the goals listed.

This chapter describes the changes in agriculture that have taken place over the past 40 years in terms of technology and input use, a range of federal government programs, the economy, and international trade.

Since the 1940s, agriculture has become more specialized and dependent on purchased off-farm inputs. Technology has facilitated specialization and constantly increasing yields, with fewer larger farms producing more food than ever before. Federal policy has responded to the farmer's needs in the context of conflicting signals such as high per acre yield goals, surplus production capacity, environmental considerations, and increased foreign competition. Although there has been some improvement in the farm economy since the recession of the mid–1980s, unprecedented levels of federal government support have financed much of this recovery. Disparities remain in productive capacities, income, and regional rural economies, even though total net farm income has reached record levels.

Farming is at the center of the food and fiber sector of the economy. Farmers are the sole consumers of agricultural inputs and the principal producers of the crops that support the multibillion dollar food and fiber industry. The production, processing, and sale of food and fiber currently represent about 17.5 percent of the gross national product (GNP) or about $700 billion in economic activity (Figure 1-1), the second largest sector of

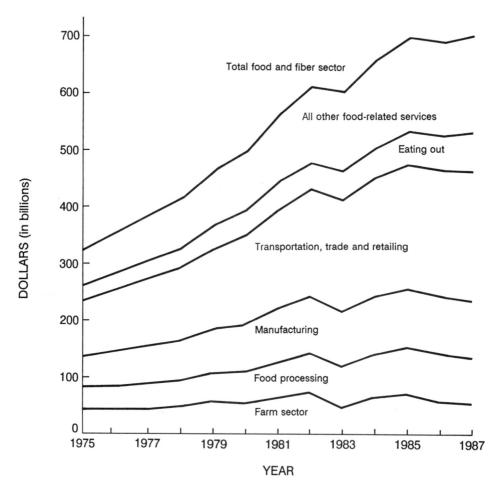

FIGURE 1-1   Food and fiber sector of the U.S. GNP. SOURCE: U.S. Department of Agriculture. 1987. Measuring the Size of the U.S. Food and Fiber System. Agricultural Economic Report No. 566. Economic Research Service. Washington, D.C.

DEFINITION OF ALTERNATIVE AGRICULTURE

Alternative agriculture is any system of food or fiber production that systematically pursues the following goals:

- More thorough incorporation of natural processes such as nutrient cycles, nitrogen fixation, and pest-predator relationships into the agricultural production process;
- Reduction in the use of off-farm inputs with the greatest potential to harm the environment or the health of farmers and consumers;
- Greater productive use of the biological and genetic potential of plant and animal species;
- Improvement of the match between cropping patterns and the productive potential and physical limitations of agricultural lands to ensure long-term sustainability of current production levels; and
- Profitable and efficient production with emphasis on improved farm management and conservation of soil, water, energy, and biological resources.

GNP next to manufacturing (U.S. Department of Agriculture, 1987f) (Figure 1-2). Farming, however, accounts for only about 2 percent of total GNP; inputs such as seed, equipment, and chemicals account for another 2 percent; and processing, marketing, and retail sales account for nearly 14 percent (U.S. Department of Agriculture, 1986e).

TRADE

Exports of agricultural commodities exploded during the 1970s, from about $7.3 billion in 1970 to $43.3 billion in 1981. Five major crops led the way: corn, cotton, rice, soybeans, and wheat (Figure 1-3). By 1981 the United States controlled 39 percent of total world agricultural trade and more than 70 percent of world trade in coarse grains, greater than 10 times the share of its nearest competitor, Argentina. During the 1970s, harvested wheat acreage increased by more than the total harvested wheat acreage of Canada (U.S. Department of Agriculture, 1986a; U.S. Office of Technology Assessment, 1986a). Economic growth in developing nations, the opening of Pacific Rim markets, grain trade with the Soviet Union, and a favorable exchange rate that fueled increased demand made this growth possible. A deliberate domestic policy designed to remove production controls helped the United States profit from these favorable conditions. The expansion of cultivated acres of wheat and feed grains, favorable tax provisions and market prices, and readily available credit helped increase the domestic supply of major commodities such as wheat, soybeans, corn, and other coarse grains. Agriculture maintained a favorable annual trade balance,

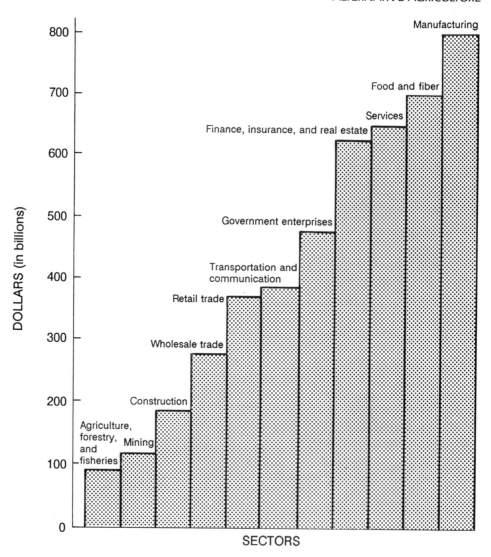

FIGURE 1-2   GNP by sector, 1985. Food and fiber sector includes farm sector; food processing; manufacturing; transportation, trade, and retailing; food; and all other nonfarm sectors. SOURCE: U.S. Department of Commerce. 1987. Survey of Current Business. Washington, D.C.

while almost all other sectors of the economy experienced growing deficits (Figure 1-4).

From 1981 to 1986, many factors contributed to a decline in agricultural exports. The loan rates in the federal commodity programs (the price that the government guarantees farmers) were rigidly set well above international market prices. This meant that most farmers sold their grain to the government at the loan rate (in practice many turn over their grain for forgiveness of the loan), instead of on the domestic or international market,

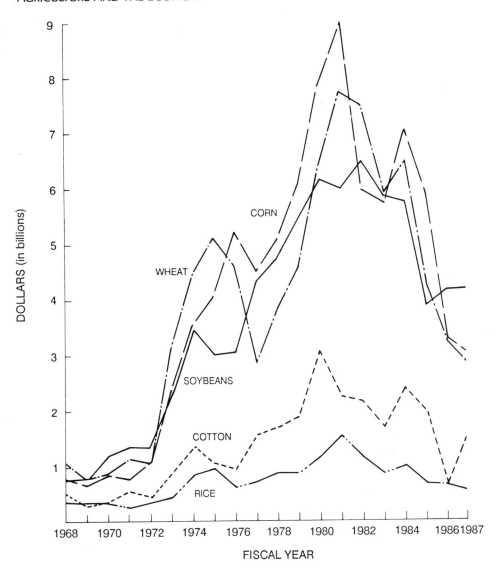

FIGURE 1-3   Value of selected agricultural exports. SOURCES: U.S. Department of Agriculture. 1983. Foreign Agricultural Trade of the United States—Annual Supplement—Fiscal Year 1982. Economic Research Service. Washington, D.C.; U.S. Department of Agriculture. 1987. Foreign Agricultural Trade of the United States—Annual Supplement—Fiscal Year 1986. Economic Research Service. Washington, D.C.; U.S. Department of Agriculture. 1988. Foreign Agricultural Trade of the United States: November/December 1987. Economic Research Service. Washington, D.C.

where prices were lower. The U.S. government ended up buying and storing the largest domestic grain surpluses in history. To compound this, the early 1980s brought global recession, increased production capacity in developing countries, an overvalued dollar, restrictive import policies and export subsidies by major competitors, foreign debt, and surpluses in major commodities. Agricultural exports fell from $43 billion in 1981 to about $26 billion in 1986 (Figure 1-5).

In 1987, the *volume* of agricultural exports increased for the first time in 7 years (Figure 1-6). The increase was largely due to a decline in the value of the dollar, falling world market prices, reduction in federal program loan rates, and implementation of the export programs of the Food Security Act of 1985 (U.S. Congress, 1985). Export programs designed to counter foreign subsidies, guarantee credit, and promote products accounted for 60 to 70 percent of wheat exports, greater than half of the vegetable oil exports, and about 40 percent of all rice exports in fiscal year (FY) 1987. Most feed grain and cotton exports were made outside these export programs (U.S. Department of Agriculture, 1988b). The *value* of agricultural exports, however,

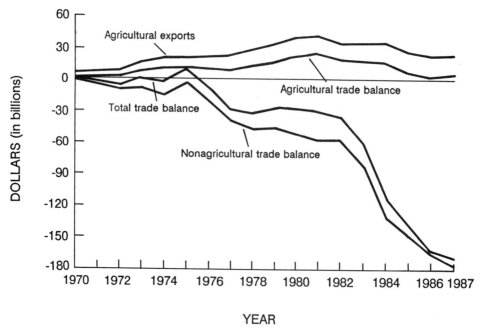

FIGURE 1-4 U.S. agricultural export trends and foreign trade balances. SOURCES: U.S. Department of Agriculture. 1988. 1988 Agricultural Chartbook. Agriculture Handbook No. 673. Washington, D.C.; U.S. Department of Agriculture. 1988. The U.S. Farm Sector: How Agricultural Exports are Shaping Rural Economics in the 1980's. Agricultural Information Bulletin 541. Economic Research Service. Washington, D.C.

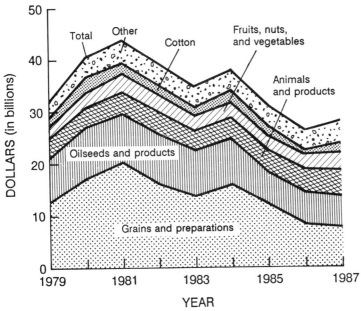

FIGURE 1-5   Value of U.S. agricultural exports by commodity. SOURCE: U.S. Department of Agriculture. 1988. 1988 Agricultural Chartbook. Agriculture Handbook No. 673. Washington, D.C.

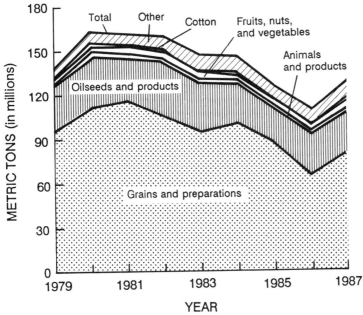

FIGURE 1-6   Volume of U.S. agricultural exports by commodity. SOURCE: U.S. Department of Agriculture. 1988. 1988 Agricultural Chartbook. Agriculture Handbook No. 673. Washington, D.C.

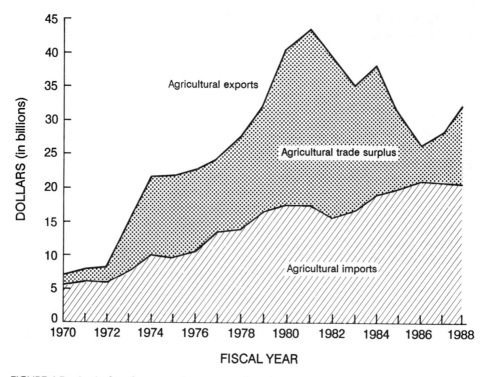

FIGURE 1-7 Agricultural exports, imports, and trade balance. Figures for 1988 are forecast. SOURCES: U.S. Department of Agriculture. 1987. National Food Review. The U.S. Food System—From Production to Consumption. NFR-37. Economic Research Service. Washington, D.C.; U.S. Department of Agriculture. 1988. 1988 Agricultural Chartbook. Agriculture Handbook No. 673. Washington, D.C.

increased only 7 percent, from $26 billion to about $28 billion in 1987. Exports are expected to continue to increase to around $33 billion in 1988. Imports, which have increased steadily since 1972, are expected to remain constant at about $20 billion, resulting in an increase in the agricultural trade surplus to about $13 billion in 1988 (Figure 1-7).

AGRICULTURAL INDUSTRIES

Mechanization and specialization increases, declining use of labor, and closer links with the input and output industries have characterized U.S. agriculture since World War II. Agricultural productivity measured as output per unit of labor has surpassed that of the nonfarm business sector for more than a decade (Figure 1-8). Adjusted for inflation, inputs purchased to produce farm output have increased from approximately $50 billion in the early 1960s to over $80 billion in the early 1980s. At no other time in U.S. history have agricultural products generated more income after they leave the farm. During the same period, economic activity in these industries rose from approximately $235 billion to about $450 billion (U.S. Department of Agriculture, 1986e).

Twenty-one million people were employed in the food and fiber economy in 1985, down from 24.5 million people in 1947 (Figure 1-9). But as a percentage of the total work force, 41 percent in 1947 were employed in the food and fiber industry compared to 18.5 percent in 1985. Increases in employment in other sectors of the economy were largely responsible for this drop. The percentage of those in the food and fiber sectors working off the farm increased from about 60 percent in 1947 to nearly 90 percent in 1985. During the same period, the size of the work force involved in farming fell from about 17 percent, or 10 million workers, to about 2 percent, or 2.5 million workers (U.S. Department of Agriculture, 1987g).

The number of farmers has declined while the total U.S. population has increased from 151.3 million in 1950 to 226.5 million in 1980. The population of employed workers increased from 56.2 million in 1950 to 97.6 million in

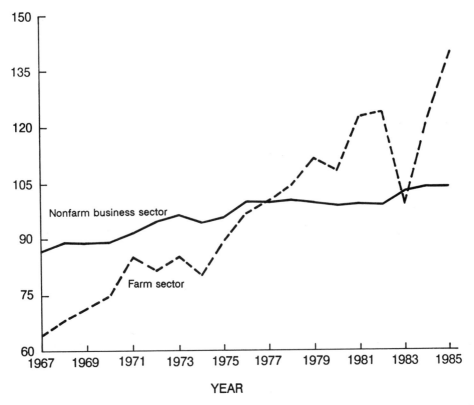

FIGURE 1-8 Agricultural productivity measured by output per unit of labor. SOURCE: U.S. Department of Agriculture. 1987. National Food Review. The U.S. Food System—From Production to Consumption. NFR-37. Economic Research Service. Washington, D.C.

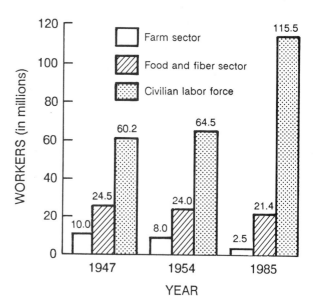

FIGURE 1-9 Distribution of food and fiber system employment in the national economy. SOURCE: U.S. Department of Agriculture. 1987. National Food Review. The U.S. Food System— From Production to Consumption. NFR-37. Economic Research Service. Washington, D.C.

1980. In contrast, farmers accounted for 6.9 million of all employed workers (or 12.2 percent) in 1950, and only 2.3 million employed workers (or 2.8 percent) in 1986 (U.S. Department of Agriculture, 1987c). Using about the same amount of cropland, fewer farmers are feeding an ever-growing population (Figure 1-10). This has been made possible by great increases in per acre yields resulting from the development and widespread adoption of fertilizers and synthetic chemical pesticides, improvements in machinery, and high-yielding varieties of major grain crops. Average yields have increased 2 percent per acre annually since 1948 (U.S. Department of Agriculture, 1986b). Average yields per acre of corn, soybeans, and wheat increased from 38.2, 21.7, and 16.5 bushels per acre in 1930 to 118, 34.1, and 37.5 bushels per acre in 1985, respectively. Cotton yields increased from 269 pounds per acre in 1950 to 630 pounds per acre in 1985 (U.S. Department of Agriculture, 1972, 1986d, 1987g). Average annual milk production per cow increased from 5,314 pounds in 1950 to 13,786 pounds in 1987 (California Department of Food and Agriculture, 1958, 1972, 1987). Poultry production rose from about 5 million birds in 1960 to nearly 20 million birds in 1987 (U.S. Department of Agriculture, 1989).

These great increases in yield and production have helped keep the price of food in the United States low as a percentage of per capita income. Americans spend only about 15 percent of their total personal disposable income on food. This figure is down from about 16.5 percent 10 years ago, largely because of the relatively rapid rise in personal income. The percentage of income spent on food varies greatly with income. Families with

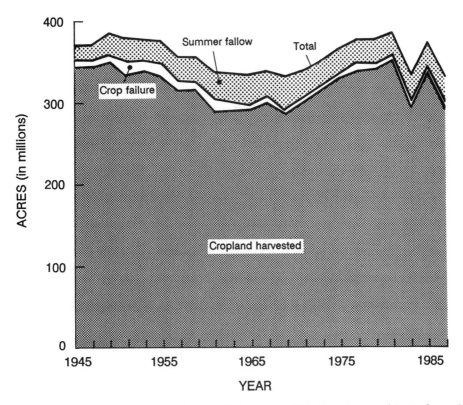

FIGURE 1-10   Cropland harvested since 1945. SOURCE: U.S. Department of Agriculture. 1988. 1988 Agricultural Chartbook. Agriculture Handbook No. 673. Washington, D.C.

before-tax annual incomes of less than $5,000 spend 49.7 percent of those incomes on food; families with incomes greater than $40,000 spend 8.7 percent (U.S. Department of Agriculture, 1986g, 1987g). Western Europeans, in contrast, spent an average of 23.8 percent of household disposable income on food in 1983. Families in many less-developed countries spend well over 50 percent (U.S. Department of Agriculture, 1986h). Since 1980, the consumer price index (CPI) for food has risen more slowly than the CPI for all other items (U.S. Department of Agriculture, 1986g) (Figure 1-11).

A decreasing amount of the total spent on food reaches farmers (Figure 1-12). This is a result of two factors: (1) the increased consumption of prepackaged foods and corresponding costs for processing, packaging, marketing, and retailing and (2) the increasing percentage of meals consumed away from home. In 1987, consumers spent about $380 billion for foods produced on farms in the United States (Figure 1-13). Preliminary 1987 data show that farmers received about $90 billion or 25 percent of the $380 billion spent on *all food*—the rest went to the food industry (Figure 1-14). As consumers spend more on food, marketers and processors have gained significant revenue. The financial returns to the farmer have remained roughly constant, but represent a shrinking piece of a growing pie. Food marketing

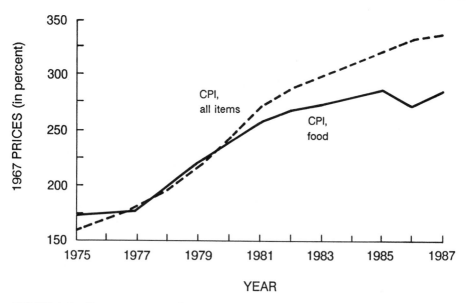

FIGURE 1-11   Consumer prices for food versus all other consumer goods. SOURCE: U.S. Department of Agriculture. 1986. Food Cost Review, 1985. Agricultural Economic Report No. 559. Economic Research Service. Washington, D.C.

FIGURE 1-12   Index of retail price for a market basket of farm foods and the value received by farmers. SOURCE: U.S. Department of Agriculture. 1986. Food Cost Review, 1985. Agricultural Economic Report No. 559. Economic Research Service. Washington, D.C.

direct labor costs represented about 50 percent of the total $260 billion accounted for by the food industry (U.S. Department of Agriculture, 1986g).

The value received at the farm for food sold in supermarkets and grocery stores declined to about 31 percent of this total, down from 34 percent in 1984 and 37 percent in 1980. Farmers receive only 16 percent of the total value of food consumed away from home. Expenditures on food consumed away from home increased from $84.3 billion in 1980 to $133.3 billion in 1986 (U.S. Department of Agriculture, 1986g).

Inputs

The scientific and technological revolution in agriculture began after World War I and accelerated after World War II. The first step in this process was the replacement of draft animals and human labor with tractors and other machinery. This conversion was virtually complete by 1960 and has continued with the introduction of larger, faster, and more powerful equipment. The power of an average tractor increased from 35 horsepower in 1963 to 60 horsepower in 1983 (U.S. Department of Agriculture, 1986f). Since 1940, the labor required to farm an acre has declined 75 percent while farm output per acre has doubled. As a result, farm labor is 8 times as productive. Farming employed 17 percent of the labor force in 1940 and about 2 percent in 1986. In contrast, employment in farm input industries such as agrochemical production, transportation, food processing, and machinery manufacturing has grown significantly during this period.

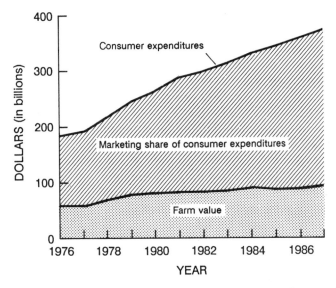

FIGURE 1-13 Marketing share of consumer expenditures, farm value, and consumer expenditures for farm foods. SOURCE: U.S. Department of Agriculture. 1988. 1988 Agricultural Chartbook. Agriculture Handbook No. 673. Washington, D.C.

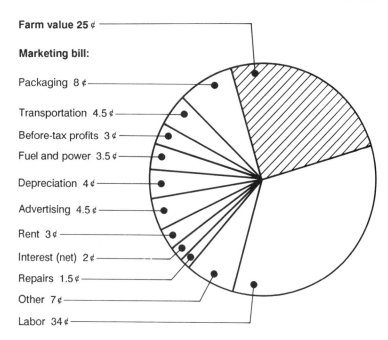

Farm value 25 ¢

Marketing bill:

Packaging 8 ¢

Transportation 4.5 ¢

Before-tax profits 3 ¢

Fuel and power 3.5 ¢

Depreciation 4 ¢

Advertising 4.5 ¢

Rent 3 ¢

Interest (net) 2 ¢

Repairs 1.5 ¢

Other 7 ¢

Labor 34 ¢

FIGURE 1-14   What a dollar spent on food paid for in 1987. SOURCE: U.S. Department of Agriculture. 1988. 1988 Agricultural Chartbook. Agriculture Handbook No. 673. Washington, D.C.

Fertilizers and pesticides currently account for a far greater share of input costs for most major crops than they did 30 years ago. This is primarily the result of widespread adoption of high-yielding seeds that are more responsive to fertilizer applications and continuous cropping that has created favorable pest habitats in certain crops. A number of federal programs and policies have encouraged the use of these seed varieties, specialized cropping practices, and fertilizer inputs. For many major commodities, fertilizer and pesticide costs far exceed other variable costs such as seeds and fuel (Figure 1-15). The national average cost of fertilizers and pesticides for corn production in 1986 was about 55 percent of variable costs and 34 percent of total costs. For soybeans, the figures were 49 and 25 percent and for wheat, 40 and 23 percent. The increasing use of these inputs has been associated with significant yield increases in major crops. For alternative systems to be successful and widely adopted, they must not result in significant overall reductions in yield or profits. The feasibility of alternative systems is discussed in Chapters 3 and 4 and in the case studies.

FIGURE 1-15   (at right) National average cost of pesticides and fertilizers, seed, and fuel as a percentage of total variable costs and total variable and fixed costs by major crop, 1986. SOURCE: U.S. Department of Agriculture. 1987. Economic Indicators of the Farm Sector: Costs of Production, 1986. ECIFS 6-1. Washington, D.C.

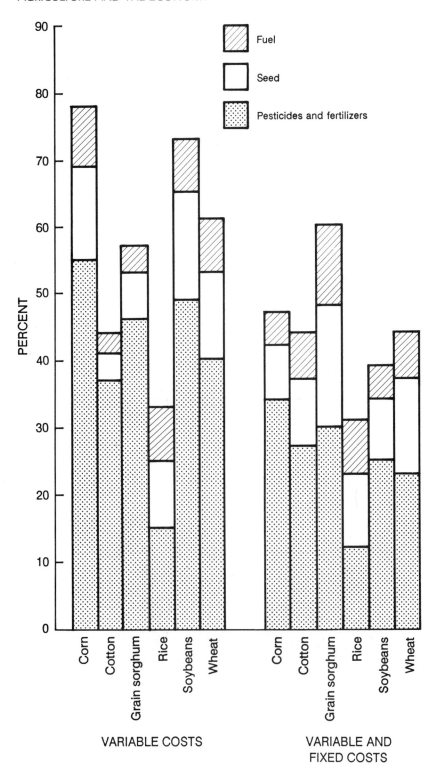

Fertilizers

Application of the three principal plant nutrients—nitrogen, phosphorus, and potassium—has increased steadily over the past 40 years. The biggest change has been in the use of nitrogen. Previously, farmers provided most nitrogen to their crops by rotating corn and small grains with leguminous crops. But as they shifted to growing nitrogen-responsive varieties of corn continuously or in short rotations with soybeans the demand for nitrogen has increased. Corn alone accounts for 44 percent of all directly applied fertilizers in agriculture, while wheat, cotton, and soybeans receive 18 percent combined (U.S. Department of Agriculture, 1987d).

The use of nitrogen, phosphorus, and potassium was roughly equivalent in 1960 at 2.7, 2.5, and 2.1 million tons, respectively (Figure 1-16). By the peak of annual fertilizer use in 1981 their respective totals were 11.9, 5.4, and 6.3 million tons. Since 1981, the total use of fertilizer has declined to slightly more than 19 million tons. This reduction primarily reflects the large number of acres currently held out of production by government programs. Nitrogen fertilizer use on a per acre basis continues to rise or remain steady for most crops (Figure 1-17).

The increased use of nitrogen fertilizer since 1964 has largely been in corn and wheat (Figure 1-18). These two crops accounted for 35 percent of all fertilizer use in 1964 and 54 percent in 1985. Fertilizer prices are sensitive to demand. They increased in the 1970s, peaked in 1981, declined through 1987, and began to increase in 1988. Total net fertilizer sales rose from $1.6 billion in 1970 to $8.6 billion in 1981, falling to $6.4 billion in 1985 (U.S. Department of Agriculture, 1987b).

Anhydrous ammonia is often applied as a source of nitrogen. The material is stored in gaseous form in pressurized tanks. It is usually released into the soil in 6- to 8-inch deep trenches formed by chisel-type blades. Being heavier than air, the gas is trapped in the soil and remains there, available for plant use. *Credit:* Soil Conservation Service, U.S. Department of Agriculture.

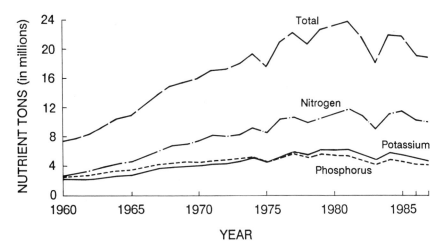

FIGURE 1-16   Total consumption of primary plant nutrients. SOURCE: U.S. Department of Agriculture. 1987. Fertilizer Use and Price Statistics, 1960–85. Statistical Bulletin No. 750. Economic Research Service. Washington, D.C.

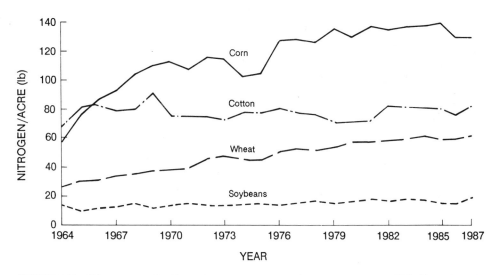

FIGURE 1-17   Nitrogen application rates per acre on major crops. SOURCE: U.S. Department of Agriculture. 1987. Fertilizer Use and Price Statistics, 1960–85. Statistical Bulletin No. 750. Economic Research Service. Washington, D.C.

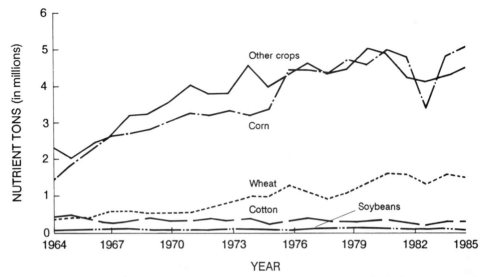

FIGURE 1-18   Total use of nitrogen fertilizer on major crops. SOURCE: U.S. Department of Agriculture. 1987. Fertilizer Use and Price Statistics, 1960–85. Statistical Bulletin No. 750. Economic Research Service. Washington, D.C.

The increased use of commercial fertilizer during the last four decades has helped to increase dramatically the per acre yields of agronomic and horticultural crops. Fertilizers make possible continuous production of major crops such as corn and wheat, decrease dependence on animal manures and leguminous nitrogen for row-crop production, and facilitate the substitution of capital for relatively more expensive inputs such as labor and land, consequently reducing labor and land requirements to produce a unit of a crop. Heavy use of most nitrogen and some other fertilizers, however, can lead to soil acidification, other changes in soil properties, and offsite environmental problems. Fertilizers are often overapplied. When this happens, the total amount of plant nutrients available to growing crops not only exceeds the need or ability of the plant to absorb them but exceeds the economic optimum as well. Estimates of crop absorption of applied nitrogen range from 25 to 70 percent and generally vary as a function of plant growth and health and the method and timing of nitrogen application. Crops are much more likely to make fuller use of properly timed applications of nitrogen. Unused nitrogen can be immobilized, denitrified, washed into streams or lakes, or leached from the soil into underground water and the subsoil (Johnson and Wittwer, 1984; Legg and Meisinger, 1982).

Pesticides

The use of synthetic organic pesticides such as dichloro diphenyl trichloroethane (DDT), benzene hexachloride (BHC), and (2,4-Dichlorophenoxy) acetic acid (2,4-D) began with great expectations in the 1940s. For the first

About two-thirds of all insecticides and fungicides are applied aerially; most herbicides in row crops are applied by spray rigs pulled by tractors. Citrus groves, such as the one shown above, may be aerially treated 10 to 20 times per season with insecticides, fungicides, and protectant oils. Helicopters are often used because the turbulence from the main rotor tends to push the pesticides down toward the crop. Fixed-wing aircraft are more commonly used in field crops such as wheat and cotton. Tractor spray rigs (bottom) are often used to apply herbicides in row crops because planting, fertilizing, and spraying can be accomplished in one pass through the field. *Credits:* U.S. Department of Agriculture (top); John Colwell from Grant Heilman (bottom).

time, satisfactory control of agricultural pests seemed possible. Substituting lower-priced chemicals for higher-priced, labor-intensive weed and insect control methods and pest-reducing practices such as rotations immediately reduced labor needs and increased the effectiveness of control and yields. Ultimately, pesticides reinforced agricultural trends such as increasing farm size and decreasing diversification.

The total pounds of pesticide active ingredients applied on farms increased 170 percent between 1964 and 1982, while total acres under cultivation remained relatively constant. Herbicide use led the way, from 210 million pounds in 1971 to a peak of 455 million pounds in 1982 (U.S. Department of Agriculture, 1984) (Figure 1-19). In 1985, 95 percent of the corn and soybean acreage was treated with herbicides, compared to about 40 percent in 1970. As a percentage of total pesticide pounds applied, herbicides rose from 33 percent in 1966 to 90 percent in 1986 (U.S. Department of Agriculture, 1970, 1986c; U.S. Environmental Protection Agency, 1986b). During that time, insecticide use declined while fungicide use held steady. Total pesticide use has declined from more than 500 million pounds of active ingredients in 1982 to about 430 million pounds in 1987. Land idled from production and the introduction of newer products that are applied at a lower rate per acre are largely responsible for this decline (U.S. Department of Agriculture, 1988a).

The total dollar value of the domestic agricultural pesticide market is about $4.0 billion. Herbicides represent the largest share of the market at about $2.5 billion, followed by insecticides at about $1.0 billion, and fungicides at about $265 million (National Agricultural Chemicals Association, 1987). Because of the size of the herbicide market and increased understanding of plant physiology and biochemistry, herbicides are the most dynamic sector of the pesticide industry. A number of new active ingredients were introduced in 1986 and 1987 (Figure 1-20). As a result, the herbicide market is currently highly competitive, particularly for use on corn, soybeans, wheat, and cotton.

The greatest volume of pesticides is applied to field crops. About 90 percent of all herbicides and insecticides are applied to just four crops: corn, cotton, soybeans, and wheat (U.S. Department of Agriculture, 1986c) (Figures 1-21 and 1-22). In 1986, corn alone accounted for 55 percent of all herbicides and 44 percent of all insecticides used on field crops. Corn replaced cotton as the leader in insecticide use in 1982. Nearly half the total pounds of pesticides applied in the nation are used in corn production (U.S. Department of Agriculture, 1986c). Soybeans receive about 25 percent of all herbicides, and cotton about 25 percent of all insecticides. Insecticides are also widely used in alfalfa, tree fruit, nut, and vegetable production.

Fungicides are primarily used as a seed treatment and to protect fruits and vegetables during production and after harvest. Fungicide use in peanuts and wheat declined between 1976 and 1986, primarily as a result of improved varieties and integrated pest management (IPM) systems, although total fungicide use remained steady (U.S. Department of Agricul-

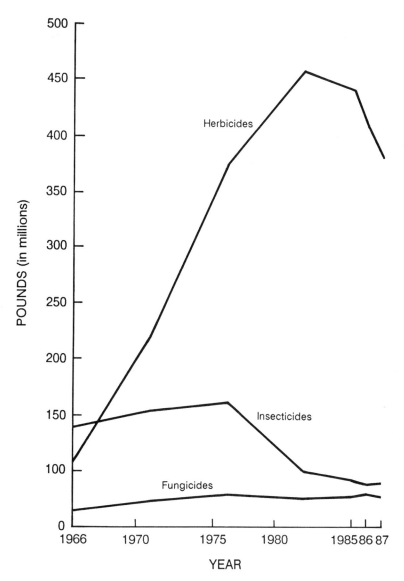

FIGURE 1-19 Herbicide, insecticide, and fungicide use estimates. SOURCES: U.S. Department of Agriculture. 1974. Farmers' Use of Pesticides in 1971. Agricultural Economic Report No. 252. Economic Research Service. Washington, D.C.; U.S. Department of Agriculture. 1978. Farmers' Use of Pesticides in 1976. Agricultural Economic Report No. 418. Economic Research Service. Washington, D.C.; U.S. Department of Agriculture. 1983. Inputs—Outlook and Situation Report. IOS-2. Economic Research Service. Washington, D.C.; U.S. Department of Agriculture. 1984. Inputs—Outlook and Situation Report. IOS-6. Economic Research Service. Washington, D.C.; U.S. Department of Agriculture. 1987. Agricultural Resources—Inputs—Situation and Outlook Report. AR-5. Economic Research Service. Washington, D.C.; U.S. Department of Agriculture. 1988. Agricultural Resources—Inputs—Situation and Outlook Report. AR-9. Economic Research Service. Washington, D.C.

ture, 1987b). Fungicides now account for less than 10 percent of all pesticides applied in agriculture.

The introduction and acceptance of new fungicides has been relatively
slow compared with those of herbicides and insecticides (see Figure 1-20).
Only five fungicides introduced since 1975 have gained more than 5 percent
of the market for any major food crop. Target pests tend to develop resistance to most new, more specific fungicides. Consequently, these fungicides
work best in combination with older broad-spectrum substances. This is
particularly true in humid areas with great pest pressure such as the South
and East. In these regions, current methods of production would not be
possible without chemical fungicides.

The adoption of IPM strategies and the increased use of synthetic pyrethroid insecticides, which are applied at about one-tenth to one-fourth the
rate of traditional insecticides, have significantly reduced the total pounds
of insecticides applied. This decreased use is particularly true for certain
crops. Reductions in insecticide use are mainly derived from IPM programs

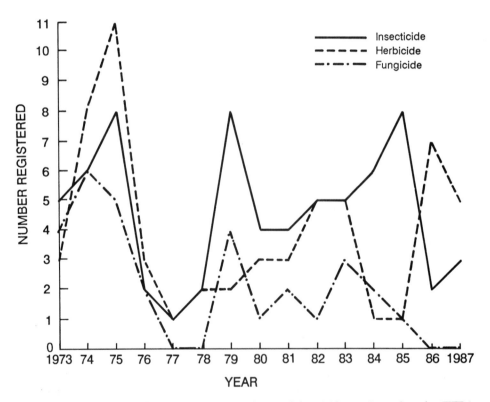

FIGURE 1-20   Number of herbicides, insecticides, and fungicides registered under FIFRA,
1973–1987. SOURCE: U.S. Environmental Protection Agency. 1988. Chemicals Registered for
the First Time as Pesticidal Active Ingredients under FIFRA (including 2 (C)(7)(A)
Registrations). Washington, D.C.

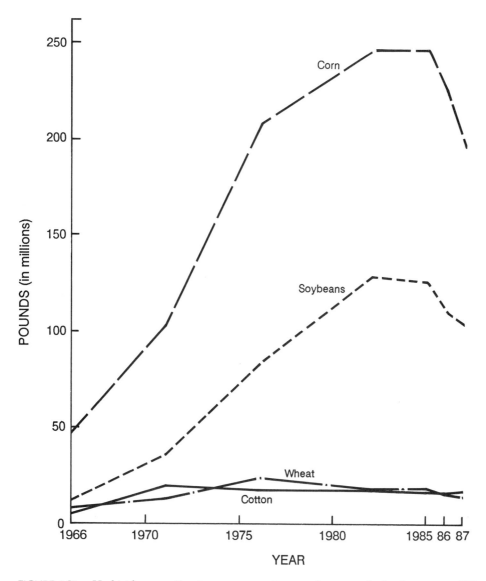

FIGURE 1-21 Herbicide use estimates on corn, cotton, soybeans, and wheat. SOURCES: U.S. Department of Agriculture. 1974. Farmers' Use of Pesticides in 1971. Agricultural Economic Report No. 252. Economic Research Service. Washington, D.C.; U.S. Department of Agriculture. 1978. Farmers' Use of Pesticides in 1976. Agricultural Economic Report No. 418. Economic Research Service. Washington, D.C.; U.S. Department of Agriculture. 1983. Inputs—Outlook and Situation Report. IOS-2. Economic Research Service. Washington, D.C.; U.S. Department of Agriculture. 1984. Inputs—Outlook and Situation Report. IOS-6. Economic Research Service. Washington, D.C.; U.S. Department of Agriculture. 1987. Agricultural Resources—Inputs—Situation and Outlook Report. AR-5. Economic Research Service. Washington, D.C.; U.S. Department of Agriculture. 1988. Agricultural Resources—Inputs—Situation and Outlook Report. AR-9. Economic Research Service. Washington, D.C.

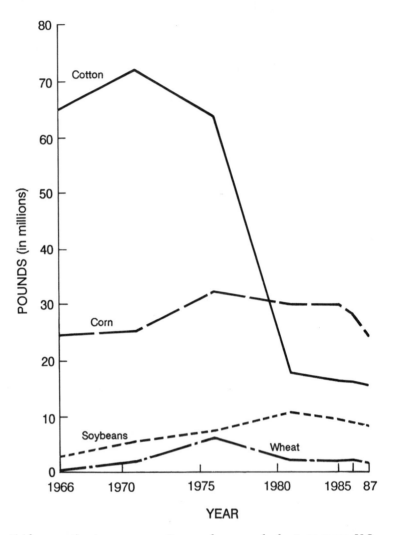

FIGURE 1-22   Insecticide use estimates on corn, cotton, soybeans, and wheat. SOURCES: U.S. Department of Agriculture. 1974. Farmers' Use of Pesticides in 1971. Agricultural Economic Report No. 252. Economic Research Service. Washington, D.C.; U.S. Department of Agriculture. 1978. Farmers' Use of Pesticides in 1976. Agricultural Economic Report No. 418. Economic Research Service. Washington, D.C.; U.S. Department of Agriculture. 1983. Inputs—Outlook and Situation Report. IOS-2. Economic Research Service. Washington, D.C.; U.S. Department of Agriculture. 1984. Inputs—Outlook and Situation Report. IOS-6. Economic Research Service. Washington, D.C.; U.S. Department of Agriculture. 1987. Agricultural Resources—Inputs—Situation and Outlook Report. AR-5. Economic Research Service. Washington, D.C.; U.S. Department of Agriculture. 1988. Agricultural Resources—Inputs—Situation and Outlook Report. AR-9. Economic Research Service. Washington, D.C.

in cotton, alfalfa, peanuts, and apples. Largely because of the success of IPM in cotton, insecticide applications to field and forage crops declined 45 percent between 1976 and 1982.

Antibiotics

Livestock and poultry producers have used antibiotics in animal production for the past 35 years. Antibiotic use in agriculture increased from 440,000 pounds in 1953 to 9.9 million pounds in 1985 (Figure 1-23). The most common agricultural use of these drugs is their subtherapeutic incorporation into animal feed. Use of antibiotics in animal feed improves the feed efficiency and growth rate of livestock. Approximately 80 percent of the poultry, 75 percent of the swine, 60 percent of the beef cattle, and 75 percent of the dairy calves raised in the United States have been fed antibiotics at some time in their lives. About 36 percent of the antibiotics produced in the United States each year are fed or administered to animals (Hays et al., 1986; Institute of Medicine, 1989).

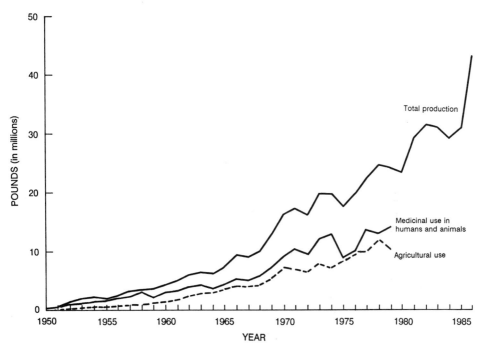

FIGURE 1-23   U.S. antibiotic production and use in animal feed. Only total production was recorded after 1979. SOURCES: U.S. International Trade Commission. 1987. Synthetic Organic Chemicals: United States Production and Sales. Washington, D.C.; National Research Council. 1980. The Effects on Human Health of Subtherapeutic Use of Antimicrobials in Animal Feeds. Washington, D.C.: National Academy Press.

The routine feeding of antibiotics to control disease has facilitated special-ization, the use of feedlots and confinement facilities, and the concentration of many animals under one manager in a small area (Council for Agricul-tural Science and Technology, 1981). While there is disagreement on the necessity of feeding antibiotics as a simple function of confinement, there is ample documentation that control of diseases more prevalent in close confinement facilities will increase animal performance (Curtis, 1983). The routine use of feed antimicrobials in confined animal units is a common practice in most regions.

Concerns that feeding animals subtherapeutic doses of antibiotics could lead to antibiotic-resistant bacteria led the U.S. Food and Drug Administra-tion (FDA) in 1977 to propose regulations revoking nearly all approved subtherapeutic uses of penicillin and the most common forms of tetracy-cline. Final action remains delayed pending the results and analysis of additional research requested by the Congress (Hays et al., 1986). Results to date show that food animals appear to be the largest single source of resistant salmonellae, although documented incidence of the development of resistant strains and their transmission to humans is rare. In the interim, subtherapeutic feeding of antibiotics has increased (see Figure 1-23). In Europe and Japan, concerns about overuse and misuse of antibiotics and the potential for bacterial resistance have led to restrictions limiting antibi-otics to use by veterinarians or by prescription. Proposals in the United States to limit drug use to the discretion of veterinarians have met with opposition from livestock producers and drug manufacturers.

Irrigation

Agriculture accounts for 85 percent of all consumptive use of water, which is use that makes water unavailable for immediate reuse because of evapo-ration, transpiration, incorporation into crops or animals, or return to groundwater or surface water sources. Ninety-four percent of agricultural water is used for irrigation, 2 percent for domestic use, and about 4 percent for livestock. On average, agricultural irrigation used about 138 billion gal-lons of water per day in 1985. During the growing season this level can exceed 500 billion gallons per day (U.S. Department of Agriculture, 1987a). From 1950 to 1978, 25 million additional acres came under irrigation (U.S. Department of Agriculture, 1986b) (Figure 1-24). Total irrigated acreage peaked at slightly more than 50 million acres in 1978, declining to just under 45 million acres in 1983. Since then, total irrigated acreage has remained steady, although the composition has changed slightly; irrigated acreage in the West has declined, while irrigated acreage in the East has increased. Ninety-four percent of all irrigated acres are in 17 western and 3 southeast-ern states. Total irrigation water withdrawals (the amount of water used for irrigation) declined in 1985 for the first time since 1950 (U.S. Department of Agriculture, 1987a).

Irrigation in some areas makes farming possible; in others it supplements

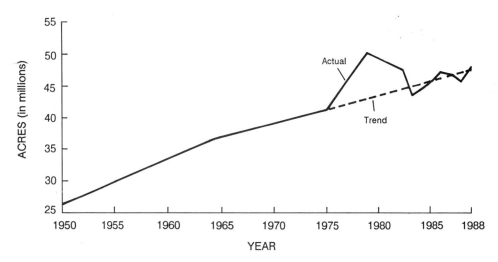

FIGURE 1-24 Irrigated agricultural land. SOURCES: U.S. Department of Agriculture. 1988. 1988 Agricultural Chartbook. Agriculture Handbook No. 673. Washington, D.C.; U.S. Department of Agriculture. 1986. Agricultural Resources—Cropland, Water, and Conservation—Situation and Outlook Report. AR-4. Economic Research Service. Washington, D.C.

rainfall. In all cases the use of irrigation results in higher and more consistent yields than regional and national average nonirrigated yields (Table 1-1). In many cases the results are dramatic (U.S. Department of Agriculture, 1986b). The 13 percent of cropland that is irrigated accounts for more than 30 percent of the value of crops produced. The high unit value of many crops produced with irrigation, as well as the high yields for irrigated grain crops, are responsible for this disparity.

The use of groundwater for irrigation increased 160 percent from 1945 to 1980. Surface water used for irrigation increased 50 percent during that time. The rapid expansion of irrigated acreage during the 1970s relied almost exclusively on the pumping of groundwater. Center pivot irrigation systems, which rely on groundwater, accounted for the largest increase of irrigated acreage of any irrigation system. These systems were used on 3.4 million acres in 1974 and 9.2 million acres in 1983. Nebraska experienced a 1 million acre increase in center pivot irrigation between 1974 and 1983 and currently accounts for 51 percent of all irrigated corn acreage (U.S. Department of Agriculture, 1985c, 1987h). Increased pumping costs due to overdrawing of aquifers and increased energy costs associated with deregulation of natural gas used for pumping in Texas and Oklahoma have caused the recent decline in irrigated acres. Energy costs for on-farm pumping of groundwater rose 352 percent between 1974 and 1983 (U.S. Department of Agriculture, 1985c).

TABLE 1-1   Average Dryland and Irrigated Yields

Crop	Dryland Yields[a] (per acre)	Irrigated Yields[b] (per acre)	Ratio of Irrigated to Dryland Yields
Corn for grain (bushels)	106.0	137.0	1.29
Wheat (bushels)	32.0	69.0	2.16
Sorghum for grain (bushels)	54.0	93.0	1.72
Barley (bushels)	48.0	81.0	1.69
Cotton (bales)	0.9	1.7	1.89
Soybeans (bushels)	31.0	36.0	1.16
Potatoes (hundredweight)	83.0	333.0	4.01

[a]In the contiguous United States, 1982.
[b]From 20 principal irrigated states with 95 percent of all irrigated acres, 1984.

SOURCES: U.S. Department of Agriculture. 1986. Agricultural Resources—Cropland, Water, and Conservation—Situation and Outlook Report. AR-4. Economic Research Service. Washington, D.C.; U.S. Department of Agriculture. 1987. U.S. Irrigation—Extent and Economic Importance. Agriculture Information Bulletin No. 523. Economic Research Service. Washington, D.C.

By 1984, irrigators obtained roughly equal amounts of water from underground and surface sources. About 44 percent of all irrigation water is from on-farm groundwater pumping, about 12 percent is from on-farm surface water supplies, and about 44 percent is from off-farm suppliers such as irrigation districts and private water companies (U.S. Department of Agriculture, 1986b). More than 85 percent of the additional irrigated acres in the past 30 years has been on land not served by the Bureau of Reclamation. The principal factors behind the increase in irrigated acres over the past two decades have been private investment stimulated by federal policies, which have included high commodity support prices, tax incentives that include investment credits and accelerated depreciation for equipment, water depletion allowances, and cheap credit. During the 1970s, about 80 percent of irrigation investment involved private funds.

Most of the recent increases in irrigation have come in crops supported by the Commodity Credit Corporation (CCC). Irrigation of these crops, primarily corn and wheat, increased by more than 8 million acres in the Great Plains between 1954 and 1982. This had a great impact on the production of these crops because irrigation generally boosts yields from 40 to 100 percent over similar nonirrigated acreage (U.S. Department of Agriculture, 1986b) (see Table 1-1). Corn and wheat have developed large surpluses in recent years. Increased productivity on irrigated lands has significantly contributed to these surpluses.

Federal efforts to reduce production are often hampered by programs or policies that encourage irrigation and its resulting high per acre yields. Between 1976 and 1985, an average of 3.7 million acres served by the Bureau of Reclamation were producing crops already in surplus. In 1986, growers producing surplus crops on the land received more than $200 million in federally subsidized water, in addition to federal income and price supports (U.S. Congress, 1987).

About 12 percent of all corn and nearly 7 percent of all wheat acres are irrigated (Table 1-2). The yield on irrigated corn acres is about 29 percent greater than national average dryland yields; for wheat, the yield is 116 percent greater. The 29 percent increase in yield on the 9.6 million irrigated acres of corn produced 298 million additional bushels of corn compared with national average yields on the same acres. Irrigated wheat acres produced nearly 170 million bushels over the average yield on the 4.6 million irrigated acres. The difference between irrigated and nonirrigated yields in regions where irrigation is common is far greater than this. Thus, the increase in production over actual nonirrigated corn and wheat production in these regions is likely to be higher than the 298 and 170 million bushel

TABLE 1-2 Harvested Irrigated Cropland and Pastureland,[a] 1982

Type of Land	Irrigated Acreage (in thousands)	Share (percent) of Crop Irrigated	Share (percent) of Total Irrigated Acres
Cropland[b]			
Corn	9,604	12.3	19.3
Sorghum	2,295	17.0	4.6
Wheat	4,650	6.6	9.3
Barley and oats	2,098	11.8	4.2
Rice	3,233	100.0	6.5
Cotton	3,424	35.0	6.9
Soybeans	2,321	3.6	4.7
Irish potatoes	812	64.0	1.6
Hay	8,507	15.0	17.1
Vegetables and melons	2,024	60.7	4.1
Orchard crops	3,343	70.4	6.7
Sugar beets	550	53.2	1.1
Other[c]	2,428	17.9	4.9
Subtotal[d]	45,289	13.4	91.0
Pastureland	4,499	0.9	9.0
Total[d]	49,788	6.1	100.0

[a]In the contiguous United States.

[b]Cropland is land on farms used for crops.

[c]Includes peanuts; dry tobacco; edible beans; and the minor acreage crops of rye, flax, sunflower, sugarcane, and dry edible peas.

[d]Figures may not add due to rounding. Irrigated cropland total includes 932,000 acres of double-cropped land.

SOURCE: U.S. Department of Agriculture. 1987. U.S. Irrigation—Extent and Economic Importance. Agriculture Information Bulletin No. 523. Economic Research Service. Washington, D.C.

figures. In many cases, these acres would not be planted without irrigation. Irrigation, in turn, often would not be profitable without government tax policies, low-cost credit, and high price supports and income support payments for these crops.

Irrigation is expected to expand in the East and other areas to supplement rainfall, increase yields, and reduce yield variability. In the arid West and Great Plains, however, irrigation will probably stabilize or decline for the remainder of the century because of the cost of water projects and competition with urban users for supply. Improved management and conservation practices will probably sustain irrigated agriculture in these areas. But declining commodity prices, changes in the tax code, and the rising demand for other uses of water in arid areas will curtail new investment in irrigated agriculture in regions where irrigated agriculture flourished in the past.

THE STRUCTURE OF AGRICULTURE

The total number of farms, which are defined as places with actual or potential sales of agricultural products of $1,000 or more, declined from 5.9 million in 1945 to slightly more than 2.2 million in 1985 (U.S. Department of Agriculture, 1987c). It is noteworthy, however, that even in the farm recession of the mid-1980s, the decline in the number of farms between 1980 and 1986 (a loss of 220,000 farms, or 11 percent of all farms) was far less than that which occurred in the 1950s (1.6 million farms, or 28 percent of all farms) or the 1960s (960,000 farms, or 24 percent of all farms) (U.S. Department of Agriculture, 1987c). Total harvested acres have remained relatively constant at approximately 340 million acres, indicating that average farm size has almost tripled. Although individuals and their families operate most farms, the growth in average farm size has been largely at the expense of the small farm with a full-time operator. Fifteen to 20 percent of all farms produce more than 80 percent of all output. Three-quarters of all farm households generate off-farm income.

Increases in farm size have been accompanied and made possible by increased specialization and substitution of purchased inputs for labor and land. At the end of World War II, most farms in the Midwest, Great Lakes, Northeast, and parts of the South were diversified crop-livestock operations. High-density animal confinement was rare. Most farmers produced forage and feed grains for their animals, which required longer crop rotations and less use of some purchased inputs, particularly fertilizers. Most farmers returned animal manure to the land. Far fewer insecticides and almost no herbicides were used. Pests were controlled through rotations, cultivation, and a variety of cultural and biological means.

Cattle and hog production clearly illustrate the relationship among increasing specialization; changing distribution of farm income among large and small producers and regions; and changing cropping patterns, farm size, and management techniques. Specialization and the increased use of feedlots and confinement rearing have made cattle rearing possible

Most beef cattle are fed or finished in feed-lots, where they eat high-energy grain ra-tions. Cattle enter feedlots between the ages of 7 and 12 months, weighing between 450 and 800 pounds. Most cattle are slaughtered between the ages of 15 and 24 months, weighing between 1,000 and 1,200 pounds. Low-cost feed grains and consumer prefer-ence for marbled (higher fat) grain-fed beef led to the proliferation of feedlots in the 1960s and 1970s. Large feedlots offer economies of scale but may also create problems with dis-ease control and manure disposal. *Credit:* Grant Heilman.

throughout the nation. Beef cattle and hog production have become concen-trated in large enterprises. Although the Corn Belt accounted for almost 50 percent of cattle feed in 1950, this fell to 22 percent in 1979. Meanwhile, the Central and Southern Plains and the Southwest increased production from 500 to 1,000 percent.

Part-time beef cattle operations with sales of $20,000 to $100,000 captured 56 percent of the market in 1969, while very large operations with sales greater than $500,000 had only 22 percent of the market (Figure 1-25). By 1982, the very large operations controlled 62 percent of the market; part-time operations accounted for only 12 percent (U.S. Office of Technology Assessment, 1986b). Hog production is showing the same trend. Part-time farm sales went from 61 percent in 1969 to 28 percent in 1982. The share of large and very large operators increased from about 5 percent to about 40

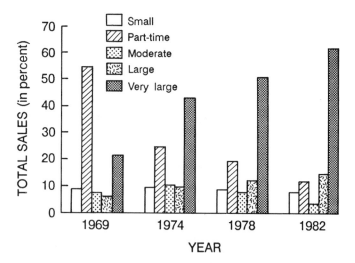

FIGURE 1-25   Percentage of cattle sales by size of operation (in 1969 dollars). Size of operation is determined by annual cattle sales: small, <$20,000; part-time, $20,000-99,999; moderate, $100,000-199,999; large, ≥$200,000-499,999; very large, ≥$500,000+. SOURCE: Office of Technology Assessment. 1986. Technology, Public Policy, and the Changing Structure of American Agriculture. Washington, D.C.

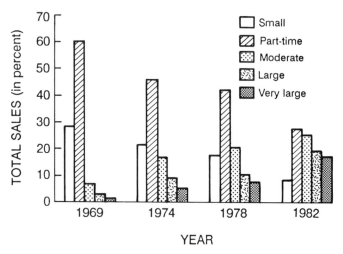

FIGURE 1-26   Percentage of hog and pig sales by size of operation (in 1969 dollars). Size of operation is determined by annual hog and pig sales: small, <$20,000; part-time, $20,000-99,999; moderate, $100,000-199,999; large, $200,000-499,999; very large, ≥$500,000+. SOURCE: Office of Technology Assessment. 1986. Technology, Public Policy, and the Changing Structure of American Agriculture. Washington, D.C.

percent of sales during that time (Figure 1-26). The dairy industry's small-farm market share declined from 66 to 41 percent between 1969 and 1982. The commercial broiler industry has moved almost entirely to vertical integration, with virtually all chickens going from egg to market without changing ownership. The egg and turkey industries are also moving in this direction.

REGIONAL DISTINCTIONS

The diversity of climatic, environmental, and economic conditions in the United States makes it essential to look beyond aggregate national agricultural trends and focus on the specifics of various regions. Needs and problems differ considerably among regions, and any effort to understand U.S. agriculture must address these differences. Types of farms and management also differ across the nation. The agricultural practices and needs of the coastal regions and southern parts of the country are quite different from those of the north and west. Dry, hot areas that depend on irrigation stand in contrast to cooler, more humid regions dependent on rainfall.

The U.S. Department of Agriculture (USDA) has identified nine representative agricultural regions based on data from the 1980 Census and the 1982 Census of Agriculture (Figure 1-27). Three basic criteria were used to identify these regions: the commodities produced and the resource base; the percentage of farms with sales between certain levels; and the degree of agricultural and nonagricultural economic integration. Although these regions omit large parts of the country, they illustrate important differences among principal agricultural areas. The USDA identified the following representative regions:

- California Metro
- Coastal Plains
- Core Corn Belt
- Delta
- Eastern Highlands
- Southeast Piedmont
- Western Corn Belt-Northern Plains
- Western Great Plains
- Wisconsin-Minnesota Dairy Area

Farm size and average sales per farm vary by region. In the Eastern Highlands, the average farm is 121 acres; on the Great Plains, the average farm is 2,334 acres (Table 1-3). Average annual farm sales are also quite different, ranging from $13,064 and $35,396, respectively, in the Eastern Highlands and the Southeast Piedmont, to $94,080 and $167,124, respectively, in the Western Great Plains and the California Metro. In the Core Corn Belt, the Wisconsin-Minnesota Dairy Area, and the Western Corn Belt-Northern Plains regions, farms reporting sales of $40,000 to $250,000 account for between 50 and 65 percent of farm sales (Table 1-4). In California, 12 percent of all farms with sales more than $250,000 account for 85 percent of farm sales. At the same time, however, in California, the greatest percentage of farms have sales between $1,000 and $9,999.

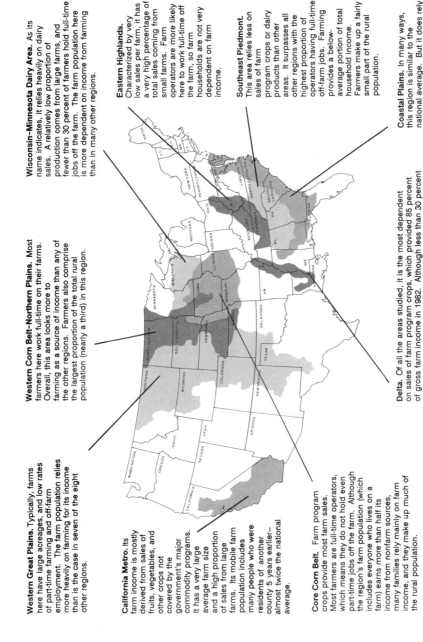

Western Great Plains. Typically, farms here have large acreages, and low rates of part-time farming and off-farm employment. The farm population relies more heavily on farming for its income than is the case in seven of the eight other regions.

Western Corn Belt–Northern Plains. Most farmers here work full-time on their farms. Overall, this area looks more to farming as a source of income than any of the other regions. Farmers also comprise the largest proportion of the total rural population (nearly a third) in this region.

Wisconsin-Minnesota Dairy Area. As its name indicates, it relies heavily on dairy sales. A relatively low proportion of production comes from large farms, and fewer than 30 percent of farmers hold full-time jobs off the farm. The farm population here is more dependent on income from farming than in many other regions.

Eastern Highlands. Characterized by very low sales per farm, it has a very high percentage of total sales coming from small farms. Farm operators are more likely here to work full-time off the farm, so farm households are not very dependent on farm income.

Southeast Piedmont. This area relies less on sales of farm program crops or dairy products than other areas. It surpasses all other regions with the highest proportion of operators having full-time off-farm jobs. Farming provides a below-average portion of total household income. Farmers make up a fairly small part of the rural population.

Coastal Plains. In many ways, this region is similar to the national average. But it does rely more heavily on sales of farm program crops and less on dairy sales. The percentage of its farmers working full-time off the farm is about average, but the area is less dependent on farm income than most of the other regions.

California Metro. Its farm income is mostly derived from sales of fruits, vegetables, and other crops not covered by the government's major commodity programs. It has a very large average farm size and a high proportion of sales from large farms. Its mobile farm population includes many people who were residents of another county 5 years earlier–almost twice the national average.

Core Corn Belt. Farm program crops provide most farm sales. Most farmers are full-time operators, which means they do not hold even part-time jobs off the farm. Although the region's farm population (which includes everyone who lives on a farm) earns more than half its income from nonfarm sources, many families rely mainly on farm income, and they make up much of the rural population.

Delta. Of all the areas studied, it is the most dependent on sales of farm program crops, which provided 85 percent of gross farm income in 1982. Although less than 30 percent of its farm operators work full-time off the farm, the percentage of the region's total farm population employed outside agriculture is about 54 percent, the national average.

FIGURE 1-27 Characteristics of nine farming regions. SOURCE: D. Martinez. 1987. Wanted: Policies to Cope with Differences in Farming Regions. Farmline 8(11):11–13.

TABLE 1-3 Diversity of U.S. Farming Regions

Region	Average Farm Size (acres)	Percentage of Farm Operators With Full-Time, Off-Farm Jobs	Average Sales per Farm (dollars)	Farm Income as Share of Total Income of Farm Population (percent)
Wisconsin-Minnesota Dairy Area	202	28	58,585	35.4
Core Corn Belt	294	28	73,944	37.4
Delta	481	29	87,042	29.5
Eastern Highlands	121	46	13,064	15.4
Western Great Plains	2,334	24	94,080	37.8
Western Corn Belt-Northern Plains	622	17	86,111	47.2
Coastal Plains	260	37	64,500	20.8
Southeast Piedmont	143	50	35,396	18.0
California Metro	362	41	167,124	25.2
United States	440	38	58,857	27.0

SOURCE: U.S. Department of Agriculture. 1988. Regional Characteristics of U.S. Farms and Farmers in the 1980's. ERS Staff Report No. AGES880128. Economic Research Service. Washington, D.C.

All regions depend significantly on income from major program crops (those for which federal price and income support programs exist) or livestock operations that rely on feed grains and oilseeds produced by crop farms (Table 1-5). Even California has a significant stake in the farm programs. Rice, cotton, wheat, corn, and dairy farmers in the state account for about 40 percent of gross agricultural income. Some regions depend on one program commodity, while others are more diversified. Only California shows diversity *and* moderate dependence on program crops.

The effect of commodity policy on regional economies, land use patterns, and farm structure is very different from region to region. These differences are accentuated by the diversity or interdependence of agricultural operations within regions. For example, the effect of a change in corn prices is quite different depending on whether a person is a corn producer, a hog farmer who is a corn consumer, or a farmer producing corn and hogs. The effect of farm policy on the overall regional economy is, in turn, a function of the importance of agriculture to the region and the nature and scope of agricultural input, food processing and marketing, and transportation industries.

In no region was the total farm income more than 50 percent of the total income of the farm population (see Table 1-3). The percentage was the highest in the Western Corn Belt-Northern Plains region at 47.2 percent, followed by the Western Great Plains at 37.8 percent, the Core Corn Belt at

TABLE 1-4 Farm Sales for Selected Agricultural Subregions, 1982 (in percent)

Item	Wisconsin-Minnesota Dairy Area	Core Corn Belt	Delta	Eastern Highlands	Western Great Plains	Western Corn Belt-Northern Plains	Coastal Plains	Southeast Piedmont	California Metro	United States
Farms with sales of:										
<$1,000	6.0	4.0	9.6	13.8	6.8	2.5	10.1	21.2	17.1	11.3
$1,000–9,999	24.9	21.7	31.0	62.7	22.6	14.4	36.1	51.2	32.9	37.7
$10,000–39,999	25.4	27.3	19.5	17.6	28.8	29.3	22.7	12.5	18.6	22.7
$40,000–99,999	27.1	24.9	15.4	3.8	23.5	30.6	14.2	5.6	10.9	14.9
$100,000–249,999	14.1	17.2	14.9	1.7	13.5	18.0	11.3	6.4	8.8	9.6
≥$250,000	2.6	4.9	9.6	0.5	4.9	5.3	5.8	3.1	11.7	3.9
Sales from farms with sales of:										
<$1,000	a	a	a	0.4	a	a	a	0.2	a	0.1
$1,000–9,999	1.9	1.4	1.5	20.0	1.1	0.8	2.3	5.1	0.8	2.6
$10,000–39,999	9.8	8.4	4.8	26.1	7.0	8.1	7.5	6.9	2.4	8.3
$40,000–99,999	30.8	22.2	11.8	18.5	16.0	23.2	14.4	10.5	4.2	16.5
$100,000–249,999	34.8	35.3	27.2	18.0	21.7	31.2	27.7	28.7	8.4	24.9
≥$250,000	22.7	32.8	54.7	17.0	54.2	36.6	48.1	48.7	84.3	47.7

^aLess than 0.05 percent.

SOURCE: U.S. Department of Agriculture. 1988. Regional Characteristics of U.S. Farms and Farmers in the 1980's. ERS Staff Report No. AGES880128. Economic Research Service. Washington, D.C.

TABLE 1-5 Major Agricultural Sources of Gross Farm Income for Selected Agricultural Subregions,[a] 1982

Sales Ranking	Wisconsin-Minnesota Dairy Area	Core Corn Belt	Delta	Eastern Highlands	Western Great Plains	Western Corn Belt-Northern Plains	Coastal Plains	Southeast Piedmont	California Metro	United States
1	Dairy (53.9)	Corn (26.7)	Soybeans (36.4)	Cattle (25.6)	Cattle (57.4)	Cattle (29.1)	Poultry (20.0)	Poultry (58.5)	Fruits (24.4)	Cattle (24.4)
2	Cattle (12.9)	Cattle (23.7)	Cotton (22.2)	Tobacco (24.0)	Wheat (14.9)	Corn (18.9)	Tobacco (17.4)	Dairy (10.6)	Vegetable (14.7)	Dairy (12.3)
3	Corn (11.1)	Hogs (21.0)	Rice (18.6)	Dairy (23.8)	Corn (5.3)	Wheat (13.0)	Soybeans (12.4)	Cattle (8.9)	Dairy (14.2)	Corn (10.5)
4	Hogs (6.4)	Soybeans (17.9)	Wheat (10.5)	Poultry (5.0)	Cotton (4.0)	Hogs (12.6)	Hogs (10.1)	Soybeans (5.5)	Cattle (11.7)	Soybeans (8.1)

[a]Numbers in parentheses represent percentages of total farm sales accruing from each commodity.

SOURCE: U.S. Department of Agriculture. 1988. Regional Characteristics of U.S. Farms and Farmers in the 1980's. ERS Staff Report No. AGES880128. Economic Research Service. Washington, D.C.

37.4 percent, and the Wisconsin-Minnesota Dairy Area at 35.4 percent. These figures do not include income from rents and interest, which average from 7 to 11 percent for all regions. They also mask the degree to which individual households within regions rely on income from farming. Approximately 46 percent of all U.S. farm households derive 50 percent or more of their income from farming. Nonetheless, these figures reflect the growing importance of off-farm income to the economic well-being of farm families. They indicate that farm families are like most other families—both adults work. Two of these four regions, the Wisconsin-Minnesota Dairy Area and the Western Great Plains, as well as the Southeast Piedmont, generate more than 50 percent of their gross farm income from the sale of one commodity (see Table 1-5). The significance of this dependence is far less in the Southeast Piedmont region because agriculture is less vital to the regional economy.

The Delta states, on the other hand, appear more diversified but nonetheless remain dependent on federal commodity programs. Soybeans, for which the government sets a support price or loan rate, and three commodity program crops—cotton, rice, and wheat—generate more than 85 percent of all agricultural gross income in the region. Prospective changes in commodity programs and market demand for these crops will directly affect growers and the region's economy because agricultural input and output industries are significant sources of income.

The Western Corn Belt-Northern Plains and Core Corn Belt show the direct and indirect influence that farm policies can have on regions dependent on the farm sector economy. Farmers in these regions derive a large part of gross farm income from beef cattle, hogs, feed corn, and soybeans. Changes affecting one commodity are felt across the entire agricultural economy. Higher feed grain prices will increase the average feed prices for livestock producers, which may influence meat prices and, ultimately, consumer demand. For many producers who raise crops and livestock, the results will be mixed. Because these regions depend on farming as a principal source of overall income, the fortunes of the agricultural economy are felt throughout the regional economy. This was clearly demonstrated in the farm recession of the mid–1980s. Feed grain producers and suppliers of machinery and inputs to these farmers were stressed; livestock producers benefited from the availability of relatively inexpensive feed.

By contrast, California farmers are less influenced by federal commodity programs, particularly farmers specializing in fruit and vegetable production. More than half of California's agricultural gross income is from nonprogram crops. These growers, however, are concerned with other federal and state policies affecting the viability of their operations, including marketing orders, trade policy, food safety regulations, and environmental policies. Federal cosmetic and grading standards for fruit and vegetables significantly influence pest management practices. Producers of these specialty high-value crops dominate the California agricultural economy, generating additional economic activity in the form of processing, packaging, marketing, and transportation. As a result of widespread production of high-value crops and significant agriculture-urban integration, average farm income and per acre value of agricultural land and buildings are high compared to those in other regions (Tables 1-3 and 1-6).

The median income of the farm population is well above the average U.S. household income in three regions: the Wisconsin-Minnesota Dairy Area, Core Corn Belt, and California Metro. In four regions, the California Metro, Core Corn Belt, Delta, and Southeast Piedmont, the median farm household income is higher than that of all other households in the region. The Delta and Eastern Highlands had the lowest median income for farm households of the nine regions. In the Delta, however, median farm income was 11 percent higher than the median income for all Delta households. Median farm income in the Eastern Highlands was only slightly less than all Eastern Highlands households.

Agriculture causes environmental problems in all nine regions. Surface water pollution from fertilizers, pesticides, sediment, and manure is the most serious problem, although not uniformly distributed throughout major agricultural regions. Contamination of groundwater with pesticides and nitrate from agricultural fertilizers appears to be the most pervasive problem, occurring in all major agricultural regions: the Core Corn Belt, Wisconsin-Minnesota Dairy Area, Western Great Plains, Western Corn Belt-Northern Plains, Delta, California Metro, and Coastal Plains (U.S. Department of Agriculture, 1987e). The most severe overall water quality problems have been identified in the California Metro, Core Corn Belt, Delta, and the Coastal Plains. Soil erosion remains a concern on some soils in some regions. Food safety concerns are affecting agricultural production practices in California and to a lesser degree in other fruit- and vegetable-producing regions. These concerns are also affecting individual beef and dairy producers in many regions. Of the nine regions considered, irrigation problems appear to be most serious in the California Metro and the Western Great Plains.

Agricultural practices and systems and the importance of alternative practices are quite different in each region. In California, agriculture is extremely diverse. Thus, the scope of alternative production practices is great and variable. California agriculture is also confronted with the greatest range of environmental and public health concerns associated with modern conventional agricultural production practices. In the Core Corn Belt, the farm sector recession and the presence of nitrates and pesticides in the groundwater are among several factors influencing farmers to adopt alternative crop nutrient and pest management practices. The relatively small number of crops produced in the Core Corn Belt, however, makes the search for alternatives easier. Problems associated with food safety, for example, are of less relevance in this predominantly feed-producing region.

Relatively few research and policy studies on regional alternative systems have been undertaken. Those that have often focus on a particular crop or policy and do not attempt to fully account for the complexity of farm management decision making. In some areas, research on and experience

TABLE 1-6 Average Value of Land and Buildings in Selected Agricultural Subregions, 1982 (in dollars)

Value	Wisconsin-Minnesota Dairy Area	Core Corn Belt	Delta	Eastern Highlands	Western Great Plains	Western Corn Belt-Northern Plains	Coastal Plains	Southeast Piedmont	California Metro	United States
Per acre	1,224	1,481	1,193	906	270	790	1,052	1,048	2,181	787
Per farm	246,962	436,103	574,384	109,512	630,975	491,263	273,083	149,560	789,633	346,071

SOURCE: U.S. Department of Agriculture. 1988. Regional Characteristics of U.S. Farms and Farmers in the 1980's. ERS Staff Report No. AGES880128. Economic Research Service. Washington, D.C.

with the implementation of alternative systems for certain crops has been significant, such as IPM and biological pest control in fruit and vegetable production in California. This is not the case for most crops and regions, however. In the future, federal research and commodity program policies will need to take into account the diversity in agricultural needs, priorities, and systems and the physical and biological limitations of different regions and farms within these regions.

THE POWER OF POLICY

Government policy influences the direction of agriculture through a variety of agricultural, economic, and regulatory programs and policies. The most important of these are the commodity price and income support programs, tax policy, credit policy, research programs, trade and domestic economic policy, soil and water conservation programs, and the U.S. Environmental Protection Agency's (EPA) pesticide and water-quality regulations. The government's major influence on agriculture is through economic policy and the setting of prices and mandates regarding how land can be used by farmers wishing to participate in government programs. Regulatory policies that influence the cost and availability of alternative technologies and science and education priorities indirectly but powerfully affect agriculture.

Recently, the impact of commodity programs on farm management decisions has become more visible. In 1986, total federal farm program outlays (including direct payments to farmers, export subsidies, storage costs, and nonrecourse loans) equaled nearly 50 percent of net cash farm income. This declined to about 40 percent in 1987 (Figure 1-28). At the same time, net farm income reached record levels of $37.5 billion in 1986 and $46.3 billion in 1987, due in part to the large subsidies paid to most growers participating in federal commodity programs (Figure 1-29). Direct payments to growers set records in 1986 and again in 1987 of $11.8 and $16.7 billion, respectively. In 1986, 50 percent of income for wheat growers was in the form of a federal producer subsidy, such as direct payment or restricted foreign access to the domestic market (U.S. Department of Agriculture, 1988c). Federal farm commodity programs will continue to play a central role in shaping farm management decisions in agricultural regions dependent on these programs.

Farm programs have enormous influence on the crops that are grown and on the choice of management practices. Prices under the commodity programs are often far above world market prices. Consequently, most farmers feel compelled to preserve or build their farm commodity program base acres—acres that determine program eligibility and future income. The land-use decisions of farmers operating about two-thirds of the harvested cropland in the United States are strongly influenced by program rules and incentives.

Price and income support programs for major commodities also influence growers not in the programs. For example, pork producers do not receive

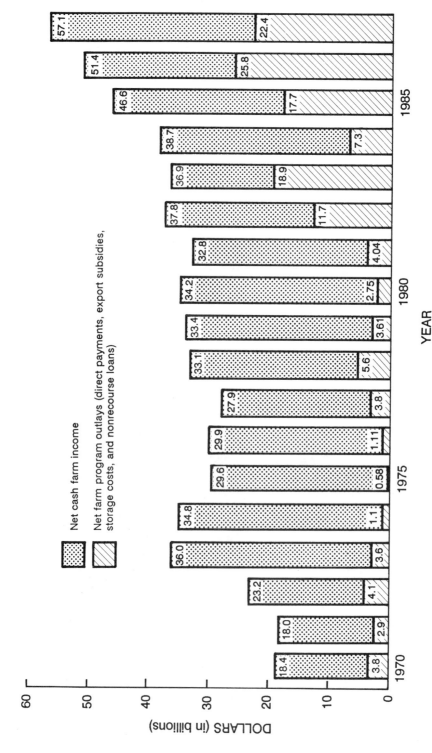

FIGURE 1-28 Net outlays for U.S. farm programs and net cash farm income (in billions of dollars). SOURCES: U.S. Department of Agriculture. 1987. Economic Indicators of the Farm Sector: National Financial Summary, 1986, Table 32. ECIFS 6-2. Economic Research Service. Washington, D.C.; U.S. Department of Agriculture. 1988. Agricultural Outlook, Table 37. Data on farm program outlays from 1970 to 1978 provided by the USDA Agricultural Stabilization and Conservation Service, Commodity Analysis Division. AO-145. Economic Research Service. Washington, D.C.

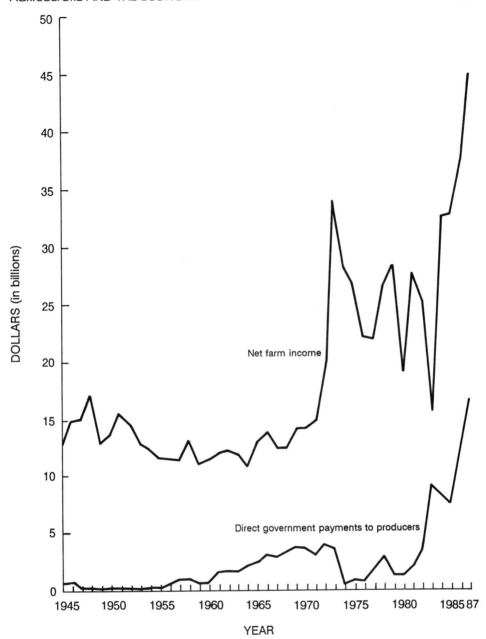

FIGURE 1-29 Net farm income and direct government payments to farmers. Net farm income includes all farm business income and expenses associated with dwellings located on farms; business income represents the profit from current production, with gross income adjusted to reflect net quantity changes in inventories. These adjustments offset sales from inventories carried over from the previous year and exclude changes in value of inventories existing on January 1. SOURCE: U.S. Department of Agriculture. 1987. Economic Indicators of the Farm Sector: National Financial Summary, 1986. ECIFS 6-2. Economic Research Service. Washington, D.C.

any government income protection. But in the past they have payed higher feed prices because of high price supports on feed grains. Recently, however, they have benefited from lower feed costs resulting from the feed grain program passed in the Food Security Act of 1985.

Policy also influences land use in indirect ways. The federal dairy termination program from 1985 to 1987 was designed to reduce overproduction of milk. Farmers were given the opportunity to leave the dairy business by selling their milking cows for slaughter or export. Almost 1 million cows— or about 9 percent of the nation's milking herd—were involved in the buy-out program. Farmers enrolled in the program were suddenly without cows to feed and had to decide on new farm enterprises. Many of these farmers decided to produce hay for local cash markets instead of for on-farm use. This decision caused a steep decline in the prices received by other established hay producers.

Lack of Long-Range National Program Goals

Federal policy evolved as a patchwork of individual programs, each created to address individual problems. No coherent strategy or national goals unite the programs, nor is there much appreciation of what the programs do or should accomplish or how they interact. Many programs, such as soil conservation and export programs, have historically had conflicting objectives. As a result, many well-intentioned policies that made sense when adopted or when viewed in isolation make less sense in the overall context of U.S. agriculture's contemporary needs. The USDA itself has recognized this failing.

> [Farm] policy has always tended to follow events and changes rather than anticipate and lead them—that is, the approach to developing policy has largely been reactive, dealing with one emergency after another.
>
> Times of a studied, deliberate approach to the design of a forward-looking farm policy, rather than adjustment of the previous statute, have been rare. . . .
>
> There is little doubt that some of the programs that have resulted from this ad hoc, crisis-oriented policymaking have subsequently exacerbated problems of farmers, or, over time, produced unintended and unwanted consequences for the farm sector as a whole (U.S. Department of Agriculture, 1981, p. 101).

In more than half a century of operation, government policy has not only affected commodity prices and the level of output, but it has also shaped technological change, encouraged uneconomical capital investments in machinery and facilities, inflated the value of land, subsidized crop production practices that have led to resource degradation such as soil erosion and surface and groundwater pollution, expanded the interstate highway system, contributed to the demise of the railway systems, financed irrigation projects, and promoted farm commodity exports. Together with other economic forces, government policy has had a far-reaching structural influence on agriculture, much of it unintended and unanticipated.

Impact of Commodity Policy on Alternative Agriculture

Federal commodity programs exist to stabilize, support, and protect crop prices and farmer income. Programs that idle land, set prices, make direct payments to farmers, and encourage and subsidize exports address these objectives. Most of the current commodity program concepts are derived from the Agriculture Adjustment Act of 1938, which established nonrecourse loans, acreage allotments, and marketing allotments for most major crops (U.S. Department of Agriculture, 1985a).

Two central components of federal commodity programs impede movement toward alternative agriculture: base acre requirements and cross-compliance provisions. All crop price and income support programs rely on the concept of an acreage base planted with a given commodity and a proven crop yield for those base acres. Generally, the crop acreage enrolled and the benefits received are related to the crop acreage planted and yield obtained in the past 5 years, although base acre *yields* are currently frozen at the 1981 to 1985 average. Most commodity program acreage is planted to maximize benefits. Farmers know that if they voluntarily reduce their planting (base acres) of a particular crop, they will not only forfeit benefits for that year, such as loan price and deficiency payments, but they will also lose future benefits by reducing their eligible acreage base (the subsequent 5-year average).

The cross-compliance provision of the Food Security Act of 1985 is designed to control government payments and production of program commodities by attaching financial penalties to the expansion of program crop base acres. It serves as an effective financial barrier to diversification into other program crops, particularly if a farmer has no established base acres for those crops. Cross-compliance stipulates that to receive *any* benefits from an established crop acreage base, the farmer must not exceed his or her acreage base for *any* other program crop. The practical impact of this provision is profound, particularly if a farmer's acreage base for other crops is small or zero. For example, a farmer with corn base acreage and *no* other crop base acres would lose the right to participate in *all* programs if *any* land on his or her farm was planted to other program crops such as wheat or rye (oats are currently exempt) as part of a rotation. If a farm had base acreage for two or more crops when cross-compliance went into effect in 1986, the farm must stay within the base acreage allotments applicable to both programs each year to retain full eligibility for commodity program payments in the future.

The conservation compliance provisions of the Food Security Act of 1985 may also complicate a farmer's adoption of alternatives. These provisions require that between 85 million and 90 million acres of highly erodible cropland have approved conservation plans or be enrolled in the conservation reserve program (CRP) by 1990. Plans must be fully implemented by 1995. About 28 million acres are currently in the CRP. For the remaining land, local soil conservation service specialists often recommend rotations in combination with conservation tillage practices as the best way to reduce erosion. Without adjustments in the cross-compliance or base acres provisions, many farmers may be forced to implement more costly, less effective conservation systems to maintain full eligibility for government program benefits.

Between the need to maintain base acres and the cross-compliance provision, farmers often face economic penalties for adopting beneficial practices, such as corn and legume or small grain rotations or strip cropping. With few exceptions, only farmers outside the programs can currently adopt these cropping systems without financial penalties. The conflict between the conservation, cross-compliance, and base acres provisions of the farm programs must be resolved to allow farmers to adopt, without economic penalty, practices and rotations that reduce erosion, input costs, and the potential for off-site environmental contamination.

Another incentive for farmers to remain enrolled in the commodity programs is the deficiency payments that farmers receive (see the boxed article, "Commodity Programs: Definition of Terms"). Since the 1940s, deficiency payments, or their previous equivalents, have been based on "proven yield," or the yield actually achieved in recent years on base acres on a particular farm. For each base acre in the program, the payment in a given year is the product of the per bushel deficiency payment times the land's proven yield in bushels per acre. The deficiency payment is the difference between the target price and the loan rate or the market price, whichever difference is less. When market prices are low, this policy rewards producers who strive for maximum per acre yield rather than maximum net return in the marketplace. The higher the farmer's established proven yield, the greater the deficiency payment received per acre.

The prospect of higher payments has encouraged heavier use of fertilizers, pesticides, and irrigation than can be justified by market forces in any given year. In effect, a high target price subsidizes the inefficient, potentially damaging use of inputs. It also encourages surplus production of the same crops that the commodity programs are in part designed to control, thus increasing government expenditures. This circumstance is illustrated by a hypothetical example from Figure 1-30: a farmer with 500 acres of wheat would produce 19,000 bushels at the market price. However, to generate additional income the farmer will produce 24,000 bushels at the target price. It costs the farmer more to produce the extra 5,000 bushels than they are worth on the market, but the taxpayer pays the difference to the farmer, in this case $10,000 (5,000 bushels × $2.00 per bushel deficiency payment).

An important change in the Food Security Act of 1985 has begun to cut the direct link of higher yields with rising program payments by freezing yield levels eligible for payments at 90 percent of the 1981 to 1985 average.

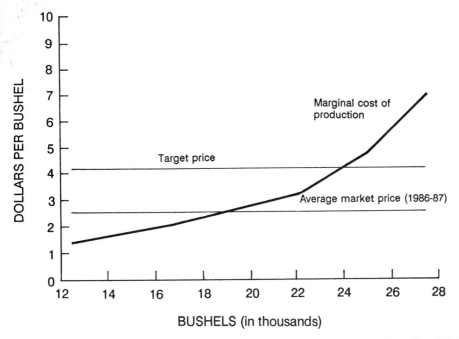

FIGURE 1-30 Hypothetical cost of production for a wheat farm. SOURCE: Agricultural Policy Working Group. 1988. Decoupling: A New Direction in Global Farm Policy. Washington, D.C.: Agricultural Policy Working Group.

Nonetheless, many farmers continue to pursue higher per acre yields in the belief that this freeze will be removed as part of the 1990 Farm Bill or some future legislation (Professional Farmers of America, 1988).

Commodity programs are a dominant force in domestic agriculture, with more than two-thirds of all U.S. cropland enrolled in these programs. The acreage enrolled in these programs has increased greatly since 1981, when export demand peaked, domestic market prices were high, and program participation was essentially zero. Program participation generally rises as market prices fall and per acre deficiency payments increase. This trend is clear for most commodities (Figures 1-31 through 1-34). With between 80 and 95 percent of the nation's corn, sorghum, wheat, cotton, and rice acreage enrolled in federal commodity programs, the chances are slim for widespread adoption of alternative practices that involve rotations with nonprogram crops, such as leguminous hay or forage crops, or the planting of other program crops for which farmers have to establish base acres. Under the current program rules, farmers simply have too much to lose.

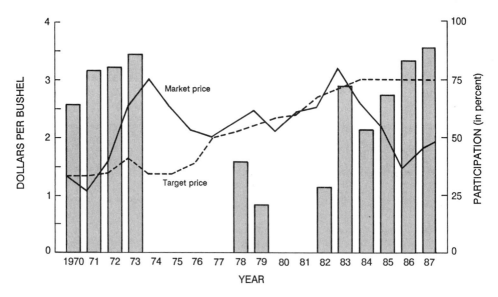

FIGURE 1-31   Market price, target price, and percentage of corn growers participating in the corn commodity program. SOURCE: Data provided by the U.S. Department of Agriculture, Agricultural Stabilization and Conservation Service, Commodity Analysis Division, 1988.

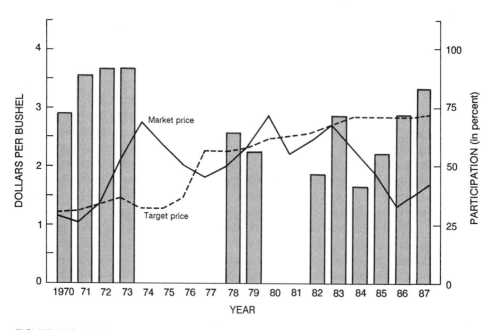

FIGURE 1-32   Market price, target price, and percentage of sorghum growers participating in the sorghum commodity program. SOURCE: Data provided by the U.S. Department of Agriculture, Agricultural Stabilization and Conservation Service, Commodity Analysis Division, 1988.

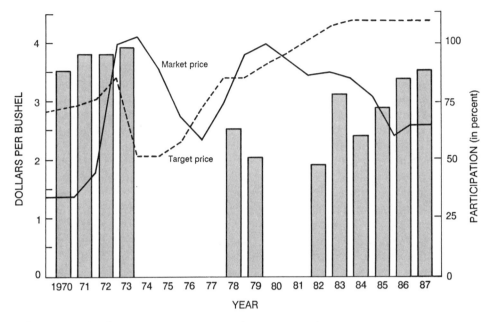

FIGURE 1-33 Market price, target price, and percentage of wheat growers participating in the wheat commodity program. SOURCE: Data provided by the U.S. Department of Agriculture, Agricultural Stabilization and Conservation Service, Commodity Analysis Division, 1988.

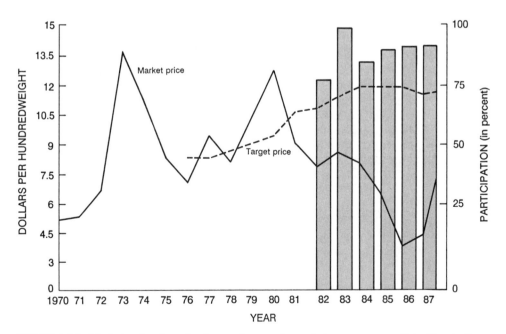

FIGURE 1-34 Market price, target price, and percentage of rice growers participating in the rice commodity program. SOURCE: Data provided by the U.S. Department of Agriculture, Agricultural Stabilization and Conservation Service, Commodity Analysis Division, 1988.

COMMODITY PROGRAMS: DEFINITION OF TERMS

The most expensive and influential government agricultural policies aim to support prices, adjust supplies, encourage exports, and maintain income for farmers producing wheat, corn, barley, sorghum, cotton, rice, sugar, tobacco, milk, and other program products. Following congressional direction, the U.S. Department of Agriculture (USDA) sets a target price and a loan rate for wheat, corn, barley, sorghum, cotton, and rice and equivalent prices for sugar, tobacco, and milk each year. If the average market price of a commodity is below the stated target price, the government pays participating farmers the difference between the target price and the loan rate or the market price, whichever difference is less. This is called a deficiency payment. It is paid to farmers in addition to income received for market sale of their crop or for placing the crop under loan with the Commodity Credit Corporation (CCC). The target price, designed to support farm income, is often set well above the market price and well above what it costs the majority of farmers to produce a crop (Figures 1-30, 1-35, and 1-36).

Deficiency payments are often a substantial share of gross farm income. The amount of the payment depends on a farmer's established per acre yield on a predetermined acreage base devoted to the crop. Direct deficiency payments may not exceed $50,000 per farmer, although Congress exempted some other types of payments in the Food Security Act of 1985 and provided separate limits on others. Many farmers have found ways to reorganize their holdings to avoid payment limitations.

In addition, the government offers nonrecourse loans with a government-set loan rate, which acts as a government-guaranteed minimum price for the commodity. If the market price falls below the loan rate, a

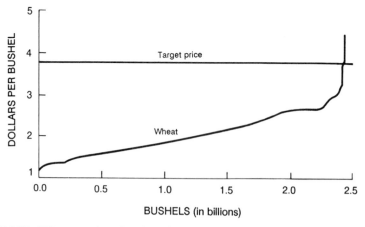

FIGURE 1-35   Wheat produced at less than the target price per bushel, 1981. SOURCE: Adapted from U.S. Department of Agriculture. 1985. Agricultural-Food Policy Review: Commodity Program Perspectives. Agricultural Economic Report No. 530. Economic Research Service. Washington, D.C. In Office of Technology Assessment. 1986. A Review of U.S. Competitiveness in Agricultural Trade—A Technical Memorandum. OTA-TM-TET-29. Washington, D.C.

farmer can forfeit crops placed under loan to the government in repayment of the loan. When the loan rate is set above the international market price, foreign producers can undercut the price of U.S. exports. In this case, most producers will take the higher price (the loan rate instead of the market price) and "sell" their crops to the government, which in turn stores them until market prices are well above loan rates or the crops are used in food aid programs. Most crops placed under CCC loan are sold at a net loss.

High and rigid loan rates were a major factor in the agricultural export decline of 1981 to 1986. Under changes in the Food Security Act of 1985, loan rates for wheat, feed grains, soybeans, upland cotton, and rice were lowered to 75 to 85 percent of the average price received by farmers over the past 5 years, dropping out the high and the low years.

Loan rates may not drop more than 5 percent from the previous year's rate, unless deemed necessary to make the U.S. crop more competitive. Using discretionary authority, the Secretary of Agriculture may not lower the loan rate more than 20 percent below the normally computed rate. Such discretionary reductions in the loan rate are not used to calculate subsequent rates.

For 1986 the secretary was required to reduce the loan rate for wheat and feed grains by at least 10 percent; the actual reduction was the maximum allowed, 20 percent. In 1986 and 1987, Congress set the soybean loan rate at $5.02 per bushel. During 1988 through 1990, with the above formula in effect, the rate is not allowed to drop below $4.50 per bushel.

Rice and cotton growers receive crop *marketing* loans that may be paid back at the loan rate or the prevailing market price, whichever is less.

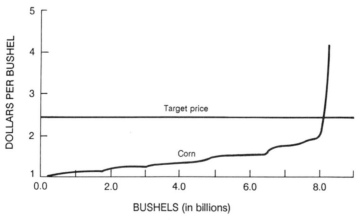

FIGURE 1-36 Corn produced at less than the target price per bushel, 1981. SOURCE: Adapted from U.S. Department of Agriculture. 1985. Agricultural-Food Policy Review: Commodity Program Perspectives. Agricultural Economic Report No. 530. Economic Research Service. Washington, D.C. In Office of Technology Assessment. 1986. A Review of U.S. Competitiveness in Agricultural Trade—A Technical Memorandum. OTA-TM-TET-29. Washington, D.C.

Tax Policy

Income tax policies over the last two decades have significantly influenced agricultural practices, even though they are not generally considered part of the farm program. The increases in irrigated acreage and animal confinement facilities are two examples. Before passage of the Tax Reform Act of 1986, agriculture received investment credit and accelerated depreciation on physical plants and equipment. Additionally, favorable capital gains treatment allowed individual farmers to exclude from taxation 60 percent of income received from the sale of assets such as land, breeder stock, and certain unharvested crops.

Favorable capital gains treatment provided incentives to purchase highly erodible fields and wetlands, rangelands, or forestlands at relatively low prices; convert these lands to cropland; sell them at a profit; and exclude 60 percent of the gain from taxation. The tax advantages of large-scale conversion of wetlands to cropland were estimated to be as much as $603 per acre, largely from the treatment of capital gains (Benfield et al., 1986). Ironically, favored tax treatment of "conservation" investments stimulated conversion of rangeland and wetlands to cropland. The Tax Reform Act of 1986 eliminated special capital gains treatment for the conversion of highly erodible land or wetlands into cropland. The act also explicitly denies the deduction of expenses associated with draining or filling a wetland. Although the effects of recent changes in the tax code will not be fully understood for several years, the similar swampbuster and sodbuster provisions of the Food Security Act of 1985 and the Tax Reform Act of 1986 eliminated many financial incentives for farming practices that contributed to soil erosion and conversion of wetlands to farmland (see the "Soil Conservation Programs" section in this chapter). Management decisions and capital expenditures profitable under the previous tax code are now less attractive.

Much of the sharp rise in farm real estate prices in the 1970s can be attributed to a speculative boom driven by tax advantages and inflation, compounded by the higher earning power of farmland, which resulted from higher commodity prices. This boom collapsed in the 1980s as lower commodity prices and rising real interest rates depressed land value (equity), in turn eroding the net worth of most farms. Lowered equity has affected not only those who had invested in land speculatively, which includes farmers and nonfarmers, but also farmers who bought land with the sole interest of farming it.

Another effect of volatile land prices is consolidation of land holdings. Capital gains taxes, which helped make farm real estate speculation profitable and thus destabilized land prices, reinforced the trend toward fewer and larger farms. When land prices rise above the value sustainable from current farm income, only buyers who have enough equity can compete for land that comes on the market. When land prices fall, farms with high debt-to-asset ratios and equity in the form of land can have greater difficulty securing capital at competitive interest rates. Those in more favorable positions can increase their holdings by buying land at "distressed" prices.

Increasing investment in physical plants and equipment, accompanied by a rapid decline in the use of labor, have been two of the most salient features of the agricultural economy for many decades. Since World War II, the value of agricultural machinery and vehicles increased sevenfold in constant dollars to more than $100 billion. Use of labor decreased by a factor of five (U.S. Department of Agriculture, 1985b). Tax advantages and shelters available through investment in agricultural facilities and equipment, such as irrigation systems, orchard plantings, and animal confinement structures, accelerated the use of certain technologies and altered the structure of many farming enterprises.

Investment tax credits spurred the use of irrigation in the Great Plains with water from the Ogallala aquifer. More than 1 million acres were brought under center pivot irrigation in Nebraska between 1973 and 1983. Converting the sandhills of Nebraska to center-pivot-irrigated corn has been estimated to generate $175 per acre in tax advantages through a combination of the water-depletion allowance, accelerated depreciation, and investment tax credits (Benfield et al., 1986). The water-depletion allowance permits farmers to claim a deduction if they can prove that they are irreversibly depleting certain groundwater reserves. Agricultural overuse is depleting the slowly recharging Ogallala aquifer in some locations. The Tax Reform Act of 1986 denies deduction of any expenses associated with preparing land for center pivot irrigation.

Accelerated depreciation and investment tax credits also motivated the rapid growth of custom beef feedlots, which started in the 1960s and increased through the early 1980s. These tax provisions are responsible for the surge of off-farm investment in hog confinement facilities in the Midwest in the early 1980s. The Tax Reform Act of 1986 lengthened the depreciation period for single-purpose agricultural structures from 5 to 7 years. Tax code changes in 1988 further lengthened this depreciation to 10 years.

Research and Education

The primary goal of agricultural research and education policy has been to increase farm production and profitability while conserving the natural resource base. Achieving higher crop yields per acre has traditionally been viewed as the best way to do both. Emphasizing the attainment of higher yields per acre makes the greatest sense if land is the most limiting factor in production or if the cost of land is high on a per acre basis relative to other costs. Some high-value per acre specialty crop operations are examples in which large investments for an irrigation system or pesticides can be justified economically. Because government programs do not support these crops, the intensity of input use is generally a function of the market demand and price paid for the crop.

Much research conducted over the past 40 years has responded to the needs of farmers operating under a set of economic and policy incentives that encourage high yields. Much of the focus has been on chemical- and drug-related technologies to support specialized, high-yield operations and simplify farm management. Until recently, research has generally not deliberately addressed the possibility of maintaining current levels of production with reduced levels of certain off-farm inputs, more intensive management, increased understanding of biological principles, or greater profitability per unit of production with reduced government support.

Yet, increased international competition, the decline in world market prices for most commodities, and the relatively high percentage of total variable costs for inputs needed to achieve current high yields warrants a reassessment of farming practices, research, and the effects of policy on farm decision making. In general, further increases in yield are an ineffective means of achieving greater profitability or international competitiveness. For many crops like corn, cotton, wheat, and small grains, higher yields are often justified in terms of profitability only in the context of government support, particularly high target prices. The added costs of purchased inputs soon become more than the free market value of the added yield. Moreover, high-yield, specialized production systems can result in more variable yields than diversified systems that also reduce per unit input costs (Helmers et al., 1986). This is especially true when rainfall or other climatic conditions deviate far from the norm (Goldstein and Young, 1987; Lockeretz et al., 1984).

Increased yield variability can also raise risk and capital costs. Farmers growing commodity program crops, however, are often willing to take this risk, because government commodity payments provide an economic safety net that does not depend on annual harvested production. Disaster relief and crop insurance benefits may also be available, further reducing the risk borne by farmers. In years when high yields are attained, farmers may have an opportunity to raise the proven yield that is used as the basis of future program benefits and insurance settlements. When high yields fail, disaster payments and insurance program mechanisms protect farmers. These programs are expensive, however. The economic, agronomic, and environmental consequences associated with these practices are leading to fundamental changes in the targets for agricultural research and education. Throughout the system, a new emphasis is being placed on identifying crops better suited to a region's natural resources and to reducing costs per unit of production, sometimes even at lower per acre yields.

Other Programs and Policies

Soil Conservation Programs

Soil conservation and other federal farm policies have been linked since the Soil Conservation and Domestic Allotment Act of 1936. This connection was politically expedient. When pictures of the Dust Bowl were a symbol of the Great Depression, the public was willing to pay farmers to shift from erosion-prone crops to soil-conserving land uses. The soil-conserving crops such as hays and forages were not in surplus, while crops that generally require more tillage and often result in higher rates of erosion, such as corn, wheat, and cotton, were in surplus. Reducing the acreage devoted to these crops provided an opportunity to reduce erosion through cover crops or other conservation measures. Since the 1940s, conservation programs have done better in times of depressed prices and surpluses and worse in periods of strong prices and expanding production.

Voluntary compliance or participation has been an underlying principle of soil conservation programs since their inception. The government has historically relied on the "carrot," such as availability of free technical assistance, cost-sharing funds, and commodity program benefits, rather than the "stick" of mandatory compliance or penalties. The price and income support aspects of farm programs have dominated environmental and conservation considerations. This was particularly apparent in the mid–1970s through the mid–1980s, when expanding production exacerbated erosion losses. As production expanded, there were no policies in place to slow the conversion of wetlands or highly erodible grasslands to cultivated crops. Nor were there policies to slow the resulting steady growth in commodity program base acreage allotments.

Congress addressed this problem in the Food Security Act of 1985 by the adoption of the so-called sodbuster and swampbuster provisions. The sodbuster provision denies all federal program benefits to farmers who plow highly erodible lands without first adopting a locally approved soil conservation plan. The swampbuster provision denies benefits to farmers who drain or otherwise convert certain wetlands to cultivated crop production.

Soil and water conservation measures often require continuous refinement, maintenance, and good management to reduce erosion significantly and protect water quality. In periods of high commodity prices and strong demand, some farmers have planted grain crops on almost all available land, with few steps taken to reduce soil and water runoff. Farmers who have continued conservation practices in boom years lost opportunities to build base acreage and, in some cases, forfeited chances to improve their farms' proven per acre yield and payments.

In response to this inequity, the Food Security Act of 1985 incorporated several mechanisms designed to simultaneously control surplus production and reduce soil erosion on the most highly erodible land. The CRP pays farmers to take their most highly erodible land out of production for 10 years. Over 60 percent of the land now in the CRP is drawn from crop base acres. Nearly half of the base acres now in the CRP are from the wheat program (Table 1-7). As of February 1988, 25.5 million acres had been idled under the CRP (Table 1-8). Five million to 10 million more acres are expected to be idled over the next 2 years. It is noteworthy that even though set-aside acreage from the commodity programs and the CRP idled nearly 70 million acres in 1987, excess production capacity of major commodities remained near its highest point at 16 percent of potential output (Figure 1-37). (Excess production capacity is defined here as the difference between potential output and commercial demand at prevailing farm prices.)

Another feature of the Food Security Act of 1985, the conservation com-

TABLE 1-7   Commodity Base Acres Enrolled in CRP Through July 1987

Crop	Million Acres		
	Total Base Acres in 1985	Base Acres Enrolled in CRP	Percentage of Base Acres Enrolled in CRP
Barley	12.4	1.8	14.5
Sorghum	18.9	1.7	9.0
Oats	9.2	0.8	8.7
Wheat	91.7	6.8	7.4
Cotton	15.4	0.9	5.8
Corn	82.2	2.7	3.3
Rice	4.1	—	—
Peanuts	1.5[a]	—	—
Tobacco	0.7[a]	—	—
Total	236.1	14.7	6.2[b]

NOTE: The dash indicates that the values were negligible.

[a]Acres harvested.
[b]This figure represents the percentage of all crop base acres.

SOURCE: U.S. Department of Agriculture. 1987. Agricultural Resources—Cropland, Water, and Conservation—Situation and Outlook Report. AR-8. Economic Research Service. Washington, D.C.

TABLE 1-8   Regional Distribution of Acres Enrolled in CRP Through February 1988

Region	Acres Enrolled (in millions)	Share (percent) of U.S. Acres Enrolled	Percentage of Region's Cropland
Northeast	0.13	0.5	0.8
Lake States	2.07	8.1	4.7
Corn Belt	3.56	13.9	3.9
Northern Plains	6.04	23.7	6.5
Appalachia	0.86	3.4	3.8
Southeast	1.25	4.9	6.8
Delta States	0.78	3.0	3.5
Southern Plains	4.10	16.1	9.1
Mountain	5.22	20.4	12.1
Pacific	1.51	5.9	6.7
United States	25.53	100.0	6.1[a]

[a]This figure represents the percentage of all crop acres.

SOURCE: U.S. Department of Agriculture. 1988. Agricultural Resources—Cropland, Water, and Conservation—Situation and Outlook Report. AR-12. Economic Research Service. Washington, D.C.

pliance provision, will require farmers wishing to retain eligibility for government program benefits to implement recommended conservation plans beginning in the 1990 growing season. To retain eligibility for any government program—diversion payments, deficiency payments, CCC commodity loans and storage payments, Farmers Home Administration (FmHA) loans, government loans for storage facilities, federal crop insurance, and conservation reserve payments—farmers must manage all highly erodible fields in accordance with an approved soil conservation plan by 1995. Between 80 million and 95 million acres will require these plans, although more than 25 million of these acres are now in the CRP.

The impact of conservation compliance on farming practices is not known, although no-tillage or conservation tillage practices are often recommended for highly erodible land. In many instances, alternative farming systems

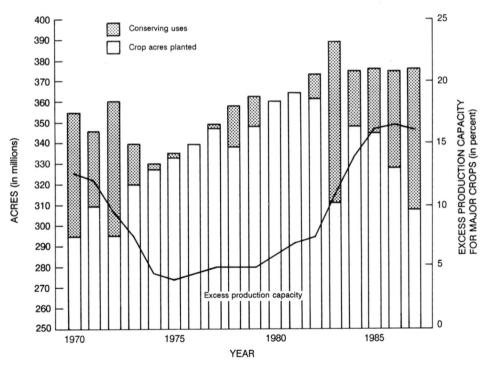

FIGURE 1-37   U.S. crop acreage in conserving uses compared with excess production capacity. Major crops include wheat, feedgrains (corn, barley, sorghum, and oats), soybeans, and cotton. Excess production capacity is the difference between potential output and commercial demand at prevailing market prices. SOURCE: U.S. Department of Agriculture. 1988. 1988 Agricultural Chartbook. Agriculture Handbook No. 673. Washington, D.C. Revised data from Economic Research Service, USDA.

may be used to sustain high levels of crop production and comply with the erosion control goals sought under conservation compliance. Future government policy may provide new incentives for farmers to develop alternative crop management systems that protect environmental quality and maintain current levels of production and farm incomes.

Pesticide Licensing

About 600 pesticide active ingredients are registered for use in the United States. Approximately 200 active ingredients, however, account for over 95 percent of all agricultural pesticide use. Congressional policy and the EPA's application of current law regulating pesticides have resulted in a slow, deliberate pesticide regulatory process. From the inception of the EPA special review program in 1975 through September 30, 1987, the agency completed 40 special reviews or risk-benefit analyses of the most hazardous pesticides. These reviews resulted in 5 cases where all agricultural uses were cancelled, 34 cases where some uses were cancelled or restrictions imposed, and 1 case where no action was taken. These cases do not include the cancellation of all food uses of aldrin, chlordane, chlordecone, DDT, dieldrin, and heptachlor, nor the voluntary cancellation of all or some uses of 21 other active ingredients that occurred outside the special review process (U.S. Environmental Protection Agency, 1987). Between 1975 and 1987, these reviews took from 2 to 7 years to complete, with some important reviews still outstanding. Recently, however, the review process has been expedited and newly initiated reviews may now take an average of $1\frac{1}{2}$ to 3 years (U.S. General Accounting Office, 1986). Legal challenges commonly delay the final resolution of regulatory actions, as does the sheer size of the task in comparison to available EPA resources.

Since the amendment of the Federal Insecticide, Fungicide and Rodenticide Act (FIFRA) in 1972, through 1987 the EPA registered 69 new insecticides, 60 new herbicides, and 31 new fungicides (U.S. Environmental Protection Agency, 1988) (see Figure 1-20). New products are generally subjected to stricter standards before they gain market entry than are existing products with which they would compete. Typically, these pesticides are safer and more biologically benign. In some cases, new compounds that are safer than the existing products they might replace have been denied registrations, while more hazardous products were left on the market (National Research Council, 1987). Current regulations are complex, sometimes inconsistent, and exceedingly difficult to implement. The Delaney Clause (1958) of the Federal Food, Drug and Cosmetic Act of 1954 offers the best example of inconsistency. This provision of the law forbids the residues of pesticides in any processed food that induce cancer in laboratory animals if those residues concentrate above the level allowed on the raw food. The Delaney Clause, however, does not apply to raw foods with no processed form or to carcinogenic pesticides that do not concentrate in processed foods. Consequently, residues of the same carcinogenic pesticides are allowed on certain fresh and processed foods, but not in processed foods where they concentrate. Further, the EPA has applied the Delaney Clause only to new pesticides, thereby maintaining registrations for many older pesticides that pose risks acknowledged by the EPA to be greater than those posed by most new substitute chemicals (National Research Council, 1987).

The benefit-assessment methods employed by the EPA are also a concern. The EPA does not, as a matter of routine procedure, incorporate alternative or nonchemical pest control methods into its assessment of pesticide benefits when carrying out a regulatory review (U.S. Congress, 1988). As a result, the benefits of currently used products are sometimes inflated, and the economic values of alternatives are not taken into account or formally recognized and acted upon. The most recent example of this is the EPA's review of the herbicide alachlor. Alachlor is the most widely used pesticide in the nation. It is used to control grassy weeds on 30 percent of the corn and 25 percent of the soybeans produced in the United States. The benefits analysis in this review was confined solely to an economic comparison of the benefits of alachlor with those of a similar herbicide, metolachlor (U.S. Environmental Protection Agency, 1986a). The comparative economic benefits and costs associated with the use of cultivation, tillage, and planting techniques that are used effectively by many farmers to control similar weeds were not considered in the analysis (see the Spray, BreDahl, Sabot Hill, Kutztown, and Thompson case studies). According to John Moore, former EPA assistant administrator for pesticides and toxic substances, the alachlor benefits assessment is representative of most EPA pesticide benefits assessments. These assessments routinely consider only the benefits of the most likely alternative pesticide, ignoring all other alternative control strategies (U.S. Congress, 1988).

Food Quality and Safety

Food safety regulations and meat inspection programs are primarily designed to prevent health risks and acute illnesses from chemical and microbial contaminants in food. These regulations, however, do not enhance food quality. For example, meat-grading standards have traditionally rewarded producers of fatty beef. Cosmetic standards for fruits and vegetables can encourage late-season pesticide use that results in higher residues in food. Certain poultry slaughter practices result in a high prevalence of microbiological contamination. Methods of producing food with fewer of these inherent risks are well known and widely practiced (see the case studies in this report; Allen et al., 1987; National Research Council, 1985, 1988b).

The EPA reviews health and safety data and establishes tolerance levels for pesticide residues in foods that are thought to present minimal health risks. Foods with pesticide residues up to these levels are then allowed in the market. The FDA then monitors food for compliance with these tolerances. For some types of risk, however, particularly cancer risk, there remains considerable debate about the certainty of the data and assumptions supporting calculations of acceptable risk. Moreover, the monitoring does not regularly check for many widely used pesticides, including a number of widely used compounds classified by the EPA as probable human carcinogens (U.S. Congress, House, 1987).

Livestock are being fed an increasing amount of various by-products from the processing of agricultural commodities. This is particularly true in states like California and Florida that produce a great variety of commodities (National Research Council, 1983). By-product feeds like citrus pulp, tomato pomace, and almond hulls are valuable livestock feeds with nutritive and economic value often comparable to that of feeds produced exclusively for animal use. Many of these feeds, however, have not been historically recognized as animal feeds. Because of this, many pesticides used on these commodities do not have tolerances for residues in by-products used as animal feed or in the ensuing animal food product. The potential for the introduction of these pesticides into food-producing animals is unknown. While animal food products may contain residues of the pesticides found in nontraditional animal feeds, the EPA has generally not examined the fate of pesticides in animals consuming these feeds or the food products derived from them (National Research Council, 1987).

There is also concern about combinations of residues on food to which people may be regularly exposed (National Research Council, 1988a). The EPA sets acceptable levels for residues in food for each pesticide separately, although many combinations of pesticides are regularly used and detected on food crops. Even though risks from pesticides are presumed to be additive, acceptable levels of exposure are calculated assuming exposure to each pesticide in isolation. Some chemicals, moreover, may act synergistically. Current regulations and standards do not assess or incorporate margins of safety reflecting the possibility of synergistic or additive effects.

The federal government also sets grading standards for farm products. Beef grading tended to equate high-fat content with high quality in Prime and Choice cuts for example, until recent changes in grading standards. Excessive consumption of animal fat is known to raise the likelihood of heart disease (National Research Council, 1988b). Similarly, the USDA grading standards and milk pricing standards reward producers for butterfat content of milk. Since the 1940s, however, butterfat consumption has declined dramatically, while consumption of low-fat and nonfat dairy products has increased. Consequently, the butterfat-based pricing system has resulted in large government-held surpluses of butter, despite the capability of producers—through genetics and management—to produce lower-fat products.

Salmonellae also remains a significant concern, particularly in poultry products. A National Research Council study reported that about one-third of all poultry sold is contaminated with salmonellae. Although salmonellae is controlled by proper cooking and sanitation, not all people follow recommended food handling procedures. The possibility of resistant strains and human health problems following infection remains a concern (Institute of Medicine, 1989; National Research Council, 1985).

SUMMARY

Agriculture produces the essential elements of the $700 billion food and fiber economy. Since World War II, agriculture has become more specialized and dependent on off-farm inputs and has substantially increased per acre yield. Machinery, pesticides, irrigation water, fertilizers, and antibiotics have replaced land, diversity, and labor as principal components of agricultural production. Fewer and larger farms produce more food and fiber than ever before. Government commodity income and price support programs, tax policy, and agricultural research heavily influence on-farm decision making in major sections of U.S. agriculture. Producing food to meet government criteria, however, often precludes farmers from responding to changing market conditions or imposes financial penalties for practices that improve food safety and environmental quality.

In the midwestern states, government programs and subsidies have reduced the risk of specialization and thus encouraged the separation of livestock operations from feed grain production. The result is a decline in two important agricultural practices: return of animal manures to the soil and rotation of cultivated crops with grass and leguminous forages. Feed grain production without livestock or legumes requires additional commercial fertilizer and often entails increased pesticide use to compensate for the lost pest control benefits of rotations. The increase in confinement livestock operations, particularly for swine and poultry, correlates with the subtherapeutic use of antibiotics to promote growth and to suppress disease incidence. Between 80 and 95 percent of program crop acreage is currently enrolled in the federal commodity programs. The base acres and cross-compliance provisions of these programs will penalize growers who want to adopt diversified crop rotations or integrated livestock feed and forage operations on this land.

There are many economic and environmental problems to be solved that are associated with current conventional agricultural practices. However, a substantial number of growers practice many systems that provide solutions, in spite of actual disincentives or little support from federal programs. Chapter 2 describes some of the major problems derived from conventional practices. Subsequent chapters describe the alternatives.

REFERENCES

Allen, W. A., E. G. Rajotte, R. F. Kazmierczak, Jr., M. T. Lambur, and G. W. Norton. 1987. The National Evaluation of Extension's Integrated Pest Management (IPM) Programs. VCES Publication 491-010. Blacksburg, Va.: Virginia Cooperative Extension Service.

Benfield, F. K., J. R. Ward, and A. E. Kinsinger. 1986. Assessing the Tax Reform Act: Gains, Questions, and Unfinished Business. Washington, D.C.: Natural Resources Defense Council.

California Department of Food and Agriculture. 1958. California Dairy Industry Statistics. Sacramento, Calif.: Crop and Livestock Reporting Service.

California Department of Food and Agriculture. 1972. California Dairy Industry Statistics. Sacramento, Calif.: Crop and Livestock Reporting Service.

California Department of Food and Agriculture. 1987. California Dairy Industry Statistics. Sacramento, Calif.: Crop and Livestock Reporting Service.

Council for Agricultural Science and Technology. 1981. Antibiotics in Animal Feeds. Report No. 88. Ames, Iowa.

Curtis, S. E. 1983. Environmental Management in Animal Agriculture. Ames, Iowa: The Iowa State University Press.

Goldstein, W. A., and D. L. Young. 1987. An economic comparison of a conventional and a low-input cropping system in the Palouse. American Journal of Alternative Agriculture 2(Spring):51–56.

Hays, V. W., D. Batson, and R. Gerrits. 1986. Public health implications of the use of antibiotics in animal agriculture: Preface. Journal of Animal Science 62(Suppl. 3):1–4.

Helmers, G. A., M. R. Langemeier, and J. Atwood. 1986. An economic analysis of alternative cropping systems for east-central Nebraska. American Journal of Alternative Agriculture 1(4):153–158.

Institute of Medicine. 1989. Human Health Risks with the Subtherapeutic Use of Penicillin or Tetracyclines in Animal Feed. Washington, D.C.: National Academy Press.

Johnson, G. L., and S. H. Wittwer. 1984. Agricultural Technology Until 2030: Prospects, Priorities, and Policies. East Lansing, Mich.: Michigan State University Agricultural Experiment Station.

Legg, J. O., and J. J. Meisinger. 1982. Soil Nutrition Budgets in Nitrogen in Agricultural Soils, F. J. Stevenson, ed. ASA Monograph No. 22. Madison, Wis.: American Society of Agronomy, Crop Science Society of America, Soil Science Society of America.

Lockeretz, W., G. Shearer, D. H. Kohl, and R. W. Klepper. 1984. Comparison of Organic and Conventional Farming in the Corn Belt. Pp. 37–48 in Organic Farming: Current Technology and Its Role in a Sustainable Agriculture, D. F. Bezdicek, and J. F. Power, eds. Madison, Wis.: American Society of Agronomy, Crop Science Society of America, Soil Science Society of America.

National Agricultural Chemicals Association. 1987. 1986 Industry Profile Survey. Washington, D.C.: Ernst and Whinney.

National Research Council. 1983. Underutilized Resources as Animal Feedstuffs. Washington, D.C.: National Academy Press.

National Research Council. 1985. Meat and Poultry Inspection: The Scientific Basis of the Nation's Program. Washington, D.C.: National Academy Press.

National Research Council. 1987. Regulating Pesticides in Food: The Delaney Paradox. Washington, D.C.: National Academy Press.

National Research Council. 1988a. Complex Mixtures: Methods for In Vivo Toxicity Testing. Washington, D.C.: National Academy Press.

National Research Council. 1988b. Designing Foods: Animal Product Options in the Marketplace. Washington, D.C.: National Academy Press.

Professional Farmers of America. 1988. Pro Farmer's Guide to Working with ASCS. Cedar Falls, Iowa: Professional Farmers of America.

U.S. Congress. 1985. The Food Security Act of 1985. Public Law 99–198. Washington, D.C.

U.S. Congress, House. Committee on Energy and Commerce. 1987. Hearing on Pesticides in Food. 100th Cong., 1st sess. Serial No. 100-7, pp. 37–39.

U.S. Congress, House. Committee on Interior and Insular Affairs, Subcommittee on Water and Power Resources. 1987. Congressman George Miller's questions and Dr. Wayne N. Marchant's responses to testimony on irrigation subsidy legislation. H.R. 100–1443. May 22. 100th Cong., 1st sess.

U.S. Congress, House. Committee on Government Operations. 1988. Hearing on Low Input Farming Systems: Benefits and Barriers. H.R. 100–1097. October 20. 100th Cong., 2nd sess.

U.S. Department of Agriculture. 1970. Quantities of Pesticides Used by Farmers in 1966. Agricultural Economic Report No. 179. Economic Research Service. Washington, D.C.

U.S. Department of Agriculture. 1972. Agricultural Statistics. Tables 1, 38, 84, and 189. Washington, D.C.

U.S. Department of Agriculture. 1981. A Time to Choose: Summary Report on the Structure of Agriculture. Washington, D.C.

U.S. Department of Agriculture. 1984. Inputs—Outlook and Situation Report. IOS-6. Economic Research Service. Washington, D.C.

U.S. Department of Agriculture. 1985a. Agricultural-Food Policy Review: Commodity Program Perspectives. Agricultural Economic Report No. 530. Economic Research Service. Washington, D.C.

U.S. Department of Agriculture. 1985b. Economic Indicators of the Farm Sector: Production and Efficiency Statistics, 1983. ECIFS 3-5. Economic Research Service. Washington, D.C.

U.S. Department of Agriculture. 1985c. Energy and U.S. Agriculture: Irrigation Pumping, 1974–1983. Agricultural Economic Report No. 545. Economic Research Service. Washington, D.C.

U.S. Department of Agriculture. 1986a. 1986 Agricultural Chartbook. Agriculture Handbook No. 663. Washington, D.C.

U.S. Department of Agriculture. 1986b. Agricultural Resources—Cropland, Water, and Conservation—Situation and Outlook Report. AR-4. Economic Research Service. Washington, D.C.

U.S. Department of Agriculture. 1986c. Agricultural Resources—Inputs—Outlook and Situation Report. AR-1. Economic Research Service. Washington, D.C.

U.S. Department of Agriculture. 1986d. Agricultural Statistics. Tables 2, 38, 81, and 166. Washington, D.C.

U.S. Department of Agriculture. 1986e. Agriculture's Links to the National Economy: Income and Employment. Agriculture Information Bulletin No. 504. Economic Research Service. Washington, D.C.

U.S. Department of Agriculture. 1986f. Cropland Use and Supply—Outlook and Situation Report. Economic Research Service. Washington, D.C.

U.S. Department of Agriculture. 1986g. Food Cost Review, 1985. Agricultural Economic Report No. 559. Economic Research Service. Washington, D.C.

U.S. Department of Agriculture. 1986h. U.S. and World Food, Beverages, and Tobacco Expenditures, 1970–1983. Economic Research Service. Washington, D.C.

U.S. Department of Agriculture. 1987a. Agricultural Resources—Cropland, Water, and Conservation—Situation and Outlook Report. AR-8. Economic Research Service. Washington, D.C.

U.S. Department of Agriculture. 1987b. Agricultural Resources—Inputs—Situation and Outlook Report. AR-5. Economic Research Service. Washington, D.C.

U.S. Department of Agriculture. 1987c. Economic Indicators of the Farm Sector: National Financial Summary, 1986. ECIFS 6-2. Economic Research Service. Washington, D.C.

U.S. Department of Agriculture. 1987d. Fertilizer Use and Price Statistics, 1960–1985. Statistical Bulletin No. 750. Economic Research Service. Washington, D.C.

U.S. Department of Agriculture. 1987e. The Magnitude and Costs of Groundwater Contamination from Agricultural Chemicals: A National Perspective. Staff Report AGES870318. Economic Research Service. Washington, D.C.

U.S. Department of Agriculture. 1987f. Measuring the Size of the U.S. Food and Fiber System. Agricultural Economic Report No. 566. Economic Research Service. Washington, D.C.

U.S. Department of Agriculture. 1987g. National Food Review: The U.S. Food System from Production to Consumption. NFR-37. Economic Research Service. Washington, D.C.

U.S. Department of Agriculture. 1987h. U.S. Irrigation: Extent and Economic Importance. Agriculture Information Bulletin No. 523. Economic Research Service. Washington, D.C.

U.S. Department of Agriculture. 1988a. 1988 Agricultural Chartbook. Agriculture Handbook No. 673. Washington, D.C.

U.S. Department of Agriculture. 1988b. Agricultural Outlook. AO-141. Economic Research Service. Washington, D.C.

U.S. Department of Agriculture. 1988c. Outlook '88 Charts: 64th Annual Agricultural Outlook Conference. Economic Research Service. Washington, D.C.

U.S. Department of Agriculture. 1989. Data from National Agricultural Statistics Service, retained in data base of the Livestock, Dairy, and Poultry Branch, Commodity Economics Division, Economic Research Service. Washington, D.C.

U.S. Environmental Protection Agency. 1986a. Alachlor: Special Review Technical Support Document. Washington, D.C.

U.S. Environmental Protection Agency. 1986b. Pesticide Industry Sales and Usage: 1985 Market Estimates. Washington, D.C.

U.S. Environmental Protection Agency. 1987. Fiscal Year 87 Report on the Status of Chemicals in the Special Review Program, Registration Standards Program, Data Call-In Program, and Other Registration Activities. Washington, D.C.

U.S. Environmental Protection Agency. 1988. Chemicals Registered for the First Time as Pesticidal Active Ingredients under FIFRA (including 3 (C)(7)(A) Registrations). Washington, D.C.

U.S. General Accounting Office. 1986. Pesticides: EPA's Formidable Task to Assess and Regulate Their Risks. GAO/RCED-86-125. Washington, D.C.

U.S. Office of Technology Assessment. 1986a. A Review of U.S. Competitiveness in Agricultural Trade: A Technical Memorandum. OTA-TM-TET-29. Washington, D.C.

U.S. Office of Technology Assessment. 1986b. Technology, Public Policy, and the Changing Structure of American Agriculture. OTA-F-285. Washington, D.C.

2

Problems in U.S. Agriculture

THE U.S. AGRICULTURAL SYSTEM has been beset by numerous economic and environmental problems in the 1980s. In the economic sphere, with storage facilities filled with surplus crops, the cost of federal farm support programs skyrocketed from $3.5 billion in 1978 to a peak of $25.8 billion in 1986, falling to $22 billion in 1987 (U.S. Department of Agriculture, 1988f). Financial stress hit tens of thousands of farmers and many rural communities. Some farmers still find it difficult to pay debt accumulated during the prosperous 1970s. Many U.S. products are no longer competitive in world markets. From 1981 through 1986, the United States' agricultural trade surplus declined substantially. Although agricultural trade performance has improved since then, this has come at considerable expense to U.S. taxpayers. Competition among nations for worldwide markets is fierce and volatile.

Agriculture is also causing serious environmental problems. Agriculture is the largest single nonpoint source of water pollutants, including sediments, salts, fertilizers, pesticides, and manures. Nonpoint pollutants account for an estimated 50 percent of all surface water pollution (Chesters and Schierow, 1985; Myers et al., 1985). Salinization of soils and irrigation water from irrigated agriculture is a growing problem in the arid West. In at least 26 states, some pesticides have found their way into groundwater as a result of normal agricultural practice. In California alone, 22 different pesticides have been detected in groundwater as a result of normal agricultural practices. Nitrate from agricultural sources (principally manures and synthetic fertilizers) is found in drinking water wells in levels above safety standards in many locations in several states.

Agriculture presents other environmental problems. Major aquifers in California and the Great Plains have been depleted because withdrawals exceeded recharge rates. Cultivation of marginal lands has caused soil erosion. The use of certain pesticides on some crops and antibiotics in animal production for disease control and growth promotion presents risks that may be avoidable.

Agricultural leaders and policymakers are currently confronting questions about contemporary production practices. These questions are the subject of this chapter. It is important to note that many problems discussed in this report are prevalent only in certain regions and under specific management practices. Almost all of these problems can be overcome. Nonetheless, problems such as groundwater contamination will likely grow if current practices are continued.

Many of these problems have developed in large part as a result of public policies and thus may be overcome through policy reform. The important link among all of these problems is that productive and profitable alternative practices are available in most cases and are already implemented in some. The benefits of alternatives in addressing these problems are presented in subsequent chapters.

Publicly and privately funded agricultural research since World War II has created a wealth of technology and information. This information and technology has led to vastly increased yields of a number of commodities and has reinforced movement toward specialization. High deficiency and disaster payments for most program crops reduced risks and further accelerated specialization. The development of specialized large farm equipment made it possible for individual farmers to grow one crop or a few related crops on more acres. Because of these trends, farmers were able to take advantage of market forces in the 1970s that stimulated demand for U.S. agricultural commodities.

THE FARM ECONOMY

In their desire to accelerate industrial growth, many developing countries neglected their agricultural sectors in the 1950s and 1960s. By the 1970s, a growing number of developing nations needed to import food to feed rapidly growing populations. Many of them imported food from the United States. Growing trade with Pacific Rim nations and trade agreements with the Soviet Union further expanded available markets for the United States. The 1970s also brought generally favorable weather for agriculture to the United States and unfavorable conditions to many other countries. Tax policies such as accelerated depreciation and capital gains preferences encouraged machinery purchases and cultivation and irrigation of previously uncultivated or erodible land. Crop prices were well above the loan rate; expanded exports were used to offset the trade deficit created by oil imports. The result was greater demand for U.S. commodities, higher prices, and an all-out effort by U.S. farmers to increase production.

Farmland prices followed the upward movement of commodity prices, inflation, and negative real interest rates. In some midwestern states, the

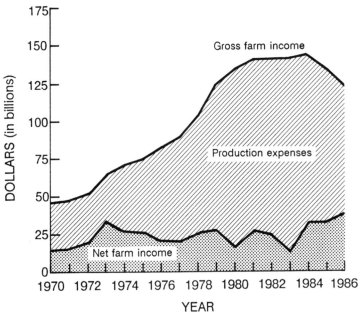

FIGURE 2-1 Gross farm income and production expenses. SOURCE: U.S. Department of Agriculture. 1988. 1988 Agricultural Chartbook. Agriculture Handbook No. 673. Washington, D.C.

price of farmland increased by 15 percent or more per year. Rising land and commodity prices led farmers to increase purchases of inputs such as fertilizers, seeds, chemicals, and equipment. Production expenses and gross farm income soared as farmers responded to a growing market (Figure 2-1). Agricultural lending organizations, responding to inflation and rising market values of farm assets, were eager to make loans to farmers. Total farm debt went from $52.8 billion in 1970 to a peak of $206.5 billion in 1983 (U.S. Department of Agriculture, 1987c).

In late 1979, events began to change the economic, political, and social environment of agricultural production. Policy changes caused increases in real interest rates and the virtual end of inflation. Prices received for crops began to level off and drop although input prices continued to rise through 1984 (Figure 2-2). Demand for U.S. agricultural commodities declined as a result of the increased value of the dollar; fixed loan rates; foreign competition from the European Community (EC), Argentina, Australia, and Brazil; foreign debt; global recession; and reduction of U.S. loans to developing countries to buy food. Commodity surpluses around the world swelled, and prices dropped. Falling commodity prices deflated land values, which fell by 1986 to less than half their 1980 value in many agricultural areas.

In a few years, prosperity turned into economic recession. Many farmers borrowed heavily in the 1970s to invest in land and machinery and take

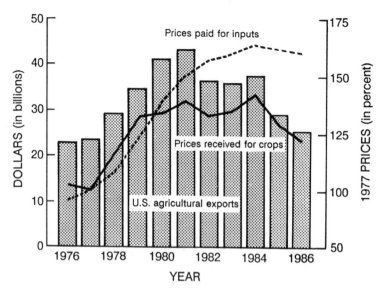

FIGURE 2-2 Input prices, crop prices, and agricultural exports. SOURCE: U.S. Department of Agriculture. 1988. 1988 Agricultural Chartbook. Agriculture Handbook No. 673. Washington, D.C.

advantage of high crop prices. The sudden change in the economic environment placed those with the greatest debt in the most vulnerable position. The debt-to-asset ratio suddenly became a major criterion for a farm's viability.

The financial plight of farmers also affected the farm credit sector. One-fourth of all farm loans—$33.7 billion—from the Farmers Home Administration (FmHA), federal land banks, production credit associations, commercial banks, and life insurance companies were nonperforming or delinquent in 1984 and 1985 (U.S. General Accounting Office, 1986a). The farm credit system lost $4.6 billion in 1985 and 1986. Agricultural banks accounted for more than half of 1985 bank failures, although they comprise only one-fourth of all banks (U.S. General Accounting Office, 1986a). New rules to implement the Agricultural Credit Act of 1987, however, will help to keep tens of thousands of farmers on their land. The act requires the FmHA, the farmers' bank of last resort, to make all feasible efforts to restructure loans, including forgiving debt. Up to $7 billion in debt and interest may be written off under this program.

Suppliers of farm inputs have also been hurt by bad debt and federal supply-control programs that have reduced sales. Farm machinery sales, for example, fell more than 50 percent from 1980 to 1985. In Nebraska and Iowa alone, hundreds of farm implement dealers have gone out of business since 1985. The industry has recovered somewhat since 1986 as farm income has risen.

As of January 1, 1988, 4 percent of farms were technically insolvent because debt exceeded assets. An additional 4.9 percent of farms had debt-to-asset ratios of 70 to 100 percent, and 10.0 percent had debt-to-asset ratios of 40 to 70 percent (U.S. Department of Agriculture, 1988d). Farms with ratios above 70 percent generally experience serious financial problems. Those with debt-to-asset ratios of 40 to 70 percent face declining equity unless commodity prices are strong or production expenses fall, which they have since 1983.

Although some farmers experienced financial hardship in the 1980s, many prospered. Total net farm income was $37.5 billion in 1986 and a record $46.3 billion in 1987 (see Figure 1-29). Off-farm income totaled a record $44.7 billion in 1986 (Van Chantfort, 1987). Table 2-1 shows that most farms had positive income in 1987, and that debt is now concentrated in farms with sales over $250,000. This record income and reduction in debt was made possible, however, only by record levels of government support.

In 1987, 44 percent of all farmers had no long-term debt. The average debt-to-asset ratio, which reached 25 percent in 1985, fell to 22 percent in 1986 and 15 percent in 1987 (Figure 2-3). Total farm debt fell from $206.5 billion in 1983 to $150 billion in 1988 (U.S. Department of Agriculture, 1988a).

Federal programs can have a great effect on the agricultural economy. In general, they are slow to alleviate the economic problems of farmers with the greatest need. This has been evident in the 1980s. Well over one-half of all major commodity producers have been enrolled in the programs since 1983. But 60 percent of direct government payments in 1985, for example, went to only 14 percent of all operators with net cash incomes averaging nearly $130,000 (Agricultural Policy Working Group, 1988). This is largely because federal payments are based on farm yields and sales. Even though Congress has limited certain categories of federal payments to $50,000 per farm, many growers have found ways to reorganize their operations to avoid this and other limitations.

TRADE

U.S. agriculture built a substantial trade surplus during the 1970s as the manufactured goods sector slipped into a deepening trade deficit. The U.S. agricultural trade balance deteriorated in the 1980s, however, falling from $27 billion in 1980 to $6 billion in 1986 (U.S. General Accounting Office, 1986b). The United States depends primarily on grain and oil seed exports; growth in this market is slowing as the U.S. share declined from 72 percent in 1979 and 1980 to 50 percent in 1986 (U.S. Department of Agriculture, 1986b).

The trade situation has improved since 1986; exports are expected to increase to about $33 billion in 1988, with the trade surplus rising to between $12 billion and $13 billion. The U.S. agricultural trade balance has increased, in part because of a drop in market prices for most export commodities. Government subsidies, credit guarantees, and product promotion

TABLE 2-1   Farm Financial Conditions by Farm Size, Region, and Commodity

Factor	Percentage of Farms in Each Financial Condition			
	Favorable (Positive Income and Favorable Solvency)	Negative Income- Favorable Solvency	Marginal Solvency- Positive Income	Vulnerable (Negative Income and Marginal Solvency)
Farm size				
≥ $250,000	59	14	20	7
$40,000–249,999	64	12	17	6
< $40,000	71	19	6	4
Region				
Northeast	68	22	7	3
Great Lakes	59	1	15	7
Corn Belt	71	12	13	5
Northern Plains	64	17	15	5
Appalachia	76	16	5	3
Southeast	73	18	6	4
Delta	72	16	8	4
Southern Plains	69	20	8	4
Mountain	64	20	10	6
Pacific	67	18	9	7
Farm type				
Cash grain	65	14	14	7
Tobacco	78	9	8	5
Cotton	65	11	15	9
Vegetable, fruit, nut	71	16	9	3
Nursery-greenhouse	80	12	6	2
Other field crops	65	17	10	7
Beef, hog, sheep	70	20	7	3
Dairy	63	12	20	5
Poultry	73	6	16	6
Other livestock	58	30	5	7

NOTE: The income measure used in these statistics is net cash farm income; marginal solvency indicates a debt-asset ratio of 40 percent or more. Favorable solvency indicates a debt-asset ratio of 40 percent or less. Adding across, numbers may not total exactly to 100 percent because of rounding.

SOURCE: U.S. Department of Agriculture. 1988. Financial Characteristics of U.S. Farms, January 1, 1988. Agriculture Information Bulletin No. 551. Economic Research Service. Washington, D.C.

also supported increased exports. The rise in export volume, however, far exceeded the increase in the value of exports in current dollars largely due to the declining value of the dollar (Figures 2-4 and 2-5) (U.S. Department of Agriculture, 1987e).

Meanwhile, the United States is increasing its imports of high-value products such as processed foods and horticultural products. The United States accounts for about 10 percent of the value of world trade in high-value markets, primarily through exports of soybean meal, tobacco, cigarettes, cattle hides, and corn-gluten feed. Imports of supplementary high-value commodities (crops also produced in the United States) have increased from

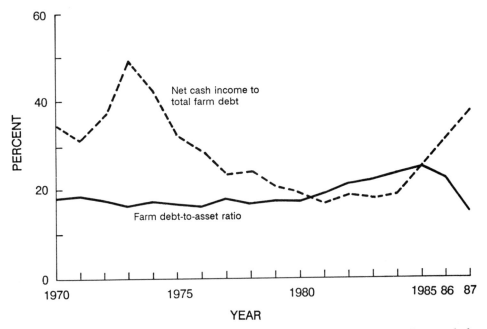

FIGURE 2-3 Farm debt-to-asset and net-cash-income-to-total-farm-debt ratio. Data exclude households. SOURCE: U.S. Department of Agriculture. 1988. 1988 Agricultural Chartbook. Agriculture Handbook No. 673. Washington, D.C.

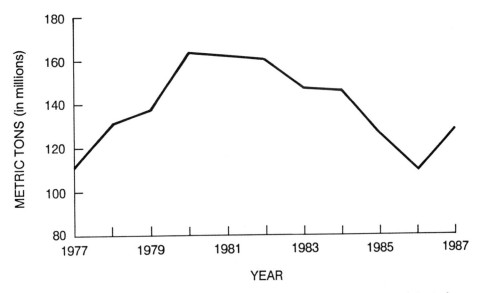

FIGURE 2-4 Volume of U.S. agricultural exports. SOURCE: U.S. Department of Agriculture. 1988. 1988 Agricultural Chartbook. Agriculture Handbook No. 673. Washington, D.C.

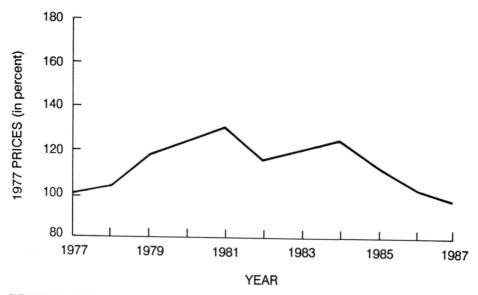

FIGURE 2-5 Value of U.S. agricultural exports. SOURCE: U.S. Department of Agriculture. 1988. 1988 Agricultural Chartbook. Agriculture Handbook No. 673. Washington, D.C.

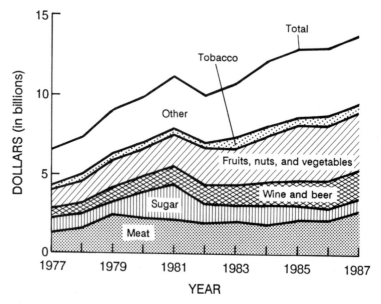

FIGURE 2-6 Value of supplementary commodity imports. SOURCE: U.S. Department of Agriculture. 1988. 1988 Agricultural Chartbook. Agriculture Handbook No. 673. Washington, D.C.

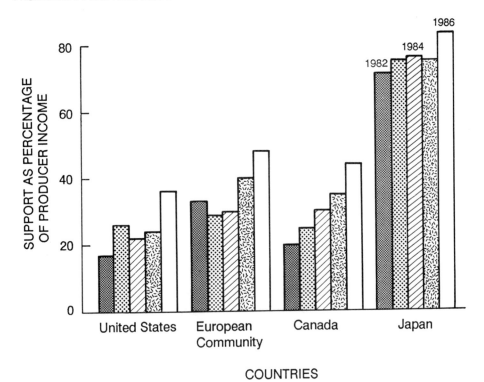

FIGURE 2-7 Average producer subsidy equivalents for grains, livestock, dairy, oilseeds, and sugar. The European Community is Belgium, Denmark, France, Greece, Ireland, Italy, Luxembourg, the Netherlands, Portugal, Spain, the United Kingdom, and West Germany. SOURCE: U.S. Department of Agriculture. 1988. 1988 Agricultural Chartbook. Agriculture Handbook No. 673. Washington, D.C.

$7 billion in 1977 to almost $14 billion in 1987 (Figure 2-6). The total value of agricultural imports reached $20 billion in 1987.

Increasing competition is also contributing to the rising cost of federal agricultural subsidies. The government spent $25.8 billion in 1986 and $22.0 billion in 1987 for price supports and related activities. Of this amount, $11.8 billion in 1986 and $16.7 billion in 1987 were direct payments to farmers (U.S. General Accounting Office, 1988). Nonetheless, U.S. agricultural subsidies as a percentage of producer income are far less than those of the EC, Canada, and Japan (Figure 2-7).

NATURAL RESOURCES

The diversity in plant and animal products produced in the United States has increased in the past three decades, but individual farms have become more specialized. Technology has contributed to a shift from multi-enterprise farming operations to those having as few as one or two income-generating crops or products. Recently, however, this trend of specialization has slowed down. Over the past decade, many farmers have adopted alternative methods more consistent with the goals of profitability with less government support and greater natural resource and human health protection.

The following section is a brief review of the adverse consequences that some current agricultural practices have on natural resources and the environment. It must be emphasized that many conventional agricultural practices are environmentally sound and are components of certain alternative strategies. The following analyses are not intended to be fully comprehensive; however, they do illustrate the factors that must be considered in any agricultural production system.

Water Quality

Surface Water

Water pollution is probably the most damaging and widespread environmental effect of agricultural production. Agriculture is the largest nonpoint source of water pollution, which accounts for about half of all water pollution (Chesters and Schierow, 1985; Myers et al., 1985). Under sections 304(f) and 305(b) of the Clean Water Act of 1972 as amended, 17 states and Puerto Rico identified agriculture as a primary or major nonpoint source of water pollution, and 27 states and the Virgin Islands identified it as a problem (Table 2-2) (U.S. Environmental Protection Agency, 1984). Surface water damage from agriculture is estimated at between $2 billion and $16 billion per year. These estimates are approximate, however, and may underestimate the long-term costs of pollution.

Precipitation- and irrigation-induced runoff carries sediment, minerals, nutrients, and pesticides into rivers, streams, lakes, and estuaries. Most experts consider erosion's effects on water resources to be greater than its potential effects on productivity (National Research Council, 1986c; Schneider, 1986). The U.S. Department of Agriculture (USDA) calculates that the economic cost of off-farm water pollution due to agricultural erosion is from two to eight times the value of erosion's effect on productivity (U.S. Department of Agriculture, 1987a). This comparison, however, is crude.

Sediment deposition and nutrient loading are major agricultural water pollution problems (Clark et al., 1985; U.S. Department of Agriculture, 1987a). Agriculture accounts for more than 50 percent of suspended sediments from all sources discharged into surface waters (U.S. Department of Agriculture, 1987a). In predominantly agricultural regions, these percentages are higher; in other regions, agriculture's contribution is less. Nationwide trends in surface water sediment deposition between 1974 and 1981 were significantly related to cropland erosion within basins. They were not closely related to estimates of total basin erosion from forestland, pastureland, or rangeland (Smith et al., 1987).

TABLE 2-2 Agriculture (Including Feedlots) as a Nonpoint Source of Water Pollution by State or Territory

Agriculture Identified as a Primary or Major Nonpoint Source of Water Pollution		Agriculture Identified as a Nonpoint Source Pollution Problem	
Delaware	Montana	Alabama	Nevada
Idaho	North Dakota	Arizona	New Jersey
Illinois	Ohio	Arkansas	New Mexico
Indiana	Oregon	California	New York
Iowa	Puerto Rico	Colorado	North Carolina
Kansas	South Dakota	Florida	Oklahoma
Kentucky	Utah	Georgia	Pennsylvania
Minnesota	Vermont	Hawaii	South Carolina
Mississippi	Washington	Louisiana	Tennessee
		Maine	Virgin Islands
		Maryland	Virginia
		Michigan	West Virginia
		Missouri	Wisconsin
		Nebraska	Wyoming

SOURCE: U.S. Environmental Protection Agency. 1984. Report to Congress: Nonpoint Source Pollution in the U.S. Office of Water Program Operations, Water Planning Division. Washington, D.C.

The principal consequence of sediment loading is increased turbidity, which causes decreased light for submerged aquatic vegetation. Species that depend on aquatic vegetation for breeding and food can thus experience stress and decline. Sediment also has direct economic consequences when it fills reservoirs, clogs navigable waterways, reduces recreational use of waters, and increases operating costs of water-treatment facilities. Between 675 million and 1 billion tons of eroded agricultural soils are deposited in waterways each year (National Research Council, 1986c; Schneider, 1986; U.S. Department of Agriculture, 1986a). The USDA (1988e) estimates that the removal from production of 30 million to 40 million acres of highly erodible land through the Conservation Reserve Program (CRP) will reduce sediment delivery to surface waters by as much as 200 million tons per year.

Phipps and Crosson (1986) and the USDA (1987a) estimate that between 50 and 70 percent of all nutrients reaching surface waters, principally nitrogen and phosphorus, originate on agricultural land in the form of fertilizer or animal waste. Nitrate, which is relatively soluble, is carried in solution by water; phosphorus is most often carried attached to sediment. From 1974 to 1981, 116 stations from the National Stream Quality Accounting Network and the National Water Quality Surveillance System reported increasing nitrate concentrations; only 27 stations reported decreases. Elevated nitrogen levels were strongly associated with agricultural activity and atmospheric deposition of nitrogen in rainfall. Phosphorus deposition has been less consistently observed because increases are closely linked to levels of suspended sediments. Nitrate moves with water; thus, nitrogen move-

Flooding and extensive water runoff carry sediment, fertilizers, and pesticides into surface water. Agriculture contributes one-half of all nonpoint surface water pollution. Flooding can also severely damage crops, land, and buildings. *Credit: Agrichemical Age.*

ment into surface waters more fully reflects the effects of agricultural activity than phosphorus movement does. Phosphorus moves as a passenger bound with sediment, much of which erodes from fields but is not deposited in surface waters (Smith et al., 1987).

Nutrient loading has had a devastating effect on many lakes, rivers, and bays throughout the country. Increased nutrient levels, particularly of phosphorus, stimulate algal growth, which can accelerate the natural process of eutrophication. In its later stages, the algal growth stimulated by nutrients will die and decay, which can significantly deplete available oxygen and reduce higher-order aquatic plant and animal populations. The accelerated eutrophication of the Chesapeake Bay is an excellent example of the consequences of nutrient loading from agricultural and municipal sources. Nutrient loading in the bay has contributed to a significant decline in the bay fisheries (Kahn and Kemp, 1985).

Sediment and nutrient runoff from agricultural land plays a part in estuary degradation. For 78 estuaries examined by the USDA, agricultural runoff supplied on average 24 percent of all nutrient loading and 40 percent of total sediment. In 22 of the 78 estuaries, agriculture contributed more than

25 percent of total nitrogen and phosphorus pollution. High rates of pesticide runoff (greater than 30 percent above the average of all coastal states) were found in 21 estuary systems. High nutrient *and* pesticide runoffs were found in 15 systems (U.S. Department of Agriculture, 1988b).

Between 450 million and 500 million pounds of pesticides are applied to row crops each year. The majority of these are herbicides, most of which are applied before planting, and many of which are incorporated into the soil. Probably less than 5 percent of all pesticides applied reach a body of water (Phipps and Crosson, 1986). The highest concentrations of pesticides are related to agricultural runoff into streams and lakes. In intensively farmed states, such as Iowa, Minnesota, and Ohio, a number of the widely used corn and soybean herbicides have been detected in rivers, many of which serve as drinking water sources. In humid areas where groundwater contributes a major proportion of stream flow, some herbicides may be delivered to surface water via groundwater (Hallberg, 1987).

It appears that many of these herbicides are not effectively removed from drinking water by conventional treatment or more sophisticated carbon filtration systems (Table 2-3). Maximum and mean levels of 10 herbicides detected in treated drinking water in Ohio and Iowa are shown in Table 2-4. In Iowa, 27 of the 33 public water supplies from surface water sources tested, or 82 percent, had 2 or more pesticides detected in treated drinking water samples; 73 percent had 3 or more; 58 percent had 4 or more; and 21 percent had 5 or more (Table 2-5). These samples were collected after rainfall between mid-April and July 30, 1986, when most herbicides are applied; consequently, they may represent a peak of exposure to those compounds. The mean detection levels were below 4.0 parts per billion (ppb). Samples collected and analyzed by Monsanto between May 1985 and March 1986, however, had only slightly lower percentages of detection for all herbicides except alachlor, which was far lower, in treated drinking water from surface water sources (Wnuk et al., 1987).

Glenn and Angle (1987) studied the effect of tillage systems on runoff of the herbicides atrazine and simazine in the Chesapeake Bay watershed. There was less runoff of water, atrazine, and simazine from the untilled fields compared to conventionally tilled fields each year that a major rain occurred during the growing season. From 1979 through 1982, total runoff from fields in the untilled watershed was 27 percent less than that from the conventionally tilled watershed. The highest levels of atrazine in runoff water were 1.332 micrograms/liter and 0.975 micrograms/liter in conventional tillage and no-tillage systems, respectively, or 1.6 and 1.1 percent of atrazine applied at 2 pounds/acre. The highest reported runoff of simazine at 2 pounds/acre was 0.456 micrograms/liter or 0.5 percent for conventional tillage and 0.210 micrograms/liter or 0.36 percent of the total applied for no tillage. Most of the runoff occurred within 2 weeks of application. In 1981 and 1982, when rainfall was delayed more than 2 weeks, levels declined substantially. Losses of up to 16 percent for atrazine and up to 3.5 percent for simazine have been reported (Glenn and Angle, 1987).

TABLE 2-3 Pesticide Detections Reported in Iowa Drinking Water Survey (in percent)

Water Type	Atrazine	Cyanazine	Metolachlor	Alachlor	Metribuzin	Trifluralin	Butylate	Carbofuran	2,4-D	Dicamba
Treated water	30/33 (91)	26/33 (79)	21/33 (64)	17/33 (52)	4/33 (12)	1/33 (3)	1/33 (3)	9/33 (27)	2/30 (7)	1/30 (3)
Untreated water	14/15 (93)	11/15 (73)	11/15 (73)	10/15 (67)	1/15 (7)	0/15 (0)	0/15 (0)	5/15 (33)	2/14 (14)	1/14 (7)
Total	44/48 (92)	37/48 (77)	32/48 (67)	27/48 (56)	5/48 (10)	1/48 (2)	1/48 (2)	14/48 (29)	4/44 (9)	2/44 (5)

NOTE: Values indicate number of positive detections/number of samples analyzed (percent).

For mean levels detected, see Table 2-4. Means for all compounds were below 4.0 μg/l. Samples were taken in the spring when the presence of compounds in water may have been highest as a result of agricultural runoff.

SOURCE: Wnuk, M., R. Kelley, G. Breuer, and L. Johnson. 1987. Pesticides in Water Supplies Using Surface Water Sources. Iowa City, Iowa: Iowa Department of Natural Resources and University Hygienic Laboratory.

TABLE 2-4 Pesticide Concentrations From Finished (Treated) Public Drinking Water Supplies Derived From Surface Waters (in micrograms/liter)

Pesticide	Maximum Reported		Mean in Iowa	Residues Detected, Iowa City, Iowa, May 18–19, 1986
	Ohio	Iowa		
Herbicides				
Alachlor	14.3	8.8	1.1	8.8
Atrazine	30.0	24.0	3.8	15.0
Butylate	—	0.3	0.27	—
Cyanazine	2.4	17.0	2.7	7.2
2,4-D	—	0.2	0.23	0.2
Dicamba	—	1.4	1.4[a]	—
Linuron	0.6	—	—	—
Metolachlor	24.2	21.0	2.9	10.0
Metribuzin	—	0.3	0.29	0.3
Simazine	1.0	—	—	—
Trifluralin	—	0.1	0.13[a]	—
Insecticide				
Carbofuran	—	14.0	—	6.0

NOTE: Maximum levels reported from studies in Ohio (D. Baker, 1985) and Iowa (M. Wnuk et al., 1987), and multiple residues in Iowa City, Iowa, tap water May 18–19, 1986.

[a]Only one detection (see Table 2-3).

SOURCES: Baker, D. B. 1985. Regional water quality impacts of intensive row-crop agriculture: Lake Erie Basin case study. Journal of Soil and Water Conservation 40(1):125–132; Hallberg, G. R. 1987. Agricultural chemicals in ground water: Extent and implications. American Journal of Alternative Agriculture 2(1):3–15; Kelley, R. D. 1987. Pesticides in Iowa's drinking water. Pp. 115–135 in Pesticides and Groundwater: A Health Concern for the Midwest. Navarre, Minn.: The Freshwater Foundation and U.S. Environmental Protection Agency; Wnuk, M., R. Kelley, G. Breuer, and L. Johnson. 1987. Pesticides in Water Supplies Using Surface Water Sources. Iowa City, Iowa: Iowa Department of Natural Resources and University Hygienic Laboratory.

The USDA predicts that pesticide and fertilizer use may be reduced by 61 million pounds and 1.4 million tons, respectively, from 1985 levels as a result of land idled under the CRP. Because idled land is highly erodible, the reduction in fertilizers and pesticides reaching surface water through runoff may be proportionally greater as a percentage of total pesticides and fertilizers applied (U.S. Department of Agriculture, 1988e).

Declining but detectable levels of many chlorinated hydrocarbon pesticides are still found in several fish, shellfish, and bird species and waterborne sediments, particularly in the Great Lakes (Hileman, 1988). Levels of dichloro diphenyl trichloroethane (DDT) in fish contributed to a dramatic decline in predatory bird populations, such as the peregrine falcon, osprey, and bald eagle in the 1960s and 1970s. The use of most persistent organochlorine pesticides has been phased out in the United States, although they continue to enter the environment as inert ingredients in a few currently used pesticides, and through their continued use worldwide. Some of these compounds are carried through the atmosphere and deposited far from

TABLE 2-5   Number of Pesticides Detected in Treated Drinking Water Samples in Iowa

Number of Individual Pesticides Detected in Treated Water Samples	Number of Supplies with the Number of Pesticides Listed in Column 1	Population (number) Served by These Supplies	Percentage of Water Supplies in the Study Containing an Equal or Greater Number of Pesticide Residues Listed in Column 1
1	3	136,725	91
2	3	2,828	82
3	5	33,222	73
4	12	115,485	58
5	3	239,386	21
6	3	15,874	9
7	1	20,000	3

SOURCE: Wnuk, M., R. Kelley, G. Breuer, and L. Johnson. 1987. Pesticides in Water Supplies Using Surface Water Sources. Iowa City, Iowa: Iowa Department of Natural Resources and University Hygienic Laboratory.

their point of application. The presence of toxaphene, chlordane, and other chlorinated hydrocarbon compounds in the Great Lakes is an example of this phenomenon (Hileman, 1988). The U.S. Environmental Protection Agency (EPA) (1987) reports that other pesticides, notably the herbicide alachlor, have been detected at up to 6.59 ppb in rainwater.

Mineralization and salinization of soils and irrigation wastewater are growing problems in irrigated agriculture, primarily in the West. Soil salinization and mineralization reduce crop yields, and, if not corrected, will ultimately leave the land unfit for agricultural purposes. The adverse reproductive effect of selenium on waterfowl in the Kesterson wildlife refuge in California is the most publicized example of the nonagricultural effects of salinization. The salinization and depletion of the Colorado River from its use as agricultural irrigation water throughout its course is perhaps the most vivid example of agriculture's effect on water quality and quantity in the West.

Amendments to the Clean Water Act in 1987 require states to report their principal nonpoint sources of water pollution and programs in place to mitigate the problem. The act does not require implementation of measures to reduce nonpoint source pollution of surface waters, however. In 1988, the USDA's National Program for Soil and Water Conservation and the Rural Clean Water Program conducted 22 water quality improvement projects around the nation. Other provisions of the Food Security Act of 1985, such as the CRP and conservation compliance, will also help reduce agricultural nonpoint surface water pollution. Incentives integrated into agricultural conservation and commodity programs will likely remain the most effective way to reduce surface water pollution from agricultural sources, in lieu of further amendments to the Clean Water Act or regulations promulgated under the act.

Groundwater

Groundwater is the source of public drinking water for nearly 75 million people. Private water wells supply water to an additional 30 million individuals. Nearly 50 percent of all drinking water, 97 percent of all rural drinking water, 55 percent of livestock water, and more than 40 percent of all irrigation water is from underground sources. Accumulating evidence indicates that a growing number of contaminants from agricultural production are now found in underground water supplies (National Research Council, 1986b; U.S. Department of Agriculture, 1987b, 1987d).

Increased use of nitrogen fertilizers and pesticides, particularly herbicides, over the past 40 years has raised the potential for groundwater contamination. Greater use of feedlots that concentrate manure production also heightens this risk. Several of the most widely used pesticides have the potential to leach into groundwater as a result of normal agricultural use. The EPA has initiated a nationwide survey of pesticides in groundwater with results anticipated in 1990. The high-priority pesticides in that survey are listed in Table 2-6 (U.S. Department of Agriculture, 1987d).

Pesticides have been detected in the groundwater of 26 states as a result of normal agricultural practices (Williams et al., 1988). The most commonly detected compounds are the herbicide atrazine and the insecticide aldicarb. Aldicarb, the most acutely toxic pesticide registered by the EPA (LD_{50}*, 0.9 milligrams/kilogram), has been found in 16 states; in many states, however, detections are isolated. Atrazine, the second most used herbicide in the nation, has been found in the groundwater of at least 5 states, usually at levels between 0.3 and 3.0 micrograms/liter. Tests show that atrazine is oncogenic in laboratory rats. The EPA is currently reviewing these studies but has not yet classified atrazine as an oncogen. The herbicide alachlor, recently banned in Canada and classified by the EPA as a probable human carcinogen, is the next most commonly detected pesticide in groundwater. It has been found in 12 states at a median concentration of 0.90 micrograms/liter. Pesticides detected in groundwater as a result of agricultural use in 26 states are listed in Table 2-7. Pesticides detected in groundwater used for drinking water in Iowa and Minnesota are listed in Table 2-8.

A survey by the U.S. Geological Survey (USGS) of 1,663 counties showed 474 counties in which 25 percent of the wells tested had nitrate-nitrogen (NO_3N) levels in excess of 3 milligrams/liter (Figure 2-8). Levels above 3 milligrams/liter are considered elevated by human activities, primarily nitrogen fertilizer use (Nielsen and Lee, 1987). In 87 of the 474 counties, at least 25 percent of the sampled wells exceeded the EPA's 10 milligrams/liter interim standard for nitrate in drinking water. Prolonged exposure to levels exceeding this standard can lead to methemoglobinemia (oxygen deficit in the blood), although reported instances of this condition have been rare. The USDA (1987d) predicted that wells in an additional 149 counties may

---

*LD_{50}, or the Lethal Dose 50, is the dose of a substance that kills 50 percent of the test animals exposed to it. The lethal dose can be measured orally or dermally.

TABLE 2-6 Priority Pesticides in EPA's National Survey of Pesticides in Groundwater (in thousands of pounds)

Pesticide	Type[a]	Estimated Use[b]	EPA Description
Acifluorfen	H	1,399	Leacher
Alachlor	H	85,015	Leacher
Aldicarb	I, N	2,271	Mobile; marginal persistence
Ametryn	H	96	Leacher
Atrazine	H	77,316	Leacher
Bentazon	H	8,410	Leacher; toxicological concern
Bromacil	H	1,234	Leacher
Butylate	H	55,095	Mobile; uncertain persistence; toxicological concern
Carbofuran	I, A, N	7,695	Leacher
Chloramben	H	6,069	Leacher
Chlordane	I	11	Persistent; possible direct contamination via termiticide use
Cyanazine	H	21,626	Leacher
Cycloate	H	52	Mobile; uncertain persistence; toxicological concern
2,4-D	H	37,217	Marginal leacher; heavy use
Dalapon	H	261	Leacher
DCPA	H	196	Leacher
Dicamba	H	4,158	Leacher
Dinoseb	H	8,835	Leacher
Diphenamid	H	698	Marginal leacher; toxicological data gaps
Disulfoton	I, A	2,105	Leacher
Diuron	H	1,861	Leacher
Fenamiphos	I, N	348	Moderate leacher; toxicological concern
Fluometuron	H	2,943	Leacher
Hexazinone	H	11	Leacher
Maleic hydrazide	H	287	Leacher; toxicological data gaps
MCPA	H	9,861	Marginal leacher
Methomyl	I	425	Leacher
Metolachlor	H	37,940	Leacher
Metribuzin	H	10,603	Leacher
Oxamyl	I, A, N	51	Leacher
Picloram	H	549	Leacher
Pronamide	H	83	Leacher
Propazine	H	1,287	Leacher
Propham	H	445	Leacher
Simazine	H	3,975	Leacher
2,4,5-T	H	204	Marginal leacher
2,4,5-TP	H	7	Marginal leacher
Terbacil	H	833	Leacher

[a]Abbreviations: A = acaricide; H = herbicide; I = insecticide; N = nematicide.
[b]Thousands of pounds of active ingredient per year used for agricultural purposes only.

SOURCE: U.S. Department of Agriculture. 1987. The Magnitude and Costs of Groundwater Contamination From Agricultural Chemicals—A National Perspective. Staff Report AGES870318. Economic Research Service. Washington, D.C.

TABLE 2-7   Confirmed Pesticide Detections in Groundwater Due to Normal Agricultural Use

Pesticide	Health Advisory Level[a] (parts per billion)	States	Median Concentration[b] (parts per billion)
Alachlor	1.5[c]	CT, FL, IL, IA, KS, LA, MA, ME, NE, PA, WI	0.90
Aldicarb	10	CA, FL, MA, NC, NY, RI, WI	9.00
Aldrin		MS, SC	0.10
Arsenic		TX	N/A
Atraton		MD	0.10
Atrazine	3.0	CA, CO, CT, IL, IA, KS, MD, ME, NE, NJ, PA, VT, WI	0.50
BHC		MS	2.70
Bromacil	80	CA, FL	9.00
Carbofuran	36	MA, NY, RI	5.30
Chlordane	0.03[c]	MS	1.70
Chlorothalonil	1.5[c]	ME, NY	0.02
Cyanazine	9.0	IA, LA, MD, NE, PA, VT	0.40
1,2-D	0.0013[c]	CA, CT, MA, NY	4.50
1,3-D	0.20[c]	NY	123.00
2,4-D	70	CT, MS	1.40
DCPA	3,500	NY	109.00
DDT		MS, NJ, SC	1.70
Diazinon	0.63	MS	162.00
Dibromochloropropane	0.02[c]	AZ, CA	0.01
Dicamba	9.0	CT, ME	0.60
Dieldrin	0.00219[c]	NE, NJ	0.02
Dinoseb	7.0	MA, ME, NY	0.70
Diuron	14	CA	N/A
Endosulfan		ME	0.30
Ethoprop		NY	N/A
Ethylene dibromide	0.0005[c]	CA, CT, GA, MA, NY, WA	0.90
Fonofos	14	IA, NE	0.10

(Table 2-7 continued on page 108)

have contaminated water based on high susceptibility to contamination and fertilizer use (Figure 2-9) (Nielsen and Lee, 1987).

The USDA calculates that 1,437 counties, or 46 percent of all U.S. counties, contain groundwater susceptible to contamination from agricultural pesticides or fertilizers (Figure 2-10). An estimated 54 million people living in these counties rely on underground sources of drinking water. The costs or benefits of decontaminating this water are not currently quantifiable. It is likely, however, that contamination in certain regions will persist for many years after remedial actions are taken (Nielsen and Lee, 1987). Several states (including California, Florida, Iowa, New York, and Wisconsin) have developed strategies for dealing with agriculturally induced groundwater contamination. But changes in agricultural practices to reduce groundwater contamination are not widespread. The EPA is also developing a national

TABLE 2-7   (Continued)

Pesticide	Health Advisory Level[a] (parts per billion)	States	Median Concentration[b] (parts per billion)
Hexazinone	210	ME	8.00
Lindane	0.026[c]	MS, NJ, SC	0.10
Linuron		WI	1.90
Malathion		MS	41.50
Methamidophos		ME	4.80
Methomyl	175	NY	N/A
Methyl parathion	2.0	MS	88.40
Metolachlor	10	CT, IL, IA, PA, WI	0.40
Metribuzin	175	IL, IA, KS, WI	0.60
Oxamyl	175	MA, NY, RI	4.30
Parathion		ND	0.03
Picloram	490	ME, ND, WI	1.40
Prometon	100	TX	16.60
Propazine	14	NE, PA	0.20
Simazine	35	CA, CT, MD, NE, NJ, PA, VT	0.30
Sulprofos		IA	1.40
TDE	0.031	MS	4.80
Toxaphene		MS	3,205.00
Trifluralin	2.0	KS, MD, MS, NE	0.40

[a]The EPA sets the Proposed Lifetime Health Advisory Level. The EPA has not set levels for all pesticides.

[b]Median of the concentration of positive detections for all confirmed studies. If multiple studies were not done on a particular chemical, the single study average is given. If the data base reports a single positive well, then the average concentration reported for that well is given.

[c]For carcinogens, the Proposed Lifetime Health Advisory Level is based on the exposure levels that present a 1 in a million risk of cancer in the exposed population.

SOURCE: Williams, W. M., P. W. Holden, D. W. Parsons, and M. N. Lorber. 1988. Pesticides in Ground Water Data Base: 1988 Interim Report. Office of Pesticide Programs. U.S. Environmental Protection Agency. Washington, D.C.

groundwater protection strategy. Once the EPA's ongoing survey of pesticides in groundwater is complete, additional time will be needed to carry out detailed risk-benefit assessments required by the Federal Insecticide, Fungicide and Rodenticide Act (FIFRA). The committee notes that opportunities exist today to reduce surface water and groundwater contamination from agricultural chemicals through modified agricultural practices. Some of the modifications include increased use of legumes as a nitrogen source, adoption of integrated pest management (IPM), or shifts in regional cropping patterns.

The Effects of Irrigation

Irrigated agricultural acreage doubled from 25 million acres in 1949 to slightly more than 50 million acres in 1978. Since then, total irrigated acre-

TABLE 2-8 Pesticides Detected in Underground Drinking Water Supplies in Iowa and Minnesota

Common Name of Active Ingredient	Maximum Concentration (micrograms/liter) in Iowa and Minnesota[a]	Mean Concentration (micrograms/liter) in Iowa and Minnesota	Percentage of Detections in Iowa and Minnesota[a]
Herbicides			
Alachlor	16.6/9.8	0.5	15/11
Atrazine	21.1/42.4	0.2	72/72
Chloramben	1.7/N.D.	—	<1/0
Cyanazine	13.0/0.10	0.5	12/1
2,4-D	0.2/4.2	0.2	<1/2
Dicamba	2.3/2.1	0.3	2/2
Metolachlor	12.2/2.1	0.5	10/2
Metribuzin	6.8/0.78	0.5	9/2
Picloram	N.D./0.13	—	0/1
Propachlor	1.7/0.52	0.3	1/3
Simazine	N.D./2.6	—	0/<1
2,4,5-TP	N.D./0.26	—	0/1
Trifluralin	0.2/N.D.	—	1/0
Insecticides (and nematicides)			
Aldicarb	N.D./30.6	—	0/<1
Carbofuran	0.06/N.D.	—	2/0
Chlorpyrifos	0.07/0.21	0.1	<1/2
Fonofos	0.90/N.D.	0.2	1/0
Phorate	0.10/N.D.	—	<1/0

NOTE: N.D. means not detected. A dash indicates insufficient data to calculate mean.

[a]The two numbers listed for each active ingredient apply to Iowa and Minnesota, respectively.

SOURCE: Hallberg, G. R. 1987. Agricultural chemicals in ground water: Extent and implications. American Journal of Alternative Agriculture 2(1):3–15.

age in production has declined to about 45 million acres (U.S. Department of Agriculture, 1986a). The growth of irrigation has been most dramatic in the western Great Plains. There, irrigation has increased the production of corn and sorghum, thus contributing to the growth of cattle feedlot operations nearby. As large quantities of water are used for irrigation, however, some water tables decline and the cost of irrigation can rise. Inefficient irrigation practices have contributed to aquifer depletion in some regions. On sandy soils, certain irrigation practices have contributed to the movement of pesticides and nitrate into groundwater.

Irrigation has made agriculture possible in areas previously unsuitable for intensive crop production, such as the sandhills of Nebraska, parts of the central valley of California, and much of the arid West. In certain regions of California, irrigation is depleting aquifers at rates up to 1.5 million acre-feet per year. Land subsidence of up to 10 feet has resulted in some areas

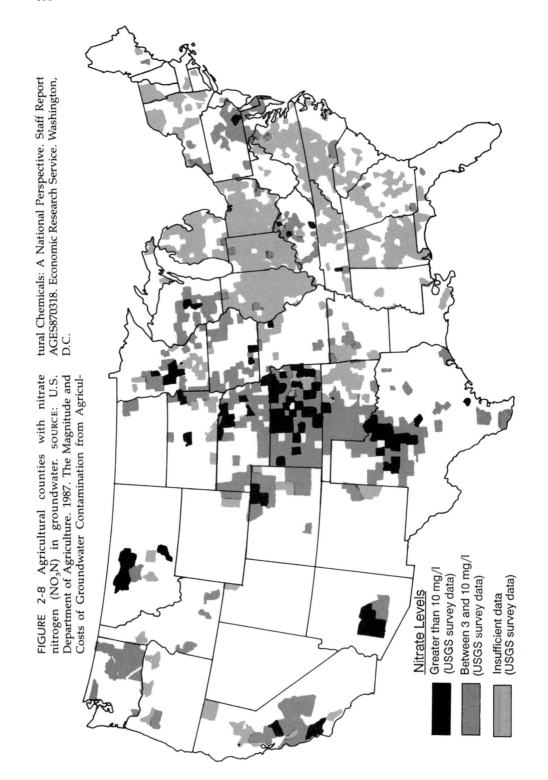

FIGURE 2-8 Agricultural counties with nitrate nitrogen (NO₃N) in groundwater. SOURCE: U.S. Department of Agriculture. 1987. The Magnitude and Costs of Groundwater Contamination from Agricultural Chemicals: A National Perspective. Staff Report AGES870318. Economic Research Service. Washington, D.C.

Nitrate Levels

Greater than 10 mg/l
(USGS survey data)

Between 3 and 10 mg/l
(USGS survey data)

Insufficient data
(USGS survey data)

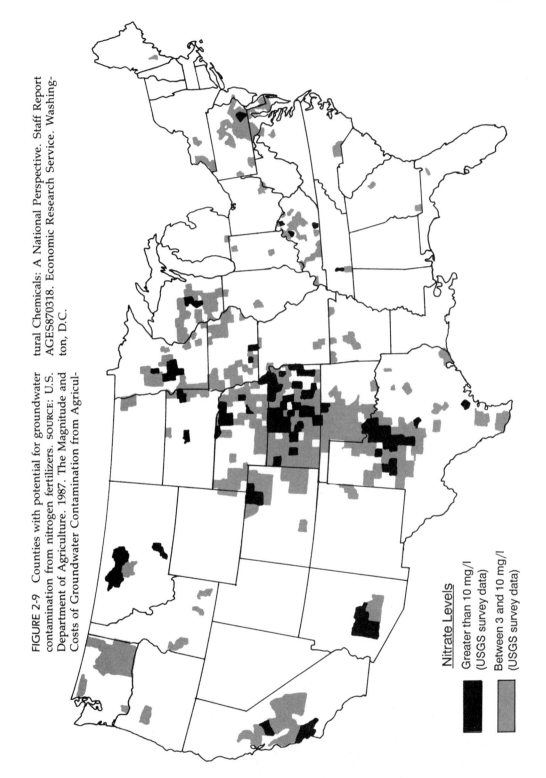

FIGURE 2-9 Counties with potential for groundwater contamination from nitrogen fertilizers. SOURCE: U.S. Department of Agriculture. 1987. The Magnitude and Costs of Groundwater Contamination from Agricultural Chemicals: A National Perspective. Staff Report AGES870318. Economic Research Service. Washington, D.C.

Nitrate Levels

Greater than 10 mg/l
(USGS survey data)

Between 3 and 10 mg/l
(USGS survey data)

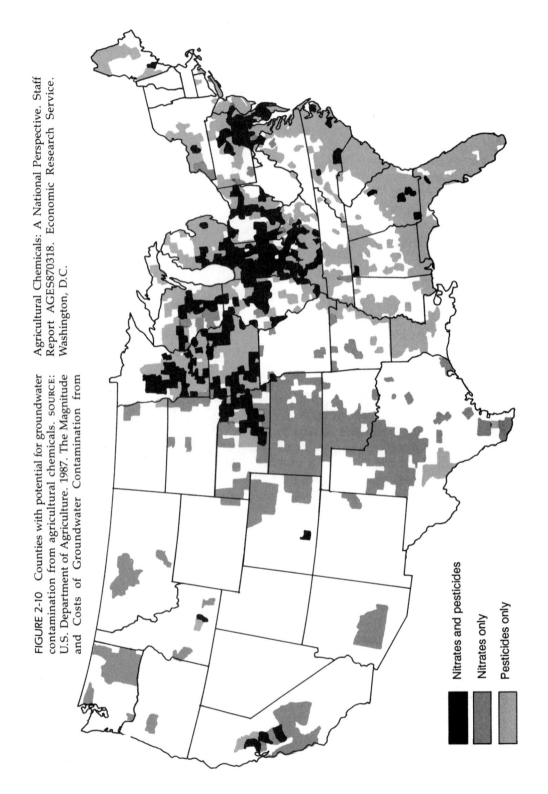

FIGURE 2-10 Counties with potential for groundwater contamination from agricultural chemicals. SOURCE: U.S. Department of Agriculture. 1987. The Magnitude and Costs of Groundwater Contamination from Agricultural Chemicals: A National Perspective. Staff Report AGES870318. Economic Research Service. Washington, D.C.

Nitrates and pesticides

Nitrates only

Pesticides only

TABLE 2-9   Acreage Irrigated in Areas With Declining Groundwater Supplies,[a] 1980–1984

State	Total Area (acres, in 1,000s) Irrigated With Groundwater	Groundwater Decline Area (acres, in 1,000s)[b]	Share (percent) of Total Area Irrigated From Declining Groundwater Aquifers	Average Annual Rate of Decline (feet)
Arizona	938	606	65	2.0—3.0
Arkansas	2,337	425	18	0.5—1.3
California	4,265	2,069	48	0.5—3.5
Colorado	1,660	590	36	2.0
Florida	1,610	250	16	2.5
Idaho	1,450	225	15	1.0—5.0
Kansas	3,504	2,180	62	1.0—4.0
Nebraska	7,025	2,039	29	0.5—2.0
New Mexico	805	560	70	1.0—2.5
Oklahoma	645	523	81	1.0—2.5
Texas	6,685	4,565	73	1.0—4.0
Total	30,924	14,032	45	

[a]In the contiguous United States.
[b]Areas with at least one-half foot average annual decline.

SOURCE: U.S. Department of Agriculture. 1987. U.S. Irrigation—Extent and Economic Importance. Agriculture Information Bulletin No. 523. Economic Research Service. Washington, D.C.

because of withdrawals in excess of recharge. In other areas of California, the groundwater table is rising, waterlogging soils and threatening agricultural production. In parts of the Great Plains, such as northern Texas and Oklahoma, where aquifer recharge is particularly slow, the Ogallala aquifer has been depleted to levels that restrict agricultural use. Between 1980 and 1984, groundwater levels declined by 0.5 to 5.0 feet per year below 14 million acres of irrigated land (Table 2-9) (U.S. Department of Agriculture, 1987d).

Total groundwater-irrigated acreage rose significantly in the 1970s and early 1980s. Center pivot irrigation alone increased from 3.4 million to 9.2 million acres between 1974 and 1983. Of the 30.9 million acres irrigated with groundwater, over 14 million acres, or 45 percent, are in areas where groundwater is declining at least 1 foot per year (see Table 2-9). California, Kansas, and Nebraska account for more than 2 million acres each of declining groundwater; Texas is responsible for more than 4 million acres.

Much of this land produces crops already in surplus. More than 10 million acres of cotton, corn, grain sorghum, and small grains are produced with water from declining aquifers (Table 2-10). More than 1.4 million acres of irrigated corn production in Nebraska are depleting groundwater between 0.5 and 2.0 feet per year. Most irrigated acres receive high levels of fertilizers and other yield-enhancing inputs to boost yields. High yields secure high per acre federal farm program payments, which help pay for the cost of irrigation. In several areas in Nebraska that produce irrigated corn, pesticides and high levels of nitrate have been detected in groundwater. This

TABLE 2-10   Irrigated Acreage of Surplus Crops in Areas of Groundwater Decline,[a] 1982 (in thousands)

State	Cotton	Corn	Grain Sorghum	Small Grains
Arizona	211	—	57	180
Arkansas	3	—	—	—
California	613	87	—	295
Colorado	—	315	56	73
Idaho	—	—	—	108
Kansas	—	664	542	683
Nebraska	—	1,456	123	44
New Mexico	72	55	96	126
Oklahoma	17	31	181	213
Texas	1,108	568	1,019	1,029
Total	2,024	3,176	2,074	2,751

NOTE: A dash indicates no irrigated crops.

[a]In the contiguous United States.

SOURCE: U.S. Department of Agriculture. 1987. U.S. Irrigation—Extent and Economic Importance. Agriculture Information Bulletin No. 523. Economic Research Service. Washington, D.C.

contamination is prevalent in areas with sandy soils, which are highly porous.

Irrigation in the arid West has been associated with mineralization and salinization of soils and water, as well as groundwater depletion and surface and groundwater contamination. The Colorado River is perhaps the most striking example of depletion of water resources. The Colorado River is so intensively used for municipal water and agricultural irrigation that in very dry years there has been virtually no water left in the river as it crosses the Mexican border. The New River in the Imperial Valley is an example of surface water pollution from irrigated cropland.

As municipalities and industry demand a greater share of available water in the West, agriculture will have to conserve. Conservation will require more prudent water use. It also may involve growing different crops and using production systems that retain more moisture in the soil. Agriculture currently uses 85 percent of available water in the West. The "use it or lose it" code of western water law encourages overuse of water based on fear of losing rights to use it in the future. With modest conservation, however, there is enough water to go around. If agriculture reduced water use through conservation by 15 percent, the amount of water available for municipal and industrial use in the region would double.

Arizona's recent decision to place urban water needs ahead of agricultural use and to demand its share of the Colorado River's water is the most dramatic example of changing western water priorities. In the future, market forces and demand for western water will continue to alter use patterns, accelerating efforts and investments in conservation practices designed to increase the efficiency of agricultural water use.

The buildup of salts on irrigated cropland can severely reduce yields. This field of grain shows heavy damage by salt. The principal method to reduce salinization of soils is to flush salts out of soils by flooding fields with water. This method, however, moves salts downstream and requires large volumes of water. *Credit:* Soil Conservation Service, U.S. Department of Agriculture.

Soil Erosion

Soil erosion remains a serious environmental problem in parts of the United States, even after 50 years of state and federal efforts to control it. Common management practices such as increased reliance on row crops grown continuously, fewer rotations involving forages, and larger farms being tilled by one operator have made it difficult to conserve soil resources in some areas. Similarly, some federal farm programs, particularly the commodity price and income support programs, have historically encouraged high levels of production that work as a disincentive for effective erosion control practices.

Soil erosion causes off-farm as well as on-farm damage. Quantifying the economic cost to society of offsite effects of erosion is difficult and estimates vary widely. Except for specific locations that have been studied intensively, it remains impossible to reach reliable judgments about the relative magnitude of on- and off-farm costs associated with erosion. The USDA calculated annual offsite damage at between $2 billion and $8 billion annually. Each year, the 350 to 400 million acres of land used for agriculture are estimated to account for more than 50 percent of suspended sediments deposited in surface waters (U.S. Department of Agriculture, 1987a). Onsite erosion damage can reduce the productivity of land, labor, and capital on the farm, and increase the need for fertilizer and other inputs. About one-fifth of U.S. cropland is subject to serious damage from erosion (Clark et al., 1985; U.S. Department of Agriculture, 1987a). The impacts of onsite erosion have been

Farm chemicals can reach dangerous levels in
rivers that drain irrigated fields. Shown here
is the New River leaving the heavily irrigated
Imperial Valley and entering the Salton Sea in
southern California. *Credit:* Richard Steven
Street.

estimated at between $1 billion and $18 billion per year, although the meth-
odology to make such estimates is complex, controversial, and of limited
value (National Research Council, 1986c; Pimentel, 1987; U.S. Department
of Agriculture, 1987a).

One part of the controversy involves what is being measured. For example, Crosson (1985) reported that about $0.5 billion is necessary to offset the
annual loss of soil nutrients by erosion; in contrast, Troeh et al. (1980)
reported that a total of $18 billion in soil nutrients is lost annually from
agriculture. Crosson estimated the nutrients directly available to crops each
growing season; Troeh et al. estimated the value of all nutrients lost, which
included those directly available and those which would have been available
after mineralization. There are many other aspects of the controversy on
the impact of soil erosion on productivity. They include water runoff, water-
holding capacity, organic matter, and soil depth (Pimentel, 1987).

It is generally recognized that soils with deep profiles are able to withstand erosion without an appreciable drop in productivity. Thin soils over
bedrock or other impermeable barriers are more vulnerable to erosion-
induced loss of productivity. Wind and water erode between 2.7 billion and

Runoff from conventionally tilled cropland in Greene County, Ohio, pollutes surface water with sediments and nutrients. The grass at the edge of the field and along the ditch filters some of the sediment from the runoff. Under new conservation policies adopted in the Food Security Act of 1985, farmers will be helped to install grassy ditches and filter strips to reduce agricultural runoff. *Credit:* Soil Conservation Service, U.S. Department of Agriculture.

3.1 billion tons of soil from the nation's cropland each year (National Research Council, 1986c; U.S. Department of Agriculture, 1986a). One ton per acre is roughly equal to 1/150 of an inch per acre. More than 118 million acres of cropland are considered highly erodible under the current federal conservation compliance provisions. According to the USDA, about 35 million of these acres are in compliance under current management. More than 25 million additional acres have been enrolled in the CRP, which pays farmers to idle highly erosive land for 10 years. When vegetation is in place on CRP land, the USDA estimates that erosion could be reduced by up to 800 million tons per year (U.S. Department of Agriculture, 1988e). The remaining 58 million acres of highly erodible land either will be placed in the CRP or, by 1990, will require approval of conservation plans to reduce erosion below tolerable levels (Table 2-11) (U.S. Department of Agriculture, 1988c). Farmers must fully implement these plans by 1995. Farmers not complying by that time will lose eligibility for nearly all federal program benefits.

The USDA projects negligible productivity loss on about 345 million acres of farmland under current practices and erosion rates. Such soils are ex-

TABLE 2-11 Cropland Affected by Conservation Compliance

Region	Cropland Requiring Compliance by 1990		Balance Requiring Compliance by 1990 or Enrollment in CRP		
	Total Acreage (in millions) Before CRP[a]	Maximum Acreage (in millions) Enrolled in CRP Through February 1988[b]	Minimum Area (in million acres)[c]	Share (percent) of U.S. Total	Share (percent) of Region's Cropland
Northeast	3.7	0.1	3.6	6.2	21
Great Lakes	3.9	2.1	1.8	3.2	4
Corn Belt	19.1	3.6	15.5	26.8	17
Northern Plains	13.8	6.0	7.8	13.5	8
Appalachia	5.9	0.9	5.0	8.6	22
Southeast	2.7	1.2	1.5	2.6	8
Delta	2.3	0.8	1.5	2.7	7
Southern Plains	13.8	2.1	9.7	16.8	22
Mountain	14.7	5.2	9.5	16.9	22
Pacific	3.5	1.5	2.0	3.5	9
United States	83.4	25.5	57.9	100.0	14[d]

NOTE: Totals may not be exact because of rounding.

[a]Includes all lands with an erodibility index (EI) equal to or greater than 8, excluding 35 million acres of land with an EI greater than 8 that is eroding at the T level or less under current use and management. This land is considered already under compliance.

[b]Maximum acreage, based on the assumption that all CRP lands through the February 1988 sign-up had an EI equal to or greater than 8.

[c]Could be slightly greater to the extent that some of the CRP acreage included lands with an EI of less than 8.

[d]This figure represents the percentage of all crop acres that will require compliance by 1990.

SOURCE: U.S. Department of Agriculture. 1988. Agricultural Resources—Cropland, Water, and Conservation—Situation and Outlook Report. AR-12. Economic Research Service. Washington, D.C.

pected to lose less than 2 percent of their productivity after 100 years. The USDA considers this loss insignificant because annual productivity gains from new technology and improved management are projected to average at least 1 percent per year. Again, the methodologies available to project long-term consequences of erosion-induced productivity losses are crude and may not fully anticipate ways in which future technologies and economic conditions could interact with soil quality. For example, if fossil fuel prices increase, nitrogen fertilizer prices could rise because of higher prices for natural gas and energy needed in fertilizer production and distribution. The future value of uneroded soils may rise appreciably because of the capacity to sustain high levels of crop yields using rotations and leguminous cover crops.

The use of conservation tillage practices has increased since 1980. Nearly 100 million acres in 1987 were farmed using some form of conservation tillage, compared with about 40 million in 1980. The main practice uses crop residue management to provide a partial mulch cover on the soil surface. This is accomplished through reduced tillage, primarily chisel plowing, which can decrease erosion rates by up to 50 percent. The use of no tillage, strip tillage, and ridge tillage, which can reduce erosion by 75 percent or more, accounts for only about 16 million acres (U.S. Department of Agriculture, 1987a).

A large body of evidence indicates that intensive tillage practices associated with continuous monoculture or short rotations may make soils more susceptible to erosion. Reganold et al. (1987) recently reported this phenomenon on two neighboring farms in the Palouse region of Washington state. The fields on one farm were worked for 38 years with conventional tillage and a shorter rotation. These fields had 6 inches less topsoil than an adjacent farm where the fields were in longer rotations. Similar but fewer tillage operations were used on the second farm. Water erosion research in the same area supports the conclusion that tillage and fertilization practices associated with longer rotations, which often use legumes to supply nitrogen, are less prone to erosion. Water erosion rates on similar fields of winter wheat were 13.1 tons per acre per year for fields not using leguminous meadows in the rotation compared to 2.4 tons per acre for fields that included them (Reganold et al., 1987).

Conventionally tilled soil with continuous intertilled crops almost always experiences a decline in organic matter and some ability to retain moisture. All other factors being equal, soils that historically receive nutrients in the form of manure or legumes tend to have higher levels of organic carbon and overall organic matter (Power, 1987). Organic matter improves soil quality by increasing granulation, water infiltration, nutrient content, soil biota activity, and soil fertility and productivity. Management systems that reduce or eliminate synthetic fertilizer applications depend on increased microbial activity to make sufficient nutrients available to sustain crop yields (Doran et al., 1987). Conventionally tilled fields without cover crops will likely have diminished organic content and be more susceptible to erosion and leaching of applied chemicals (Hoyt and Hargrove, 1986; Reganold et al., 1987).

Genetic Diversity

U.S. agriculture produces a diverse array of crops and livestock. The genetic diversity of these crops has been substantially redistributed in recent years, however. A limited number of improved varieties of crops resistant to certain diseases and pests and responsive to fertilizers, management, and other inputs are now widely used. Increased genetic potential and improved cultural practices share about equal credit for past productivity gains. As a result, yield increases of 100 percent or more per acre have been recorded for most major commodities (Fehr, 1984). Vast areas are now planted with wheat, corn, and soybean varieties that are closely related and very uniform. Recent isozyme electrophoresis and zein chromatography analysis of 88 corn hybrids identified 49 as genetically unique; the remaining 39 fell into 6 categories that these techniques were unable to distinguish as different (Smith, 1988).

Concern for the genetic resources of domestic animals has been limited in spite of the decline in the populations of many breed populations and increasing uniformity within dominant breeds. The situation for dairy cattle is particularly acute. With the exception of Holsteins, most breed populations are decreasing. This may eventually result in a decline in genetic variability within breeds. The Holstein breed is predominant; more than 90 percent of U.S. dairy cows are Holsteins (Niedermeier et al., 1983). Only 400 to 500 artificial insemination sires impregnate about 65 percent of the 6 million to 7 million dairy cows bred each year in the United States. And of the approximately 1,000 performance-tested dairy bulls used for artificial insemination in a given year, nearly half are the sons of the 10 best bulls of the previous generation (U.S. Office of Technology Assessment, 1987). With practices like this, the genetic base will become narrower.

A few major producers control the production of eggs and poultry. The genetic diversity of the breeding lines of chickens, which have undergone extensive selection for production in a controlled, high-input environment, is likely to be quite limited. A similar trend is beginning to emerge in the swine industry, with the development of specialized inbred lines raised under controlled conditions. There is a basis for concern regarding the loss of potentially important variation in these species.

In the short term, genetically uniform plant and animal varieties can be resistant to certain pests and therefore be very productive. They can, however, be susceptible to other pests. The number of pests to which the variety or breed is susceptible can increase rapidly, resulting in vulnerability to devastating epidemics (National Research Council, 1972). Genetic diversity within a crop variety provides some buffering against environmental extremes, including pressures of diseases and insects. Likewise, planting several varieties of the same crop, which differ genetically for resistance to diseases, together or in different fields decreases the likelihood of losses due to a particular disease. This contributes to stability in yields and therefore stability in income for a farmer. In the long term, unless genes are preserved and maintained in germplasm banks or in the wild, crop and livestock species will suffer an irretrievable loss of genetic variability and, thus, reduce our ability to respond to specific stresses such as diseases (Duvick, 1986).

The value of maintaining a diverse plant genetic resource base is illustrated in the case of the use of classical plant breeding to control greenbugs in grain sorghum. The rapid increase in greenbugs caused an estimated $100 million loss to the U.S. sorghum crop in 1968. In the following year, about $50 million was expended for chemical insecticides on about 8 million acres. By 1976, however, resistance to the greenbug was found in a sorghum variety and incorporated into hybrids, which were grown on about 4 million acres. This example shows how common plant-breeding techniques, drawing on the genetic resource base for sorghum, could reduce chemical dependence, pest control costs, and pest damage.

The subsequent emergence of a new biotype of greenbug illustrates another point. Biotype E of the greenbug emerged in 1980 and attacked the previously resistant sorghum hybrids. Again, researchers have found another resistant variety of sorghum, which is now in general use by sorghum hybrid producers. Because insecticides to control greenbugs were not used for the length of time and in a manner that produced resistant greenbugs, the chemical could be used later when an emergency arose.

Genetic resistance to pests in plants is widespread. The reservoir of plant genetic resources for biological pest control is an extremely valuable pest control component. There are many other examples (including wheat stem rust and European corn borer) where host plant resistance genes have continuously controlled a once-serious pest for several decades after a devastating crop loss.

Effects of Pesticides

One of the consequences of widespread genetic uniformity in crops and livestock is that when pests appear in epidemic numbers, they can have devastating effects on productivity. Existing pesticides are effective in controlling many serious threats to production and assuring unblemished products for market. They have helped maintain pest damage at between 5 and 30 percent of potential production in many cropping situations, including highly uniform and often continuous monocultures that would otherwise be highly susceptible to severe pest damage. However, the adverse effects of some pesticides are a serious problem in U.S. agriculture.

Although the data are not conclusive, evidence suggests that pesticide use creates several immediate health hazards on the farm. There is growing evidence that pesticide use may pose serious health problems for farmers and farmworkers. A 1986 study by the National Cancer Institute found that Kansas farmworkers who were exposed to herbicides for more than 20 days per year had a 6 times higher risk of developing non-Hodgkin's lymphomas (NHL) than nonfarm workers (Hoar et al., 1986). Follow-up work in Nebraska found that exposure to the herbicide (2,4-Dichlorophenoxy) acetic acid (2,4-D) more than 20 days per year increased the risk of developing NHL threefold (Hoar et al., 1988). Other studies have suggested a link between pesticide use and increased incidence of NHL and multiple myeloma among farmers (Pearce et al., 1985; Weisenburger, 1985). In addition to the risk of developing cancer, pesticides also increase the incidence of acute illness.

> Acute exposure to pesticides may result in systematic or local disease. With systemic poisonings the clinical picture reflects the known toxicology of the compound and occurs shortly after exposure. Cholinergic illness due to cholinesterase inhibition from excessive organophosphate and carbamate exposure is the commonest type of systemic poisoning. Other less common, but equally life threatening examples of acute poisonings include the gastrointestinal, hepatic, renal, and pulmonary phases of paraquat poisoning, the metabolic stimulation that follows excessive exposure to the nitrophenol group of pesticides, . . . and the seizure disorders that herald the excessive chlorinated hydrocarbon exposures (Davies, 1985).

A number of recent studies have documented farmer applicator exposure to organophosphate and carbamate insecticides through residues in urine and cholinesterase reduction (American Farm Bureau Federation, 1988; McDonald, 1987). Many farmworkers and their families, particularly migrant farmworkers, live and work in close and regular proximity to pesticides and are exposed to far greater amounts of these compounds than the average consumer. There is, however, no systematic monitoring of the health or exposure to pesticides of the more than 2 million farmworkers, applicators, harvesters, irrigators, and field hands who work around pesticides. Industrial workers who produce these pesticides receive the benefits of such monitoring.

Widespread and heavy use of pesticides in this country has severely stressed some animals, including honeybee and wild bee populations (Brown, 1978). Honeybees and wild bees are vital to the production of about $20 billion worth of fruits, vegetables, and forage crops. The large number of honeybees killed by pesticides resulted in the Bee Indemnity Act of 1970 to compensate apiarists for such losses. The act was repealed in 1980. But honeybees killed by pesticide use, loss of honey, and reduced crop yields account for at least $135 million in losses each year (Pimentel et al., 1980).

Because ecological interactions are extremely complicated and have generally not been studied by the EPA, the effect of pesticides on the environment is not well understood. The decline of predatory birds in the 1960s and 1970s because of chlorinated hydrocarbon pesticide use, however, is well documented, as is their recovery since the cancellation of these compounds. This recovery, however, is an anomaly. Although the ecological effects of pesticides are thought to be substantial, human health risks have traditionally been given priority. The EPA's special review of the insecticide carbofuran in October 1985 was the first time that an agricultural pesticide

Insecticides and fungicides are often applied to tree crops using blasters. Here, herbicides also have been used and have killed all grass and weed growth beneath the trees. As a result, there is a limited habitat for beneficial insects. Erosion may also become a problem. *Credit: Agrichemical Age.*

had been so treated solely on the basis of its effects on wildlife. The EPA is currently reviewing a number of pesticides for ecological effects in conjunction with human health effects, including the widely used ethylene bisdithiocarbamate (EBDC) group of fungicides and the insecticides dicofol and diazinon. The EPA has canceled other pesticides, based partially on their effects on the ecology and wildlife. They include the insecticides DDT, endrin, and toxaphene. The EPA, the Department of the Interior, and states are currently in the process of implementing restrictions on more than 100 major pesticides to protect between 250 and 300 endangered plant and animal species on croplands, rangelands, and forestlands in more than 900 counties, pursuant to the Federal Endangered Species Act of 1973. When implemented, these restrictions could benefit plant and wildlife species remaining on or around these lands.

Other unintended effects of pesticides include the resurgence of pests after treatment, occurrence of secondary pest outbreaks, and development of pesticide resistance in target pests. When insecticides or other pesticides are employed against one pest, its natural enemies or those of another pest may be reduced or eliminated. The control of insects by broad-spectrum insecticides also destroys beneficial insect populations. Populations of many previously innocuous species may then increase rapidly and cause major economic damage.

After heavy applications of pesticides over many years, Colorado potato beetles are now resistant to most registered insecticides and can cause severe damage in some crops, notably potatoes. *Credit:* Mycogen Corporation.

In the early 1900s, for example, the major pests of cotton were the boll weevil and cotton leafworm (Newsom, 1962). Since 1945 and the extensive use of toxaphene, DDT, methyl parathion, and other insecticides on cotton, the cotton bollworm, tobacco budworm, cotton aphid, and spider mite have become more serious pests than they were previously (National Research Council, 1975). In particular, the cotton bollworm and tobacco budworm populations have grown because pesticides destroyed their natural enemies. In 1978, it was estimated that in California 24 of the 25 top agricultural pests were secondary pests. The pesticides that wiped out their predators created or aggravated their role or dominance as pests (Van den Bosch, 1980).

More than 440 insect and mite species and more than 70 fungus species are now known to be resistant to some pesticides (National Research Council, 1986a). The committee expects that the problem will worsen. Pest populations already resistant to one or more pesticides generally develop resistance to other chemicals more rapidly, especially when the compounds work in the same way as previously used pesticides (National Research Council, 1986a). To counteract this, increased pesticide resistance in insect, mite, and fungus populations, larger doses and more frequent applications of the previously used pesticides become necessary. It often becomes necessary to combine pesticides or substitute a different type of pesticide to achieve control. In some cases, more expensive, toxic, or ecologically hazardous pesticides have to be used. This starts a cycle of shifting resistance and increased use of pesticides. For these reasons, increasing levels of pesticide resistance in pest populations have significant environmental and economic costs.

Pesticides can also cause crop losses. This can occur when the usual dosages of pesticides are applied improperly; when herbicides drift from a treated crop to nearby, susceptible crops; when herbicide residues prevent chemical-sensitive crops from being planted in rotation or inhibit the growth of subsequent crops; and when excessive residues of pesticides accumulate on crops, causing the harvested products to be destroyed or devalued in the marketplace.

Beetles have seriously damaged potato plants in the foreground, despite insecticide treatments. A new biological insecticide that controls the Colorado potato beetle and is less toxic than routinely used insecticides protects the healthy plants. *Credit:* Mycogen Corporation.

Food Safety

Many of the chemical agents introduced into the food supply, including pesticides, fertilizers, plant-growth regulators, and antibiotics can be harmful to humans at high doses or after prolonged exposure at lower doses. Although cancer-causing chemicals have attracted the most concern, agricultural chemicals can also have behavioral effects, alter immune system function, cause allergic reactions, and affect the body in other ways.

Concern about the adverse effects of synthetic chemical pesticides on human and animal health began in the 1950s when it was discovered that organochlorine pesticides such as DDT are very persistent in the environment and can damage animal systems. In the following years, the use of pesticides increased dramatically, largely because of their affordability, effectiveness, ability to cut labor costs, and a variety of economic incentives for higher yields. Pest resistance also led to more applications per growing season. Increased use placed a growing burden on regulatory agencies to ensure the safety and proper use of the compounds, and set the stage for subsequent dietary exposure and environmental problems.

The two major problems facing policymakers attempting to regulate pesticides are the lack of data on the health hazards of pesticides and a lack of accurate exposure data. A National Research Council (NRC) panel estimated that data to conduct a complete assessment of health effects were publicly available for only 10 percent of the ingredients in pesticide products, mainly because of a lack of testing of older, widely used pesticides (National Research Council, 1984). Pesticide producers and the EPA held more confidential data at that time, however. And since 1984, more data have been generated on the chronic health effects of these compounds. To date, insecticides accounting for 30 percent, herbicides accounting for 50 percent, and fungicides accounting for 90 percent of all agricultural use have been found to cause tumors in laboratory animals (National Research Council, 1987). There is still much scientific debate, however, over the extrapolation of the results of these studies to adverse effects in humans. Lack of accurate human exposure data further complicates the problem. A recent NRC report found little data on the actual levels of pesticides present in the human diet (National Research Council, 1987). Although residue studies are being conducted, a complete picture of residue patterns in the food supply is still lacking.

Based on available data, pesticide residues in the average diet do not make a major contribution to the overall risk of cancer for humans (National Research Council, 1982, 1987). The risk, however, may not be insignificant and in most cases can be substantially reduced. Fungicides pose a particularly difficult chronic health problem. They account for more estimated oncogenic risk than herbicides and insecticides combined, but few effective alternatives are available or under development (National Research Council, 1987). Further complications in risk assessment are that fungicides are often used in combinations, and residues of several oncogenic fungicides and other pesticides are commonly detected on the same crop.

Although little research has been done, there is evidence of synergistic interactions among pesticides and their contaminants with other compounds and with each other (DuBois, 1972; Knorr, 1975). In 26 percent of 15 fruits and vegetables tested by the Florida Department of Agriculture, residues of two or more pesticides (including DDT, which was banned for agricultural use in 1972) were detected. This may understate actual residues, however, because the analytical method used cannot detect some compounds widely used on these crops. Although several pesticides are often present on a given food, pesticides continue to be regulated individually (Florida Department of Agriculture, 1988; Mott, 1984).

Organic fertilizers (manures and sewage sludge) and some inorganic fertilizers present health hazards if used inappropriately. These hazards include increased nitrate levels in some foods and water, which pose a health problem when they are converted to nitrite through the action of bacteria and enzymes in the stomach. Nitrate can also be further metabolized during digestion to form nitrosamines, which are strongly carcinogenic. The potential accumulation of nitrate in parts of some crops is generally greater when nitrogen is supplied in the synthetic chemical form because there is usually more nitrate available for uptake (Hodges and Scofield, 1983).

Nitrate percolation to groundwater and runoff from fields and feedlots are major water contamination problems. Organic and inorganic fertilizers can cause these problems. Some sewage sludges, particularly those from industry, can contain high levels of heavy metals. These metals, which include cadmium, chromium, lead, and others, are toxic to most life forms and can accumulate in soil and in plant and animal tissues. The EPA has established guidelines for the agricultural application of sludges that contain heavy metals to avoid toxic accumulations in soil, forages, and vegetables. Additionally, sludges that are not dried and/or completely composted can result in contamination of the soil with human pathogens (Maga, 1983; Poincelot, 1986; Vogtmann, 1978).

In addition, a wide variety of food-borne illnesses constitute a significant health problem in the United States. It is estimated that all types of food-borne illnesses are responsible for 33 million human illnesses and 9,000 human deaths in the United States each year (Young, 1987). A significant percentage of these can be attributed to pathogenic bacteria of animal origin. The bacterial pathogens listeria and salmonellae, found in contaminated dairy products, and salmonellae and campylobacter, found on some meat and poultry, have taken a significant disease toll in recent years. According to the Centers for Disease Control, bacteria from animal products account for approximately 53 percent of all outbreaks of food-borne illness for which a source was determined (Tauxe, 1986).

Antibiotics

There has been scientific debate and concern about the subtherapeutic use of antibiotics in animal feed for nearly 20 years (Ahmed et al., 1984;

Council for Agricultural Science and Technology, 1981; Jukes, 1973; Kennedy, 1977; National Research Council, 1980). The focus of concern is the frequent development of antibiotic resistance in pathogenic bacteria as a consequence of antibiotic use in animals and in humans. Because many antibiotics used in animal feed are also used in human medicine, antibiotic-resistant pathogenic bacteria, particularly salmonellae, could develop and cause infections in animals and humans. The effectiveness of antibiotics for disease therapy would thus be diminished (Institute of Medicine, 1989; Murray, 1984).

Hirsch and Wigner (1978) demonstrated the transmission of resistant pathogens from animals to humans. This has been the subject of a thorough review (Feinman, 1984). But there are still few studies that document the incidence of human disease caused by antibiotic-resistant pathogens of animal origin. Disease in humans due to antibiotic-resistant salmonellae of animal origin is difficult to confirm and appears to be rare. Holmerg et al. (1984), however, demonstrated that antibiotic-resistant salmonellae caused disease in humans who consumed meat from animals harboring salmonellae. In a study of 542 human cases of salmonellosis in 1979, 28 percent of the bacteria isolated were resistant to at least 1 antibiotic. Resistance to 2 or more antibiotics was found in 12 percent of the salmonellae strains (Tauxe, 1986).

In addition to an apparent increase in the incidence of salmonellosis in humans, there are data to show that antibiotic resistance in the bacteria in animal intestinal microflora can be transmitted to humans because the same antibiotic-resistant bacteria are found in the human intestinal tract (Institute of Medicine, 1989). This increases the concern that antibiotic resistance in animal pathogens might spread from animals to humans. The risk that this transmission poses to human populations is a matter of intense scientific debate. Meanwhile, antibiotic use continues to increase.

A recent report by an Institute of Medicine (IOM) committee assessed human health risks resulting from the subtherapeutic use of penicillin and tetracyclines in animal feeds (Institute of Medicine, 1989). Although the IOM committee recognized that there is little direct evidence implicating subtherapeutic use of antimicrobials as a potential human health hazard, the committee found substantial indirect or circumstantial evidence indicating a potential human health risk from subtherapeutic use of antibiotics in animal feeds. This evidence includes the following:

- The use of antimicrobials in a variety of doses generates a strong selective pressure for the emergence of drug-resistant bacteria.
- Antimicrobial resistance among isolates of salmonellae from farm animals is prevalent because of extensive antimicrobial use on farms.
- Animal and poultry carcasses in meat-processing plants are often found to be contaminated with intestinal pathogens resistant to antimicrobials.
- Human infections from salmonellae or other enteric bacteria may follow handling and ingestion of improperly cooked meat or food products from animals contaminated with these organisms.

In assessing human health risk, the committee used a risk model that estimated the number of deaths from salmonellosis attributable to use of antimicrobials in animal feeds for prophylaxis and growth promotion and concluded that the likeliest estimate was in the range of 40 deaths per year (Institute of Medicine, 1989). Further, it found that increased difficulty of treatment probably led to 20 additional deaths per year. The committee estimated that less than half of these deaths were from the use of antimicrobials in growth promotion. It recognized, however, that the distinction between the use of these antimicrobials for growth promotion and prophylaxis may not be great. The committee did not estimate incidences of morbidity because even fewer data were available. For the same reason, it did not estimate deaths due to other infectious organisms that cause food-borne illnesses and are known to develop resistance to the antimicrobials. The committee's conclusions suggested that reductions in subtherapeutic antibiotic use would lessen the severity of human disease complications following infection with salmonellae. Because data are limited, it is not possible to predict accurately the magnitude of public health gains that would result from a reduction of antimicrobial use in livestock agriculture.

Human health concerns from antibiotic use go beyond bacterial resistance. Drug residues in food may also present risks. Many types of animal drugs are available to lay persons or farmers without the necessity of a veterinarian's prescription. Furthermore, it appears that even antibiotics limited to veterinary prescriptions are also widely available to lay persons (U.S. Congress, 1985). An example of the inappropriate use of antibiotics is the use of chloramphenicol. Chloramphenicol was never approved for any use in food-producing animals; however, residues of chloramphenicol have been detected in animal food products (U.S. Congress, 1985). The drug's sale in large containers, which was designed for the treatment of dogs, was banned by the U.S. Food and Drug Administration (FDA) in 1986 in an attempt to discourage the mixing of chloramphenicol with animal feed (U.S. Food and Drug Administration, 1986). Chloramphenicol nonetheless continues to be used in food-producing animals. Recent surveys of milk in New Jersey, New York, Oregon, and Pennsylvania found residues of chloramphenicol in 15 to 20 percent of the samples analyzed (Brady and Katz, 1988). Its only FDA-approved use is for pet animals under veterinary care.

University- and government-sponsored studies have found sulfamethazine residues in meat and milk (Brady and Katz, 1988). Sulfamethazine is available over the counter only in combination with other antibiotics, for use in swine and cattle. It is not allowed for use in lactating dairy animals. Surveys of commercial milk, however, revealed that in certain parts of the country, greater than 50 percent of the samples had detectable sulfamethazine residues. The human health hazard from these residues is not clear, although the compound may be carcinogenic in rodents (U.S. Food and Drug Administration, 1988). Further, approximately 3 percent of the human population is allergic to sulfamethazine and many other antimicrobial drugs that may contaminate food products (Bigby et al., 1986).

The FDA surveillance programs for the detection of violative residues of all veterinary antibiotics and chemicals are limited. Field investigations into tissue residue violations have revealed areas where the FDA may need to concentrate its enforcement activities. Dairy cows culled from herds had the highest rate of violative residues, followed by Bob veal calves (calves slaughtered at less than 4 weeks of age). In addition, 18 percent of the violative tissue residues in meat were from intramammary medication. Of these residues, 85 percent were derived from gentamicin, a drug not approved by the FDA for intramammary use and legally available only through veterinarians (Paige and Kent, 1987). These problems point out the need to improve the effectiveness of the FDA's regulation of animal drugs.

SUMMARY

Many economic and environmental factors have converged in the 1980s to make alternative farming practices more appealing. Exports have declined since 1981. Although the situation is improving, sectors of the agricultural economy continue to experience hardships. Despite the fact that net farm income has reached record levels, federal programs support an unprecedented percentage of total net farm income.

Nonpoint surface water pollution and contamination of groundwater by agricultural chemicals are recognized as environmental problems. Soil erosion remains serious in certain regions. In subhumid and arid regions, irrigation practices continue to deplete aquifers and cause salinization of agricultural land and water. Antibiotic and pesticide residues in food present risks that, while difficult to quantify and evaluate, can be reduced through alternate management systems. The ecological effects of certain pesticides are considered to be significant in some regions, although they remain largely unstudied.

In response to these factors, some farmers are beginning to implement a range of alternative practices. The scientific bases for the major components of alternative agricultural systems are presented in Chapter 3.

REFERENCES

Agricultural Policy Working Group. 1988. Decoupling: A New Direction in Global Farm Policy. Washington, D.C.: Agricultural Policy Working Group.

Ahmed, A. K., S. Chasis, and B. McBarnette. 1984. Petition of the Natural Resources Defense Council, Inc., to the Secretary of Health and Human Services requesting immediate suspension of approval of the subtherapeutic use of penicillin and tetracyclines in animal feeds. New York: Natural Resources Defense Council.

American Farm Bureau Federation. 1988. Nine show pesticide exposure in health test. Maryland Agriculture 19(4):5.

Bigby, M., S. Jick, H. Jick, and K. Arndt. 1986. Drug-induced cutaneous reactions: A report from the Boston collaborative drug surveillance program on 15,438 consecutive inpatients, 1975 to 1982. Journal of the American Medical Association 256(24):3358–3363.

Brady, M. S., and S. E. Katz. 1988. Antibiotic/antimicrobial residues in milk. Journal of Food Protection 51(1):8–11.

Brown, A. W. A. 1978. Ecology of Pesticides. New York: Wiley.

Chesters, G., and L. J. Schierow. 1985. A primer on nonpoint pollution. Journal of Soil and Water Conservation 40:14–18.

Clark, E. H., II, J. A. Haverkamp, and W. Chapman. 1985. Eroding Soils: The Off-Farm Impacts. Washington, D.C.: The Conservation Foundation.

Council for Agricultural Science and Technology. 1981. Antibiotics in Animal Feeds. Report No. 88. Ames, Iowa: Council for Agricultural Science and Technology.

Crosson, P. 1985. National Costs of Erosion on Productivity. Pp. 254–265 in Erosion and Soil Productivity: Proceedings of the National Symposium on Erosion and Soil Productivity. St. Joseph, Mich.: American Society of Agricultural Engineers.

Davies, J. E. 1985. Health Effects of Global Pesticide Use. Miami, Fla.: World Resources Institute.

Doran, J. W., D. G. Fraser, M. N. Culik, and W. C. Liebhardt. 1987. Influence of alternative and conventional agricultural management on soil microbial processes and nitrogen availability. American Journal of Alternative Agriculture 2(3):99–106.

DuBois, K. D. 1972. Interaction of chemicals as a result of enzyme inhibition. Pp. 97–107 in Multiple Factors in the Causation of Environmentally Induced Disease, D. H. K. Lee and P. Kotin, eds. New York: Academic Press.

Duvick, D. N. 1986. Plant breeding: Past achievements and expectations for the future. Economic Botany 40:289–297.

Fehr, W. R., ed. 1984. Genetic Contributions to Yield Gains of Five Major Crop Plants. Special Publication No. 7. Madison, Wis.: Crop Science Society of America.

Feinman, S. E. 1984. The transmission of antibiotic-resistant bacteria to people and animals. Pp. 151–171 in Zoonoses, Vol. I, J. H. Steele and G. W. Beran, eds. CRC Handbook Series. Boca Raton, Fla.: CRC Press.

Florida Department of Agriculture. Residue Testing Laboratory. 1988. Data compiled from the 1986–1987 growing season, Florida Department of Agriculture, Tallahasse. Available from Environmental Health Research, Vero Beach, Fla.

Glenn, S., and J. S. Angle. 1987. Atrazine and simazine to runoff from conventional and no-till corn watersheds. Pp. 273–280 in Agriculture Ecosystems and Environment. Amsterdam: Elsevier.

Hallberg, G. R. 1987. Agricultural chemicals in groundwater: Extent and implications. American Journal of Alternative Agriculture 2(1):3–15.

Hileman, B. 1988. The Great Lakes cleanup effort: Much progress, but persistent contaminants remain a problem. Chemical and Engineering News, February 8, pp. 22–39.

Hirsh, D. C., and N. Wigner. 1978. The effect of tetracycline upon the spread of bacterial resistance from calves to man. Journal of Animal Science 46:1437.

Hoar, S. K., A. Blair, F. F. Holmes, C. D. Boysen, R. J. Robel, R. Hoover, and J. F. Fraumeni, Jr. 1986. Agricultural herbicide use and risk of lymphomas and soft-tissue sarcoma. Journal of the American Medical Association 256(9):1141–1147.

Hoar, S. K., D. D. Weisenburger, P. A. Babbitt, R. C. Saal, K. P. Cantor, and A. Blair. 1988. A case-control study of non-Hodgkin's lymphoma and agricultural factors in eastern Nebraska. American Journal of Epidemiology 128(4):901.

Hodges, R. D., and A. M. Scofield. 1983. Effect of agricultural practice on the health of plants and animals produced: A review. Pp. 3–33 in Environmentally Sound Agriculture: Selected Papers from the Fourth International Conference of the International Federation of Organic Movements, W. Lockeretz, ed. New York: Praeger.

Holmberg, S. D., M. T. Osterholm, K. A. Senger, and M. L. Cohen. 1984. Drug-resistant salmonella from animals fed antimicrobials. New England Journal of Medicine 311:617–622.

Hoyt, G. D., and W. H. Hargrove. 1986. Legume cover crops for improving crop and soil management in the southern United States. Horticultural Science 21:397–402.

Institute of Medicine. 1989. Human Health Risks with the Subtherapeutic Use of Penicillin or Tetracyclines in Animal Feed. Washington, D.C.: National Academy Press.

Jukes, T. 1973. Public health significance of feeding low levels of antibiotics to animals. Advanced Applications of Microbiology 16:1–30.

Kahn, J. R., and W. M. Kemp. 1985. Economic losses associated with the degradation of an ecosystem: The case of submerged aquatic vegetation in Chesapeake Bay. Journal of Environmental Economic Management 12(3):246–263.

Kennedy, D. 1977. Antibiotics used in animal feeds. HEW News. Rockville, Md.: U.S. Food and Drug Administration. April 15.

Knorr, D. 1975. Tin resorption, peroral toxicity and maximum admissable concentration in foods. Lebensmittel—Wissenschaft Technologie 8:51–57.

Maga, J. 1983. Organically grown foods. Pp. 305–350 in Sustainable Food Systems, D. Knorr, ed. Westport, Conn.: AVI Publishing Co.

McDonald, D. 1987. Chemicals and your health: What's the risk? Farm Journal 3(2):8–11.

Mott, L. 1984. Pesticides in Food: What the Public Needs to Know. San Francisco: Natural Resources Defense Council, Inc.

Murray, B. E. 1984. Emergence of diseases caused by bacteria resistant to antimicrobial agents. In Zoonoses, Vol. I, J. H. Steele and G. W. Beran, eds. CRC Handbook Series. Boca Raton, Fla.: CRC Press.

Myers, C. F., J. Meek, S. Tuller, and A. Weinberg. 1985. Nonpoint sources of water pollution. Journal of Soil and Water Conservation 40:14–18.

National Research Council. 1972. Genetic Vulnerability of Major Farm Crops. Washington, D.C.: National Academy Press.

National Research Council. 1975. Cotton Pest Control. Washington, D.C.: National Academy Press.

National Research Council. 1980. The Effects on Human Health of Subtherapeutic Use of Antimicrobials in Animal Feeds. Washington, D.C.: National Academy Press.

National Research Council. 1982. Diet, Nutrition, and Cancer. Washington, D.C.: National Academy Press.

National Research Council. 1984. Toxicity Testing: Strategies to Determine Need and Priorities. Washington, D.C.: National Academy Press.

National Research Council. 1986a. Pesticide Resistance: Strategies and Tactics for Management. Washington, D.C.: National Academy Press.

National Research Council. 1986b. Pesticides and Groundwater Quality: Issues and Problems in Four States. Washington, D.C.: National Academy Press.

National Research Council. 1986c. Soil Conservation: Assessing the National Resources Inventory, Vols. 1 and 2. Washington, D.C.: National Academy Press.

National Research Council. 1987. Regulating Pesticides in Food: The Delaney Paradox. Washington, D.C.: National Academy Press.

Newsom, L. D. 1962. The boll weevil problem in relation to other cotton insects. Pp. 83–94 in Proceedings of the Boll Weevil Research Symposium. State College, Miss.: Mississippi State University.

Niedermeier, R. P., J. W. Crowley, and E. C. Meyer. 1983. United States dairying: Changes and challenges. Journal of Animal Science 57(Suppl. 2):44–57.

Nielsen, E. G., and L. K. Lee. 1987. The Magnitude and Costs of Groundwater Contamination from Agricultural Chemicals. Staff Report AGES87. Economic Research Service. U.S. Department of Agriculture. Washington, D.C.

Paige, J. C., and R. Kent. 1987. Tissue residue briefs. FDA Veterinarian 2(6):10–11.

Pearce, N. E., A. H. Smith, and D. O. Fisher. 1985. Malignant lymphoma and multiple myeloma linked with agricultural occupations in a New Zealand cancer registry-based study. American Journal of Epidemiology 121(2):225–237.

Phipps, T. T., and P. R. Crosson. 1986. Agriculture and the environment: An overview. Pp. 3–31 in Agriculture and the Environment: Annual Policy Review 1986, T. T. Phipps, P. R. Crosson, and K. A. Price, eds. Washington, D.C.: The National Center for Food and Agricultural Policy, Resources for the Future.

Pimentel, D. 1987. Soil erosion effects on farm economics. Pp. 217–241 in Agricultural Soil Loss: Processes, Policies, and Prospects, J. M. Harlin and A. Hawkins, eds. Boulder, Colo.: Westview Press.

Pimentel, D., D. Andow, R. Dyson-Hudson, D. Gallahan, S. Jacobson, M. Irish, S. Kroop, A.

Moss, I. Schreiner, M. Shepard, T. Thompson, and B. Vinzant. 1980. Environmental and social costs of pesticides: A preliminary assessment. Oikos 34:127–140.

Poincelot, R. 1986. Towards More Sustainable Agriculture. Westport, Conn.: AVI Publishing Co.

Power, J. F. 1987. Legumes: Their potential role in agricultural production. American Journal of Alternative Agriculture 2(2):69–73.

Reganold, J. P., L. F. Elliott, and Y. L. Unger. 1987. Long-term effects of organic and conventional farming on soil erosion. Nature 330:370–372.

Schneider, K. 1986. Erosion Is Called Small Threat to Crop Yields. The New York Times, 16 May 1986.

Smith, J. S. C. 1988. Diversity of United States hybrid maize germplasm; Isozymic and chromatographic evidence. Crop Science 28:63–69.

Smith, R. A., R. B. Alexander, and M. G. Wolman. 1987. Water-quality trends in the nation's rivers. Science 235:1607–1615.

Tauxe, R. V. 1986. Antimicrobial resistance in human salmonellosis in the United States. Journal of Animal Science 62(Suppl. 3):65–73.

Troeh, F. R., J. A. Hobbs, and R. L. Donahue. 1980. Soil and Water Conservation for Productivity and Environmental Protection. Englewood Cliffs, N.J.: Prentice-Hall.

U.S. Congress, House. Committee on Government Operations. 1985. Human Food Safety and the Regulation of Animal Drugs. Union Calendar No. 274. Washington, D.C.

U.S. Department of Agriculture. 1986a. Agricultural Resources—Cropland, Water, and Conservation—Situation and Outlook Report. AR-4. Economic Research Service. Washington, D.C.

U.S. Department of Agriculture. 1986b. World Agriculture—Situation and Outlook Report. WAS-45. Economic Research Service. Washington, D.C.

U.S. Department of Agriculture. 1987a. Agricultural Resources—Cropland, Water, and Conservation—Situation and Outlook Report. AR-8. Economic Research Service. Washington, D.C.

U.S. Department of Agriculture. 1987b. Agricultural Resources—Inputs— Situation and Outlook Report. AR-5. Economic Research Service. Washington, D.C.

U.S. Department of Agriculture. 1987c. Economic Indicators of the Farm Sector: National Financial Summary. 1986 ECIFS 6-2. Economic Research Service. Washington, D.C.

U.S. Department of Agriculture. 1987d. U.S. Irrigation: Extent and Economic Importance. Agriculture Information Bulletin No. 523. Economic Research Service. Washington, D.C.

U.S. Department of Agriculture. 1987e. World Agriculture—Situation and Outlook Report. WAS-49. Economic Research Service. Washington, D.C.

U.S. Department of Agriculture. 1988a. Agricultural Income and Finance—Situation and Outlook Report. AFO-31. Economic Research Service. Washington, D.C.

U.S. Department of Agriculture. 1988b. Agricultural Outlook. Special Reprint: Agricultural Chemicals and the Environment. Economic Research Service. Washington, D.C.

U.S. Department of Agriculture. 1988c. Agricultural Resources—Cropland, Water, and Conservation—Situation and Outlook Report. AR-12. Economic Research Service. Washington, D.C.

U.S. Department of Agriculture. 1988d. Financial Characteristics of U.S. Farms, January 1, 1988. Economic Research Service. Washington, D.C.

U.S. Department of Agriculture. 1988e. A National Program for Soil and Water Conservation: The 1988–97 Update. Washington, D.C.

U.S. Department of Agriculture. 1988f. Outlook '88 Charts: 64th Annual Agricultural Outlook Conference. Economic Research Service. Washington, D.C.

U.S. Environmental Protection Agency. 1984. Report to Congress: Nonpoint Source Pollution in the U.S. Washington, D.C.

U.S. Environmental Protection Agency. 1987. Alachlor; Notice of Intent to Cancel Registrations; Conclusion of Special Review. Office of Pesticides and Toxic Substances. Washington, D.C.

U.S. Food and Drug Administration. 1986. Chloramphenicol Oral Solution; Withdrawal of Approval of NADA's. Federal Register 51(8):1441.

U.S. Food and Drug Administration. 1988. Proposed Removal of Regulation Regarding Sulfonamide-Containing Drugs for Use in Food-Producing Animals. Federal Register 53(179):35,833–35,836.

U.S. General Accounting Office. 1986a. Farm Finance: Financial Condition of American Agriculture as of December 31, 1985. Washington, D.C.

U.S. General Accounting Office. 1986b. U.S. Agricultural Exports: Factors Affecting Competitiveness in World Markets. Washington, D.C.

U.S. General Accounting Office. 1988. Farm Programs: An Overview of Price and Income Support, and Storage Programs. GAO/RCED-88-84BR. Washington, D.C.

U.S. Office of Technology Assessment. 1987. Technologies to Maintain Biological Diversity. Washington, D.C.: U.S. Government Printing Office.

Van Chantfort, E. 1987. Farm financial profile details improvement, diversity. Farmline 8(11):4–7.

Van den Bosch, R. 1980. The Pesticide Conspiracy. Garden City, N.Y.: Anchor Books.

Vogtmann, H. 1978. Ecologically sound preparation of farm yard manure and slurry. In Towards a Sustainable Agriculture, Proceedings of International Federation of Organic Agriculture Movements Conference, J. M. Besson and H. Vogtmann, eds. Switzerland: IFOAM.

Weisenburger, D. D. 1985. Lymphoid malignancies in Nebraska: A hypothesis. The Nebraska Medical Journal 70(8):300–305.

Williams, W. M., P. W. Holden, D. W. Parsons, and M. N. Lorber. 1988. Pesticides in Ground Water Data Base: 1988 Interim Report. Office of Pesticide Programs. U.S. Environmental Protection Agency. Washington, D.C.

Wnuk, M., R. Kelley, G. Breuer, and L. Johnson. 1987. Pesticides in Water Supplies Using Surface Water Sources. Iowa City, Iowa: Iowa Department of Natural Resources and University Hygienic Laboratory.

Young, F. E. 1987. Food safety and the FDA's action plan, phase II. Food Technology (Nov.):116–124.

3

Research and Science

ALTERNATIVE AGRICULTURE is a systems approach to farming that is more responsive to natural cycles and biological interactions than conventional farming methods. For example, in alternative farming systems, farmers try to integrate the beneficial aspects of biological interaction among crops, pests, and their predators into profitable agricultural systems. Alternative farming is based on a number of accepted scientific principles and a wealth of empirical evidence. Some of both are presented in this chapter. The specific mechanisms of many of these phenomena and interactions need further study, however. In general, much is known about some of the components of alternative systems, but not nearly enough is known about how these systems work as a whole.

Examples of practices or components of alternative systems that the committee has considered are listed below. Some of these practices are already part of conventional farming enterprises. These practices include:

- Crop rotations that mitigate weed, disease, and insect problems; increase available soil nitrogen and reduce the need for synthetic fertilizers; and, in conjunction with conservation tillage practices, reduce soil erosion.
- Integrated pest management (IPM), which reduces the need for pesticides by crop rotations, scouting, weather monitoring, use of resistant cultivars, timing of planting, and biological pest controls.
- Management systems to improve plant health and crops' abilities to resist pests and disease.
- Soil-conserving tillage.
- Animal production systems that emphasize preventative disease management and reduce reliance on high-density confinement, costs associated with disease, and need for use of subtherapeutic levels of antibiotics.

ADVOCATES AND PRACTITIONERS OF ALTERNATIVE FARMING SYSTEMS

• *Individuals who adhere to philosophies that advocate nonconventional farming practices.* Some farmers never changed to the chemically intensive, specialized approach to crop and animal production that currently dominates U.S. agriculture. These farmers include followers of traditional organic farming movements, such as biodynamic agriculture and the systems advocated by Albert Howard and Eve Balfour (Balfour, 1976; Howard, 1943). These individuals also include farmers who farm organically because of religious beliefs, such as some Amish and Mennonite farmers of Pennsylvania and the Midwest. Others have practiced a generic form of organic farming not associated with any of the established organic movements (Harwood, 1983).

• *Farmers looking for new ways to reduce production costs.* Throughout the United States, individual farmers have recognized that heavy purchases of off-farm inputs can put them in a less competitive economic position. These farmers have modified their farming practices, often in innovative ways, to reduce production costs. Examples include a wide variety of conservation tillage systems; the use of legume-fixed nitrogen through rotations; interplanting; the substitution of manures, sewage sludges, or other organic waste materials for purchased inorganic fertilizers; and the use of IPM systems and biological pest control.

• *Farmers responding to consumer interest in chemical-free organic produce.* Many enterprising farmers producing agronomic and horticultural crops, milk, eggs, poultry, beef, and pork without synthetic chemical inputs have taken advantage of the fact that many consumers and businesses are willing to pay higher prices for these sorts of products. In response to market demand, several commercial supermarket chains have recently begun to market produce grown with no or very low levels of certain synthetic chemical pesticides at prices roughly comparable to those of conventionally grown produce.

• *Farmers responding to concerns about the adverse impact of many conventional farming practices on the environment.* Environmental groups and soil conservation organizations have raised public awareness of the environmental hazards of conventional agricultural practices. As a result of these hazards and personal concern for the environment, some farmers have adopted alternative farming practices that are helping to reduce the deterioration of our nation's soil and water resources.

• *University research scientists.* Critics have attacked the colleges and schools of agriculture in the land-grant universities and the U.S. Department of Agriculture (USDA) for not researching farming systems that protect the environment and reduce dependence on synthetic chemical inputs. But many individuals at these institutions have been investigating for years practices and systems that have alternative agricultural applications. Examples include integrated pest management (IPM), biological controls of pests, rotations, nitrogen fixation, timing of fertilizer applications, disease- and stress-resistant plant cultivars, conservation tillage, and use of green manure crops. These research efforts have fostered some important changes in U.S. agriculture. As greater effort is made toward implementing the results of this research, more progress can be expected in the future. Much of the scientific knowledge of alternative practices summarized below is the result of research at the land-grant universities and the USDA.

• *Alternative agriculture organizations.* Groups such as Practical Farmers of Iowa, the Land Stewardship Project, the Institute for Alternative Agriculture, the Regenerative Agriculture Association, the Center for Rural Affairs, the Land Institute, and many others have worked to provide farmers with information on alternatives. They have organized research and demonstration projects, lobbied state legislatures and Congress for research and demonstration support, and produced numerous technical publications and reports with information designed to help and encourage farmers to adopt alternatives.

• Genetic improvement of crops to resist pests and diseases and to use nutrients more effectively.

Many alternative agricultural systems developed by farmers are highly productive (see the boxed article, "Advocates and Practitioners of Alternative Farming Systems," and Part Two). They typically share much in common, such as greater diversity of crops grown, use of legume rotations, integration of livestock and crop operations, and reduced synthetic chemical use. Although many practices show great promise, the scientific bases for many of them are often incompletely understood.

During the last four decades, agricultural research at the land-grant universities and the USDA has been extensive and very productive. Most of the new knowledge has been generated through an intradisciplinary approach to research. Scientists in individual disciplines have focused their expertise on one aspect of a particular disease, pest, or other agronomic facet of a particular crop. Solving on-farm problems, however, requires more than an intradisciplinary approach. Broadly trained individuals or interdisciplinary teams must implement the knowledge gained from those in individual disciplines with the objective of providing solutions to problems at the whole-farm level. This interdisciplinary problem-solving team approach is essential to understanding alternative farming practices.

Agricultural research has not been organized to address this need except in a few areas, such as IPM, the use of organic residues as an alternative nutrient source, and the use of leguminous green manure crops and rotations for erosion control and as a nitrogen source. Even this research has not significantly contributed to the adoption of alternative agricultural *systems* for two principal reasons. First, most research has focused on individual farming practices in isolation and not on the development of agricultural *systems*. This is because of the high expense of farming systems research, the intradisciplinary nature of university research, and lack of resources. Second, most research results have been implemented under policies that encouraged ever-increasing per acre yields as the best way to increase farm profits and the world food supply.

In contrast, alternative farming research must include the interaction and integration of all farm operations and must consider the more comprehensive goals of resource management, productivity, environmental quality, and profitability with minimal government support. Only a limited amount of research has taken this comprehensive approach. Nevertheless, the scientific literature about specific farm practices and the empirical evidence from individual operators illustrate the efficacy and potential of alternative farming methods and provide the foundation on which to build a program of alternative farming research.

Important elements of the scientific knowledge base relevant to further development of alternative agricultural systems are briefly reviewed in the following sections. Knowledge of biological systems and the management of their interactions throughout agricultural ecosystems are emphasized.

CROP ROTATION

Crop rotation is the successive planting of different crops in the same field. A typical example would be corn followed by soybeans, followed by oats, followed by alfalfa. Rotations are the opposite of continuous cropping, which involves successively planting the same field with the same crop. Rotations may range between 2 and 5 years (sometimes more) in length and generally involve a farmer planting a part of his or her land to each crop in the rotation. Rotations provide many well-documented economic and environmental benefits to agricultural producers (Baker and Cook, 1982; Heady, 1948; Heady and Jensen, 1951; Heichel, 1987; Kilkenny, 1984; Power, 1987; Shrader and Voss, 1980; Voss and Shrader, 1984). Some of these benefits are inherent to all rotations; others depend on the crops planted and length of the rotation; and others depend on the types of tillage, cultivation, fertilization, and pest control practices used in the rotation. When rotations involve hay crops, on-farm livestock or a local hay market are generally required to make the hay crop profitable.

Much of the literature on crop rotations refers to the rotational effect (Heichel, 1987; Power, 1987). This term is used to describe the fact that in most cases rotations will increase yields of a grain crop beyond yields achieved with continuous cropping under similar conditions. This rotational effect has been shown to exist whether rotations include nonleguminous or leguminous crops. Corn following wheat, which is not a legume, produces greater yields than continuous corn when the same amount of fertilizer is applied (Power, 1987). The increase in crop yields following a leguminous crop is usually greater than expected from the estimated quan-

Between 40 and 45 percent of the U.S. corn crop is grown in continuous monoculture. Corn grown continuously generally requires greater use of fertilizers and pesticides than corn grown in rotation. This corn field is 10 miles from Kearney, Nebraska, which can be seen on the horizon. *Credit:* U.S. Department of Agriculture.

tity of nitrogen supplied (Cook, 1984; Goldstein and Young, 1987; Heichel, 1987; Pimentel et al., 1984; Voss and Shrader, 1984). In fact, yields of grains following legumes are often 10 to 20 percent greater than continuous grain regardless of the amount of fertilizer applied.

Many factors are thought to contribute to the rotational effect, including increased soil moisture, pest control, and the availability of nutrients. It is generally agreed, however, that the most important component of this effect is the insect and disease control benefits of rotations (Cook, 1984, 1986). The increase in soil organic matter, particularly in sod-based rotations, may

Contour strip cropping can reduce erosion and pest infestation. When a legume is included in a rotation, such as the corn-wheat-alfalfa rotation shown here, nitrogen fertilizer needs can be decreased. *Credit:* Grant Heilman.

be the basis for the improved physical characteristics of soil observed in rotations. This may account for some yield increase. Certain deep-rooted leguminous and nonleguminous crops in rotations may use soil nutrients from deep in the soil profile. In the process, these plants may bring the nutrients to the surface, making them available to a subsequent shallow-rooted crop if crop residue is not removed.

Another benefit common to all rotations is the control of weeds, insects, and diseases, particularly insects and diseases that attack the plant roots (Cook, 1986). This pest control is achieved primarily through the seasonal change in food source (the crop), which usually prevents the establishment of destructive levels of pests. As root disease and insect damage are reduced, the healthy root system is better able to absorb nutrients in the soil, which can reduce the rates of fertilizers needed (Cook, 1984). Healthy root systems also take up nutrients more effectively, thus reducing the likelihood of nutrient leaching out of the root zone.

Rotations with particular crops or crop combinations can provide additional benefits. Legumes in rotations will fix nitrogen from the atmosphere into the soil. The amount of nitrogen fixed depends on the legume and the management system; however, without any additional nitrogen fertilizers, leguminous nitrogen can support high grain yields (Heichel, 1987; Voss and Shrader, 1984). The length of the rotation and yield expectations of the farmers, however, influence the level and acceptability of these yields.

Hay and forage crops and closely sown field grain crops, such as wheat, barley, and oats, can provide some soil erosion control benefits in rotations. In some eroding areas with steep terrains, the practice of strip cropping corn (a row crop) with wheat (a closely sown crop) or a hay crop, such as alfalfa, is a common use of rotations to slow erosion. It must be stressed, however, that tillage practices greatly influence the erosion control benefits of crops planted in rotations (Elliott et al., 1987). For example, a rotation of corn, soybeans, and wheat is excellent for disease control but not for erosion control unless no tillage or reduced tillage is used.

An indirect but important benefit of all rotations is that they involve diversification. The benefits of diversification are described in more detail later in this chapter. In general, however, diversification provides an economic buffer against price fluctuations for crops and production inputs as well as the vagaries of pest infestations and the weather.

Rotations may have their disadvantages, however, particularly in the context of current government subsidies and requirements for federal program participation (see Chapters 1 and 4). Rotations that involve diversifying from cash grains to crops such as leguminous hays with less market value involve economic tradeoffs (see Chapter 4). Adopting the use of rotations may also require purchasing new equipment. As with all sound management practices, rotations must be tailored to local soil, water, economic, and agronomic conditions.

PLANT NUTRIENTS

Soil, water, and air supply the chemical elements needed for plant growth. Photosynthesis captures energy from the sun and converts it into stored chemical energy by transforming carbon dioxide from the air into simple carbohydrates. This stored chemical energy becomes the fuel for all life on earth. Water is also needed to provide essential elements, transport nutrients and sugars within plants, serve as a medium for essential chemical reactions, and provide structural form and strength by exerting turgor pressure from inside plant cells. Nutrient elements essential to the chemical reactions that occur within the plant are taken up from the soil through the roots. If nutrient elements or water are not adequately available at the time they are needed, plant growth and development will be affected. Growth and yield will be reduced or the plant may die.

Plants need three soil-derived nutrient elements in large amounts—nitrogen, phosphorus, and potassium. These elements are frequently not available in adequate amounts from soil. Nitrogen is a constituent of all proteins and a part of chlorophyll, the pigment that reacts to light energy. Nitrogen is a component of nucleic acids and the coenzymes that facilitate cell reactions. Phosphorus, as a component of adenosine triphosphate (ATP), is critical to the development and use of chemical energy within the cell. Phosphorus is also a constituent of many proteins, coenzymes, metabolic substrates, and nucleic acids. Unlike nitrogen and phosphorus, potassium does not have a clear function as a constituent of chemical compounds within the plant. It is important in regulatory mechanisms affecting fundamental plant processes, such as photosynthesis and carbohydrate translocation. In addition to these three nutrients, other soil-supplied nutrients are essential to plant growth and development: boron, calcium, chlorine, cobalt, copper, iron, magnesium, manganese, molybdenum, sulfur, and zinc. These elements are needed in small amounts that are often available in soil.

Soil Properties and Plant Nutrients

Soil Texture

The mineral particles that make up the soil are classified on the basis of their size. Clay particles are the smallest, silt is intermediate, and sand particles are the largest. The relative proportions of clay, silt, and sand determine soil texture. Soil texture has a critical influence on water and nutrient retention and movement through the soil. The large pores among grains of sand in a sandy soil allow water to pass through with relative ease, whereas the small pores formed in clay soils slow the flow and retain water.

Soil particles can exist separately or they can be bound together in larger aggregates. Organic colloids and clays play a critical role in binding soil particles into soil aggregates, which increase pore space and water and air movement.

Cation Exchange

The molecular surfaces of clays and organic colloids have a net negative charge that interacts with the polar charge of surrounding water molecules. This causes the colloids to bind with positively charged ions of elements (cations). Because cations have differing abilities to bind with soil colloids, one cation may displace another; this is referred to as cation exchange. Displacement depends on relative bond strength and relative concentration. The cation exchange capacity of a soil is an expression of the number of cation-binding sites available per unit weight of soil (Foth, 1978). This capacity has a significant effect on nutrient movement and availability and binding of pesticides in different soils. Because hydrogen ions are cations that compete with nutrient cations for exchange sites, soil acidity, which is a measure of hydrogen ion concentration, has a marked effect on which nutrient elements are bound and which are displaced.

Soil Quality

The quality of agricultural soils is derived from their effectiveness as a medium that provides essential nutrients and water. Mineral elements in soil required for plant growth exist in soluble and insoluble forms, which affects their availability for plant uptake. For example, under acidic or alkaline soil conditions, phosphorus fertilizer is rapidly converted into less soluble compounds that may be nearly unavailable for plant nutrition. Even available forms of phosphorus are bound to clay, and organic soil compounds and are relatively immobile in the soil profile except as a passenger during soil erosion. In contrast, potassium and the ammonium and nitrate forms of nitrogen are more soluble than phosphorus. Nitrate ions are not held by negatively charged soil and are readily leached. Because of their positive charges, potassium and ammonium nitrogen are held on the cation exchange and will not leach appreciably except through sandy soils.

Organic matter in soils influences plant growth in a number of ways. The greatest benefits of organic matter in soil are its water-holding capacity; the manner in which it alters soil structure to improve soil tilth; its high exchange capacity for binding and releasing some mineral nutrients; its presence as a food source for soil microbiota that recycle soil nutrients; and its mineralization to nitrogen, phosphorus, and sulfur. The cycling of mineral nutrients between living organisms and dead organic components of the soil system provides an important reservoir of the elements needed in plant growth.

Nutrients are lost from soil through removal by crops, leaching, and soil erosion. Nitrate nitrogen can also be lost from the soil by conversion to nitrogen gases (denitrification) or by volatilization of ammonia. Gaseous loss of sulfur can also occur. Some farming practices help to mitigate the loss of nutrients and in some cases replace nutrients. For example, crop rotations that include nitrogen-fixing legumes benefit the soil in several ways. Legumes, in symbiotic relationships with microbes, fix atmospheric nitrogen into nitrogen compounds available for plant nutrition. When legumes are plowed under as green manures, they add nitrogen and organic matter to the soil. Cover crops help hold nitrogen in the root zone during the winter.

The accumulated scientific knowledge on the role and fate of mineral elements, organic matter, and water in crop growth provides some indication of why some alternative farming practices succeed and others fail. Characteristics of a particular crop or farming system that yield maximum efficiency are not well understood, however. The task remains to assemble the interdisciplinary expertise needed to analyze and understand the complex relationships that contribute to the relative efficiencies of different farming systems.

Nutrient Management

The adequate supply of nutrients—particularly nitrogen, phosphorus, and potassium—and maintenance of proper soil pH are essential to crop growth.

Ideally, soil nutrients should be available in the proper amounts at the time the plant can use them; this avoids supplying an excess that cannot be used by plants and may become a potential source of environmental contamination. The current conventional approach is to apply nutrients in the form of fertilizers at levels needed for maximum profitability. Profitability in the context of current government programs has generally been achieved, however, through maximum yield per acre, often in continuous cropping or short rotations that require significant amounts of fertilizer. The nutrients in any excess fertilizer or high levels of decomposing organic matter are subject to leaching or runoff.

An alternative, more environmentally benign approach to nutrient management is to reduce the need for fertilizer through more efficient management of nutrient cycles and precise applications of fertilizer. Such practices include application of organic waste residues from animals and crops, crop rotations with legumes, improved crop health that may result in better use of nutrients, and banded or split applications of fertilizers. In mixed crop and livestock operations, for example, many of the nutrients contained in the grain and residue from crops grown on the farm can be returned to the soil if the manure and crop residues are incorporated into the soil. Crop rotations that include legumes can also play an essential role in nutrient cycling, particularly for replenishing the nitrogen supply. Plant residues and manure can release nitrogen more continuously throughout the growing season than can common commercial fertilizers. However, nitrogen from organic sources may be released when crops are not actively absorbing it. In contrast, inorganic fertilizer nitrogen is relatively quickly converted to the soluble and leachable nitrate form.

Efforts to provide adequate nutrition to crops continue to be hindered by inadequate understanding and forecasting of factors that influence nutrient storage, cycling, accessibility, uptake, and use by crops during the growing seasons. Soil testing and plant tissue analysis can provide the farmer with information to assure adequate nutrition for all agronomic and horticultural crops. But variable soil and climatic conditions that influence nutrient uptake and loss make it difficult to predict the most profitable and environmentally safe levels of nutrients. As a result, farmers often follow broad guidelines that lead to insufficient or excessive fertilization. For example, studies of fertilizer recommendations revealed that some commercial soil testing services consistently recommended the use of far more fertilizer than was needed (Olson et al., 1981; Randall and Kelly, 1987). Additionally, some farmers apply more nitrogen than is recommended.

Nitrogen

Nitrogen is the soil-derived plant nutrient most frequently limiting grain production in the United States. This is ironic because the atmosphere is 79 percent nitrogen by volume. Atmospheric nitrogen is in the form of inert nitrogen gas, however, which higher-order plants cannot use. Converting atmospheric nitrogen to ammonia and other forms that plants can use requires a high energy input. This is true for biological nitrogen fixation as well as industrial synthesis. The biological process is fueled by photosynthates; the synthetic industrial process is fueled by natural gas, petroleum, coal, or hydroelectric power. The predominant process for producing synthetic nitrogen fertilizers involves combining hydrogen from methane gas and atmospheric nitrogen at high temperature and pressure to form ammonia. Ammonia can then be converted to nitric acid or combined with other elements to form a number of nitrogen fertilizers, including ammonium nitrate, ammonium sulfate, ammonium phosphate, and urea. A significant amount of energy is required to synthesize ammonia. Consequently, energy and methane gas costs can affect the availability and cost of synthetic nitrogen fertilizers.

Neutral ammonia molecules gain a hydrogen ion when added to moist soil and become stable ammonium ions with a net positive charge. Most of the ammonium ions in soil undergo biological nitrification, in which oxidation results in the formation of a nitrate ion as well as hydrogen ions that acidify the soil. Because ammonium ions have a positive charge, they are adsorbed and held on the soil cation exchange. Nitrate ions, because of their negative charge, are not adsorbed on the soil exchange complex. While readily available for plant use, the nitrate freely moves through soil in water unless it is absorbed by the plant.

Although these basic processes are understood, there is a need to know much more about nutrient cycling and the behavior of nitrogen under various environmental conditions. To accomplish this, progress is needed in estimating the rates of biological reactions that control nitrogen transformation in soil.

Legumes as a Source of Nitrogen

Nitrogen can be provided by growing legumes in rotation with grains. For alternative farming, legumes are an effective and often profitable way to supply nitrogen. Leguminous nitrogen is consistently released throughout the growing season when temperatures are high enough to permit microbial decomposition. Combined with the rotational effect, leguminous nitrogen can support high yields of corn and wheat (Holben, 1956; Koerner and Power, 1987; Voss and Shrader, 1984). The overall contribution of legumes, however, depends on the management system and climate. For example, forage legumes are most effective in humid and subhumid regions (Meisenbach, 1983; U.S. Department of Agriculture, 1980). In regions with less than 20 inches of rain a year, deep-rooted, nonirrigated legumes may decrease subsoil moisture and lead to reduced corn yields the following year (Meisenbach, 1983). The profitability of leguminous hay crops is strongly influenced by the presence of on-farm livestock or a local hay market.

Legumes supply substantial nitrogen to the soil, but the amount of nitrogen fixed is highly variable. Different species and cultivars fix different

These fields, currently producing corn and soybeans, have received no nitrogen, phosphorus, or potassium fertilizer for 18 years. A corn, soybean, small grain, and red clover crop rotation and the application of manure supply nutrients. *Credit:* Rex Spray, the Spray Brothers Farm.

amounts of atmospheric nitrogen. A number of physical and managerial factors, including soil acidity, temperature, drainage, the timing of harvest, and whether foliage is turned under as green manure, influence the amount of nitrogen fixed as well as the amount of fixed nitrogen subsequently incorporated into the soil. Nitrogen fixation by soybeans, for example, was found to vary from 0 to 277 pounds per acre depending on management practices, soil characteristics, and water availability.

The amount of nitrate in the soil also affects nitrogen fixation. Soil rich in nitrate inhibits nitrogen fixation. In the Midwest, soybeans are managed for grain production and are commonly grown after corn in soils with residual nitrate. Where there is residual nitrate in the soil, soybean production can result in a net export of nitrogen. For example, nitrogen budget analyses on midwestern soybeans show that 40 percent of nitrogen in the crop is derived from nitrogen fixation and 60 percent is from residual nitrogen in the soil. Typically, the nitrogen removed in the soybeans at harvest exceeds the amount of nitrogen fixed, leading to a net nitrogen loss of about 70 pounds per acre. Thus, under these circumstances, soybeans may be depleting the soil of nitrogen and increasing nitrogen fertilizer needs for the subsequent crop, rather than enriching soil nitrogen as had been previously thought. In contrast, when soybeans can be managed to fix 90 percent of their nitrogen needs, the result is a 20 pound per acre nitrogen gain (Heichel, 1987).

Management systems also influence nitrogen made available by leguminous hay crops. Leguminous hays are commonly grown for their value as hay and for their ability to fix nitrogen from the atmosphere into the soil and provide nitrogen in the form of crop residue. The timing of harvest, however, dramatically affects the amount of nitrogen available for subsequent crops. After they are cut, leguminous hays first use the reserve of nitrogen in the crown, roots, and soil to support their own growth. As the growing plant increases leaf area and photosynthesis, the energy is again available for nitrogen fixation. The nitrogen-rich leaves and stems of the plant are then removed when the crop is harvested (Heichel, 1987). Heichel (1987) reported results of studies with alfalfa showing that one harvest followed by moldboard plowing of lush, late August regrowth resulted in a net nitrogen gain of 48 pounds per acre. In contrast, harvest of the August regrowth followed by plowing under of October regrowth resulted in an insignificant net nitrogen loss of 4 pounds per acre. The slight loss occurred because most of the nitrogen fixed by the lush August regrowth was removed during harvest. Harvest of the October regrowth followed by plowing under only the roots and crowns resulted in a net nitrogen loss of 38 pounds per acre.

Management also affects the amount of legume-fixed nitrogen that drains to groundwater. Results of an unpublished Minnesota experiment (G. Randall) showed that over 4 years, continuous soybean (45 bushels/acre) contributed two-thirds as much nitrate to drainage water as heavily fertilized corn (165 bushels/acre). Unpublished experiments in Michigan (B. Ellis) found more than twice the concentration of nitrates below crop root systems when alfalfa was plowed down than under irrigated or nonirrigated corn. Few measurements have been made of the contribution of legumes to groundwater contamination or, if necessary, how to minimize it.

Tillage practices also influence the amount and availability of nitrogen supplied by legumes (Dabney et al., 1987; Heichel, 1987). No-tillage systems may reduce the nitrogen available to the subsequent crop compared with a conventional tillage system such as moldboard plowing, which more thoroughly incorporates plant matter into the soil (Varco et al., 1987). Koerner and Power (1987) reported increased corn yield following double-disking of vetch. Corn yields were lower when vetch was left standing throughout the corn growing system or when the vetch was killed with herbicides.

In a greenhouse experiment, the relative nitrogen fixation varied widely depending on species of legume, duration of growth, and temperature (Zachariassen and Power, 1987) (Table 3-1). Some species performed best under low temperatures; others fixed more nitrogen at higher temperatures. For example, fava beans were found to fix 54 percent more nitrogen than hairy vetch early in the growing season (42 days) at a temperature of 10°C. But at 30°C the nitrogen fixation of the fava beans declined by 86 percent.

Using legumes in a rotation, or as winter cover crops in the South, can reduce and, in some cases, eliminate the need for nitrogen fertilizers (Dabney et al., 1987; Goldstein and Young, 1987; Neely et al., 1987). Cultivars

TABLE 3-1   Nitrogen Fixation of Legume Species as Affected by Soil
Temperature and Time Under Greenhouse Conditions

Species and Temperature (°C)	Nitrogen Fixation (percent) at[a]:			
	42 Days	63 Days	84 Days	105 Days
Hairy vetch				
10	100	108	147	223
20	46	88	67	122
30	14	8	12	30
Sweetclover				
10	21	46	67	66
20	52	86	128	122
30	42	46	50	104
Fava bean				
10	154	131	135	122
20	136	124	122	134
30	22	12	11	4
Lespedeza				
10	0	4	3	0
20	9	29	93	145
30	16	22	60	176
Field pea				
10	36	48	29	51
20	38	21	12	8
30	12	10	6	0
White clover				
10	20	54	88	162
20	39	78	108	153
30	2	0	40	38
Nodulated soybeans				
10	31	33	37	34
20	116	215	315	415
30	94	163	260	291
Crimson clover				
10	43	43	72	107
20	54	79	86	56
30	9	2	14	5

[a]Values are expressed as a percentage of the nitrogen fixation found in hairy vetch at 10°C for 42 days, which was arbitrarily selected as the basis for comparisons. To translate these percentages to the originally reported data (in milligrams/pot) multiply by 1.1339.

SOURCE: Zachariassen, J. A., and J. F. Powers. 1987. Soil temperature and the growth, nitrogen uptake, dinitrogen fixation, and water use by legumes. Pp. 24–26 in The Role of Legumes in Conservation Tillage, J. F. Power, ed. Ankeny, Iowa: Soil Conservation Society of America.

are being developed that fix more nitrogen than their predecessors. Bacteria in the *Rhizobium* and *Bradyrhizobium* genera that fix atmospheric nitrogen in the roots of legumes are being studied extensively. Current work focuses on the mechanism of nitrogen fixation itself, the infection process that leads to a successful symbiosis, the genetic determinants and biochemical processes that make plants receptive, and the bacteria capable of sustaining the asso-

TABLE 3-2 Reported Quantities of Dinitrogen Fixed by Various Legume Species

Species	N Fixed (pounds/acre/year)	Species	N Fixed (pounds/acre/year)
Alfalfa	70–198	Hairy vetch	99
Alfalfa-orchardgrass	13–121	Ladino clover	146–167
Birdsfoot trefoil	44–100	Lentil	149–168
Chickpea	21–75	Red clover	61–101
Clarke clover	19	Soybean	20–276
Common bean	1.8–192	Sub clover	52–163
Crimson clover	57	Sweet clover	4
Fava bean	158–223	White clover	114
Field peas	155–174		

SOURCE: Adapted from Heichel, G. H. 1987. Legume nitrogen: Symbiotic fixation and recovery by subsequent crops. Pp. 63–80 in Energy in Plant Nutrition and Pest Control, Z. R. Helsel, ed. Amsterdam, The Netherlands: Elsevier Science Publishers B. V.

ciation between plants and nitrogen. Improvements have already resulted from selecting for crop varieties and naturally occurring *Rhizobium* strains that fix large amounts of nitrogen. In a 2-year rotation with corn in Minnesota, a new annual cultivar of alfalfa, Nitro, developed by the U.S. Department of Agriculture's (USDA) Agricultural Research Service and the University of Minnesota Agricultural Experiment Station, fixed 94 pounds of nitrogen per acre between the last harvest in early September and the death of the plant at the first frost in October (Barnes et al., 1986). This was 59 percent more nitrogen than was fixed after the September harvest of the commonly grown perennial cultivars used as controls in the study. About 11 percent of this increase was from the improved nitrogen-fixing capability of the legume. The remaining 48 percent was from this variety's greater productivity at the end of the growing season.

Nitro alfalfa was bred as an annual crop for use in a 2-year rotation with corn. It is grown for 1 year and continues to grow and fix nitrogen until it is killed by the first frost, usually in mid-October. Commonly used alfalfa varieties, in contrast, are usually grown for 2 or more years, and (in Minnesota) begin to go dormant and stop fixing nitrogen in early September of each year.

Research conducted in northern California showed that vetch can be an economical source of nitrogen for rice. The study examined the effect of cultivar selection and time of planting. Aerial broadcast of purple vetch or Lana woolypod vetch seeded 2 days before or after field drainage produced an excellent stand. Vetch fixed between 30 and 60 pounds of nitrogen per acre in rice stubble—up to 100 pounds under ideal conditions (Williams and Dawson, 1980). The nitrogen-fixing capabilities of various legume species are listed in Table 3-2.

There is a general knowledge of the factors that affect nitrogen fixation by legumes, but little is understood about the interaction of these variables

Incorporating manure into the soil soon after application can keep nutrient losses to a minimum. The manure truck pictured is followed by a moldboard plow and a leveling harrow to incorporate manure into the soil. Between the time of its application and the emergence of the first crop, this manure tends to be vulnerable to erosion and runoff from rainfall. *Credit:* Rodale Press.

and their effect on total available nitrogen. This information is essential for determining nitrogen fertility credits of legumes and assessing the nitrate that legumes contribute to groundwater. More information is needed on nitrogen cycling in agricultural systems; the yield-boosting effects of rotations; the effects of tillage practices; and how fixed nitrogen is affected by other sources of nitrogen, soil organic matter, compost, and crop residues.

Manure as a Source of Nutrients

Animal wastes can make a substantial contribution to nitrogen, phosphorus, potassium, and other nutrient needs. Total supply, however, depends on the nature and size of animal enterprises and the methods used in storing and spreading the manure (Young et al., 1985). The potential nutrient contribution from manure is very high in some regions (Van Dyne and Gilbertson, 1978).

Most animal manure is returned to the land. Its nutrients, however, are often inefficiently used as a result of poor storage and application practices (Smith, 1988; U.S. Department of Agriculture, 1978). Runoff, volatilization, and leaching losses of plant nutrients in stored animal manure may be so high that only a fraction of the original nutrients remain to be applied to cropland. Poor manure hauling and spreading practices add to these losses. However, practices that increase the efficient use of nutrients may be eco-

TABLE 3-3   Nitrogen Losses in Manure Affected by Application Method

Method of Application	Type of Manure	Nitrogen Loss (percent)
Broadcast without incorporation	Solid	15–30
	Liquid	10–25
Broadcast with incorporation	Solid	1–5
	Liquid	1–5
Injection (knifing)	Liquid	0–2

NOTE: These numbers do not include losses from storage.

SOURCE: Sutton, A. L., D. W. Nelson, and D. D. Jones. 1985. Utilization of Animal Manure as Fertilizer. Extension Bulletin AG-FO-2613. St. Paul: University of Minnesota.

nomically costly. The cost of proper application, for example, may exceed the value of the increase in available nutrients compared with inefficient application methods. The effect of manure application methods on nitrogen loss, not including loss during storage, is shown in Table 3-3. Table 3-4 indicates the range of nutrient loss possible through different storage and handling systems. In many cases, manure is not applied at a time in the growing or fallow season that results in optimal use of the manure as fertilizer. For example, winter application of manure can result in significant nutrient loss. Little use is being made of animal manure in aerobic composting, even though composting may offer advantages of increasing nutrient concentration and reducing the volume of material to be applied (Granatstein, 1988). Anaerobic fermentation of manure to produce the biogas methane is not economical compared to the cost for other fuels. In addition,

TABLE 3-4   Nitrogen Losses in Manure Affected by Handling and Storage

Method	N loss (percent)
Solid systems	
Daily scrape and haul	15–35
Manure pack	20–40
Open lot	40–60
Deep pit (poultry)	15–35
Liquid systems	
Anaerobic deep pit	15–30
Above-ground storage	15–30
Earthen storage pit	20–40
Lagoon	70–80

NOTE: These numbers do not include losses due to application.

SOURCE: Sutton, A. L., D. W. Nelson, and D. D. Jones. 1985. Utilization of Animal Manure as Fertilizer. Extension Bulletin AG-FO-2613. St. Paul: University of Minnesota.

fermentation creates residue that must be disposed of or otherwise used (Smith, 1988).

Although systems are available to handle wastes in slurry form, other systems are essentially designed to dispose of the animal waste as an undesirable by-product. If animal waste is to be used more efficiently, systems and related equipment for profitably storing, handling, and spreading it are needed. Research is needed to devise low-cost systems of producing biogas from animal manures, make efficient application systems more economical, and educate farmers about the beneficial aspects of manure.

About 110 million tons (dry weight) of manure were voided by livestock and poultry in 1974. The total estimated amounts of nitrogen, phosphorus, and potassium in this manure were 4.1, 1.0, and 2.4 million tons, respectively. An estimated 40 percent of the total, or 1.3, 0.5, and 1.2 million tons of the nitrogen, phosphorus, and potassium voided, was estimated to be available and economically recoverable for use elsewhere. Cattle provided about 62 percent, or 800,000 tons, of the economically recoverable nitrogen from livestock manure in 1974 (Van Dyne and Gilbertson, 1978). The total nutrients economically recoverable from manure contained the equivalent of about 15 percent of the total nitrogen, 9.9 percent of the total phosphorus, and 24.2 percent of the total potassium fertilizer applied on farms in the United States during 1974 (U.S. Department of Agriculture, 1987). Similar data are not available for trace nutrients and organic matter supplied by manure.

The amount of nutrients available from manure largely depends on how it is stored and handled. Nitrogen is most readily lost; in fact, some loss is inevitable no matter how the manure is stored or applied. Phosphorus and potassium losses are less likely except through direct runoff and leaching from open storage lots or as a result of settling in open lagoons. Table 3-5 lists the approximate nutrient content of several types of manures as a result of different storage and handling techniques. Good management in manure handling is essential to successful use of manure as a nutrient source.

Farms in Lancaster County, Pennsylvania, have a high ratio of livestock per acre of cropland. Manure applications average over 40 tons per acre per year, supplying far more than the nutrient needs of the crops grown in the region. In addition, many farmers apply about 100 pounds of commercial nitrogen per acre, bringing the total nitrogen applied to corn to 350 pounds per acre (Young et al., 1985). Many of the county's wells have nitrate levels two to three times the U.S. Environmental Protection Agency's (EPA) standards for safe drinking water.

Phosphorus

The amount of phosphorus in solution in soil water determines the availability of phosphorus for plants. There is often a substantial amount of phosphorus in agricultural soils, but it is in a form that releases phosphorus to the surrounding soil water in a slow equilibrium reaction. In soil, the soluble phosphorus in fertilizer quickly reacts with aluminum and iron

TABLE 3-5   Approximate Nutrient Content of Several Manures

Type of Livestock	Storage/Handling[a]	Nutrient Content (pounds/ton)			
		Total N	Ammonium (NH_4^+)	Phosphate (P_2O_5)	Potash (K_2O)
Swine	Solid NB	10	6	9	8
	Solid B	8	5	7	7
	Liquid P	36	26	27	22
	Liquid L	4	3	2	4
Beef cattle	Solid NB	21	7	14	23
	Solid B	21	8	18	26
	Liquid P	40	24	27	34
	Liquid L	4	2	9	5
Dairy cattle	Solid NB	9	4	4	10
	Solid B	9	5	4	10
	Liquid P	24	12	18	29
	Liquid L	4	2.5	4	5
Turkeys	Solid NB	27	17	20	17
	Solid B	20	13	16	13
Horses	Solid B	14	4	4	14

[a]NB = No bedding; B = bedding; P = pit; L = lagoon.

SOURCE: Sutton, A. L., D. W. Nelson, and D. D. Jones. 1985. Utilization of Animal Manure as Fertilizer. Extension Bulletin AG-FO-2613. St. Paul: University of Minnesota.

oxide and with calcium to form compounds that are relatively insoluble and slowly available to plants. Moreover, phosphate quickly binds to clay colloids (Foth, 1978). Consequently, phosphate does not readily leach but neither does it remain in a form readily available to plants in either acid or alkaline soils.

Some organic farmers apply rock phosphate to their fields instead of acidulated phosphate. Rock phosphate found in the United States varies in solubility but generally has very low immediate availability to plants even when finely ground (Council for Agricultural Science and Technology, 1980). Chemical treatment, or acidulation, with sulfuric or phosphoric acids significantly enhances its availability. One 24-year study showed rock phosphate to be only one-sixth as effective as acidulated phosphate in a corn-oats-alfalfa rotation in slightly acid soils. Rock phosphate is even less effective on a less acid soil (Webb, 1982, 1984).

Farmers have built up the phosphorus content of U.S. soils over the past three decades. According to the USDA (1980), growers could eliminate the use of acidulated phosphates for several years in some regions without yield loss on well-managed soils. How long the resulting mining of phosphorus from such soils could continue is not known and will vary. Application of manures and organic wastes can replenish some phosphorus but removal by crops is inevitable. Because phosphorus does not leach, replacement applications to meet crop needs are advisable. Farms cannot be self-sufficient in phosphorus (Eggert and Kahrmann, 1984).

Phosphorus bound to sediment in runoff water is often implicated in eutrophication and decline in surface water quality. Although most farmers could reduce applications of phosphorus with little effect on crop yield, reductions are best made based on soil tests. Properly managed phosphorus applications and reduced soil erosion will be accompanied by improved water quality in watersheds with highly erodible cropland.

Potassium

Weathering of minerals has supplied many soils in subhumid or arid regions with adequate levels of available potassium. Potassium fertilizer applications, however, are especially required in humid regions and highly organic soils. Potassium has a net positive charge, so it is bound to the soil's cation exchange complex. Potassium is available in soil water solution in an equilibrium reaction with exchangeable and fixed potassium in the soil. It is more soluble and readily available than phosphorus but has little mobility and leaches only in sandy soils.

If there is excess potassium in the soil, plants will absorb more than is needed for normal physiological functioning. Forage crops such as alfalfa and clovers absorb large amounts of potassium; thus, harvest of hay or fodder (silage) leads to rapid depletion of readily available potassium in soils. Grain crops deplete potassium in the soil less rapidly, provided only the grain is removed. Potentially high levels of potassium in leguminous forages emphasize the need for conservation of manure from animals consuming these forages. Most of the potassium ingested by animals passes through the intestinal and urinary tracts. If all of the manure is conserved and uniformly redistributed to the land, little additional potassium will be needed for soils that begin with adequate levels. Leaching of exposed manure by rain can cause large losses of potassium, and this must be avoided to make recycling fully effective.

Amending Soil Reaction

One of the basic principles of soil management is to maintain a soil reaction appropriate to the crops produced. Soil reaction is measured by hydrogen ion activity and reported as pH. A neutral pH is 7.0. Most agronomic crops perform best between pH 6.0 and pH 7.5. The leguminous crops tend to perform better toward the upper end of the range while grain crops are usually not as responsive.

Figure 3-1 illustrates the relative availability of 12 essential plant nutrients as a function of the pH of the soil (University of Kentucky, 1978). The three primary fertilizer nutrients—nitrogen, potassium, and phosphorus—have the greatest availability within the pH 6.0 to pH 7.5 range. Hence, it is understandable why crops might grow well over this range of reaction. The availability of magnesium, sulfur, copper, and boron are also well represented within this range, although their availability tends to taper off as pH 7.5 is approached.

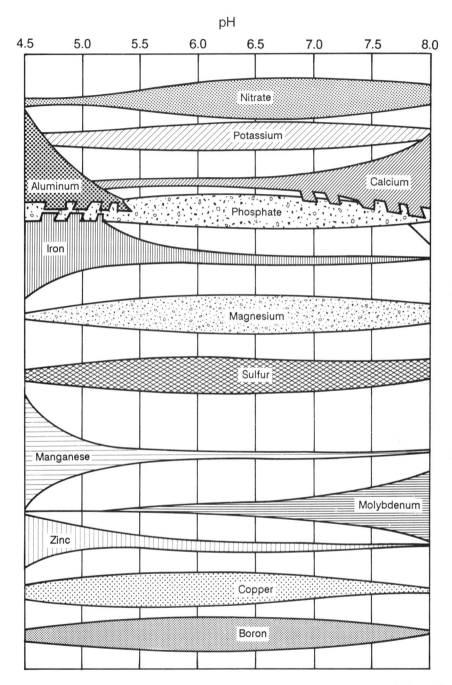

FIGURE 3-1 The relative availability of 12 essential plant nutrients in well-drained mineral soils in temperate regions in relation to soil pH. (Aluminum is not an essential nutrient for plants, but it is shown because it may be toxic below a soil pH of 5.2.) A pH range between 6.0 and 7.0 is considered ideal for most plants. The thirteenth essential plant nutrient from the soil, chlorine, is not shown because its availability is not pH-dependent. The saw-toothed pattern in the figure represents precipitation. SOURCE: University of Kentucky. 1978. Liming Acid Soils. Leaflet AGR-19. Lexington, Ky.

The availability (solubility) of aluminum, iron, and manganese increases rapidly below about pH 5.3. The increase in the solubility of these three elements, especially aluminum, creates a toxic environment for plant roots. Growth of most plants is severely limited at low pH values. Poor plant performance, however, is not necessarily linked to low pH. Deficiencies in calcium and magnesium or toxic levels of aluminum or manganese may also affect plant growth. At low values of pH, aluminum and iron cause precipitation of phosphate (Figure 3-1). This can cause phosphorus deficiencies. At the upper end of the pH scale in the presence of large amounts of calcium, the calcium precipitates phosphates, but usually not to as great an extent as aluminum and iron do at the lower end.

The application of agricultural limestone—either calcitic limestone containing mostly calcium carbonate or dolomitic limestone containing a mixture of calcium and magnesium carbonates—can amend soils with pH values and increase the pH to the favorable range. The procedure for determining the amount of neutralization required is a simple one involving the measurement of the pH of a soil sample suspended in a buffered solution. This pH value, compared to the pH of a suspension of the soil in water, provides a basis for measuring the total amount of acidity that needs to be neutralized and thus the amount of agricultural limestone needed.

Excess calcium carbonate usually dominates soil systems with pH values up to 8.0 or 8.2. This is a consequence of a soil development process that occurs mostly in drier climates. Additionally, in some crops, there may be a deficiency of iron or other micro nutrients that need to be corrected. Soils with pH values above 8.3 are likely to contain sodium carbonate or potassium carbonate. These alkaline soils also have poor physical characteristics because of the presence of sodium. They are generally found in arid or semiarid climates. The application of gypsum (calcium sulfate) and leaching with adequate quantities of high-quality irrigation water can mend these soils. Calcium replaces sodium on the exchange complex of the soils, displacing the sodium that is leached out of the root zone. This results in improved soil properties.

Tillage

Farmers have adopted a wide range of tillage practices in the past three decades. Most of these conservation tillage practices were developed by researchers to slow soil erosion and conserve water through decreased soil disturbance. Nearly 100 million acres are farmed using some form of conservation tillage. Most of this is in the form of mulch tillage or reduced tillage; no tillage, strip tillage, and ridge tillage are practiced on about 16 million acres (U.S. Department of Agriculture, 1987). While these practices may advance some of the goals of alternative farming, such as increasing organic matter in soil and reducing soil erosion, some conservation tillage practices may increase the need for pesticides, particularly herbicides (Gebhardt et al., 1985).

Conservation tillage generally leaves a layer of crop residue on top of the

A ridge-tillage planter places seeds in a narrow band of soil scraped off the top of the ridges. This method preserves the residue cover between the rows, reducing soil and water runoff. Planted last fall as a cover crop, hairy vetch helps to meet the nitrogen needs of the corn crop now being planted. *Credit:* Dick Thompson, the Thompson Farm.

Ridge tillage is an effective alternative tillage system that can reduce erosion and costs and help control weeds without herbicides. Here, seeds are planted in a field with the residue of last year's corn crop, which protects the soil surface from erosion. *Credit:* Randall Reeder, Ohio State University.

Chisel plowing, a form of conservation tillage, mixes crop stubble into the soil, leaving the ground partially protected from wind and rain. Chisel plowing can reduce erosion by 30 to 50 percent from levels expected with moldboard plowing. *Credit:* John Deere Company.

soil (Hendrix et al., 1986; House et al., 1984). This residue may provide a favorable habitat to some pests. Some plant diseases overwinter in crop residues left on the soil, above and below ground insects survive, and perennial weeds may establish a foothold. The effects of these pest populations are more severe if the same crop is planted the next year but may be inconsequential or minimized in a rotation.

Conservation tillage changes soil properties in ways that affect plant growth (Phillips et al., 1980). Researchers are studying trophic interactions in croplands with no tillage and conventional tillage. Plant nutrients in no-tillage soils are more stratified than those in soils under reduced or conventional tillage. Nutrients also tend to concentrate in the upper portion of the soil profile. Soil under conservation tillage practices, which leaves a surface mulch, is often 3 to 4°C cooler in late spring than soil under conventional tillage. In the spring, the cooler temperatures can slow early season plant growth at higher latitudes. With no tillage, the soil is also more likely to be compacted, which can also reduce plant growth.

In summer, however, the mulched soil is cooler and the soil surface under the residue is moister. As a result, many conservation tillage systems have been very successful. The concentration of soil microbes and earthworms is greater in conservation tillage systems. Inadequate research on the range of conservation tillage practices, however, makes it impossible to draw general conclusions for most crops, soils, or climates. A number of the committee's case studies demonstrate the effectiveness of various tillage practices in

No tillage is another effective way to reduce erosion. Soybeans are planted into barley stubble as barley is harvested. Farmers in the South and mid-Atlantic are often able to harvest two crops in one season. *Credit:* Soil Conservation Service, U.S. Department of Agriculture.

Corn grows on top of the ridges in a soybean field. *Credit:* Soil Conservation Service, U.S. Department of Agriculture.

One conventional tillage method is moldboard plowing. The steel blades of the plow turn over furrows of soil. This method helps to control some diseases and insect pests by disrupting their environment and burying weed seeds. But it also leaves the soil surface fully exposed to rain and wind and increases soil erosion. *Credit:* Soil Conservation Service, U.S. Department of Agriculture.

controlling weeds and as a component of viable alternative systems (see the Spray, BreDahl, Kutztown, and Thompson case studies).

Conventional moldboard plowing, in contrast to reduced tillage, contributes to pest control by destroying some perennial weeds, disrupting the life cycle of some insect pests, and burying disease inoculum. But conventional tillage may also disrupt the life cycle of beneficial organisms, contribute to soil erosion, and require more energy and larger tractors. Chisel plowing, the most widely used form of reduced tillage, requires large tractors but uses less energy than moldboard plowing. Conventional tillage creates more bacterial activity and has a "boom-and-bust" effect on nutrient cycling processes (Holland and Coleman, 1987). No tillage or other conservation tillage provides a slower but more even rate of nutrient release. Legumes are an effective source of nitrogen in some conservation tillage systems, although different tillage methods can influence the amount of nitrogen available to subsequent crops (Heichel, 1987).

Conservation tillage can reduce water runoff from fields. Data reported by Hall et al. (1984) showed that compared with conventional tillage, no-tillage systems reduced runoff by 86.3 to 98.7 percent, soil losses by 96.7 to 100 percent, and the herbicide cyanazine losses by 84.9 to 99.4 percent. Glenn and Angle (1987) reported 27 percent less total runoff of water and the herbicides atrazine and simazine with no-tillage versus conventional tillage systems.

Some forms of conservation tillage, however, may increase the concentration of broadcast nitrogen and phosphorus fertilizers and pesticides in the

An offset disk is a conventional tillage tool used for further soil preparation (top). Final field preparation is accomplished with a field cultivator, which in this case is also incorporating a preplant herbicide into the soil (bottom). At the left, a planter follows. These conventional tillage operations work the soil into fine particles and leave almost no crop residue on the soil surface. This field could experience high soil losses if hard spring rains occur before the growing crop begins to protect the soil surface. *Credits:* Soil Conservation Service, U.S. Department of Agriculture (top); Grant Heilman (bottom).

runoff. However, the total amount lost is generally reduced because runoff is usually less (Andraski et al., 1985; Sauer et al., 1987; Springman et al., 1986). With conservation tillage, fertilizer loss via runoff can be further reduced by drilling the fertilizer in the row. The effects of conservation tillage practices on pesticide runoff are largely unknown; simulation models generally show a decrease in runoff (Crowder et al., 1984). This may be accompanied by an increase in percolation to groundwater, however.

Ridge tillage is a form of conservation tillage with significant erosion control benefits that overcomes some of the soil temperature, weed control, and soil compaction problems associated with untilled systems. In the spring the tops of the ridges are tilled for planting (Figure 3-2). This removes residue from the top of the ridges and disturbs the soil enough to create a seedbed. Soil on the ridges is also generally warmer than that between ridges or in fields without ridges. Warmer soil facilitates crop germination. Tilling only the top of the ridges disturbs fewer weed seeds, reducing weed germination. Erosion is slowed because soil and crop residue between the ridges is not disturbed. Weeds that emerge later in the growing season tend to be between the ridges. Cultivation easily controls these weeds and reduces soil compaction in crop rows, thereby enhancing plant root growth and water infiltration.

The use of reduced tillage is expected to increase primarily as a means to meet conservation compliance provisions of the Food Security Act of 1985. As its use increases, farmers and researchers will need to better understand its effects on soil structure, soil biota, plant growth, and pest populations. Research is needed to reduce the use of herbicides to control weeds in untilled fields and increase the use of conservation tillage in vegetable and other specialty crops.

Effect of Soil Biota on Nutrient Availability

Numerous free-living but plant-associated microorganisms aid in nutrient uptake (Hendrix et al., 1986). Mycorrhizal fungi are important in the uptake of nutrients from soil and in the establishment of vigorous seedling growth in many crop and nursery species (Gerdeman, 1976). Little is known about the genetics of these or other similar organisms or how their association benefits plants. Improvements in their use and establishment of beneficial microorganisms in the rhizosphere could make crop plants more efficient in their use of soil nutrients (Holland and Coleman, 1987). In particular, work under way on the genetic basis of specific root associations may make it possible for free-living nitrogen-fixing bacteria to attach to the roots of cereal grasses or establish dominant relationships in the surrounding rhizosphere, thereby improving nitrogen use (Baldani et al., 1987).

The harmful effects of insect, nematode, and fungus pests in soil on crop growth and yields are well known and receive substantial research attention. Less studied is another group of organisms that benefit crop production by decomposing leaf litter and root residues, playing an integral role in

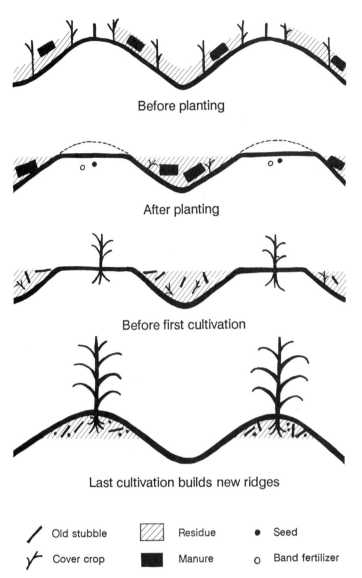

Before planting

After planting

Before first cultivation

Last cultivation builds new ridges

/ Old stubble   ▨ Residue   ● Seed

Y Cover crop   ■ Manure   ○ Band fertilizer

FIGURE 3-2 Ridge tillage advantages in alternative production systems. The planter tills 2 to 4 inches of soil in a 6-inch band on top of the ridges. Seeds are planted on top of the ridges, and soil from the ridges is mixed with crop residue between the ridges. Soil on ridges is generally warmer than soil in flat fields or between ridges. Warm soil facilitates crop germination, which slows weed emergence. Crop residue between the ridges also reduces soil erosion and increases moisture retention. Mechanical cultivation during the growing season helps to control weeds, reduces the need for herbicides, and rebuilds the ridges for the next season. SOURCE: Dick Thompson, The Thompson Farm.

making inorganic nutrients available for subsequent plant growth (Coleman et al., 1983, 1984a, 1984b; Ingham et al., 1985). The activities of these biota could influence crop nutrient needs, although their relationships with plants are not precisely understood. It is not known whether pesticides used to control soil insects and plant pathogens destroy other soil organisms. If they do, however, pesticide use may be a factor in nutrient management. Research is needed to know how pesticides affect soil biotic activity and thus influence nutrient cycles and crop nutrient requirements.

Bacteria and fungi, the primary decomposers, break down sugars, polysaccharides, and proteins in organic matter. They also assimilate mineral nutrients such as nitrogen and phosphorus into their tissues. When these organisms die, they release mineral nutrients, such as ammonium, nitrate, phosphates, and sulfates. This step is known as mineralization. A considerable portion of mineralization occurs when various members of the soil fauna prey on dead bacteria and fungi (Coleman et al., 1983, 1984a, 1984b; Mitchell and Nakas, 1986). Much of this microbial and faunal activity occurs in the top few inches of the soil. A typical plow layer of 6 inches on 1 acre of farmland, which equals about 1,000 tons, contains approximately 4 tons of flora (2 tons of bacteria and an equal amount of fungi) and about 1 ton of all fauna combined (Ingham et al., 1985).

A majority of soil fauna is found in the plant rhizosphere (Ingham et al., 1985). Current agroecosystem research on microbial flora, fauna, and organic matter sources is directed toward discovering how microbial production and turnover can best be understood and managed to synchronize nutrient supply with the nutrient uptake needed for plant growth (Hendrix et al., 1986). As a result, some farmers in the northern Great Plains are switching to spring wheat, because its growth is more synchronous with organic matter decomposition and nutrient cycles. Several ecologists and agronomists argue that farming practices matched to soil microbe activity are more energy and nutrient efficient than conventional practice (Coleman et al., 1984a, 1984b; Hendrix et al., 1986). Such factors as soil moisture, temperature, and texture must be included in decisions on timing of cultivation, planting, and general tillage management. The relationships among soil microorganisms, nutrient cycling, pest pressure, plant growth, yield, and many other factors of crop production need further study.

LIVESTOCK

Livestock play an important role in many alternative farming systems in terms of nutrient cycling and their ability to make crop rotations economically feasible through the consumption of forage crops (see the Spray, BreDahl, Sabot Hill, Kutztown, and Thompson case studies). Converting marginal cropland to pasture for grazing also helps promote soil conservation and reduce water runoff. Nutrients not retained by the animal can be readily returned to the soil in the form of manure. Manure provides soil nutrients and enhances organic content and tilth. Research in this area has been neglected in recent years because agriculture has focused on monocrop and monospecies systems and has increasingly separated crop and livestock activities. Much could be done to further understand and enhance crop-livestock interactions.

Research is needed in the following areas to achieve the full benefits of crop-livestock systems:

- Crop and forage rotations and forage handling, harvest, and storage;
- Ruminant digestion of lignocellulose;
- Quality and digestibility of pastures and forages;
- Animal muscle-to-fat ratio;
- Animal health systems; and
- Manure handling.

Obstacles to Greater Use of Hay and Forage

Impressive advancements have been made in the harvesting, storage, and processing of forages for hay and silage, but major difficulties remain in integrating hay and silage into crop-livestock systems (Wedin et al., 1980). The principal obstacles to further adoption of hay or forage crops are weather constraints and the time and labor required to harvest forage crops, particularly when cash crops require attention. Weather can interfere with harvesting; rain can seriously damage hay crops. Harvest of forage crops often occurs at peak demand times for labor and equipment and is often a lower priority than taking care of grain crops. These factors can significantly increase production costs per unit of digestible energy produced. Consequently, many livestock producers choose to maximize use of grains and silage and to minimize the area devoted to pastures and forages.

Because pastures and forages have some advantages over row crops, such as reduced soil erosion and the potential to supply nitrogen, they are an important element in many alternative farming systems. Convincing more farmers to incorporate or expand acreage devoted to pastures and forages, however, will require animal systems that are profitable and that save time and labor and an agricultural policy that encourages their adoption.

Lignocellulose Digestion

In many alternative crop-livestock systems, forages and crop residues comprise a large portion of the total diet for beef cattle, dairy cattle, and sheep, particularly in contrast to more intensive systems that rely on purchased feed grains. The availability of the energy stored in forages thus becomes critical to the viability of these operations. The gross energy content of these cellulosic plants and crop residues is equal to that of grain and root crops. But animals do not use this energy efficiently because the cellulose structure is highly ordered and often associated with indigestible lignin

(Council for Agricultural Science and Technology, 1986; Oltjen et al., 1980; Wedin et al., 1980).

Forage or hay crops will often be produced on land that could otherwise be in grain or other high-value crops. The economic viability of producing these crops, and ultimately the viability of incorporating livestock into an operation, therefore, will depend on the efficiency of ruminants in transforming these forages, hays, and crop residues into animal products.

Unless corn stalks are harvested prior to maturity, as with silage, little sugar is in the stalk. When left in the field, feeding value of corn stalks is further reduced by weathering. Methods for improving the feeding value of lignocellulose materials, such as corn stalks and small grain straw, have been studied. These methods include treatment with sodium hydroxide (Klopfenstein, 1978), a mixture of sodium and calcium hydroxides (Klopfenstein and Owen, 1981), ammonia, urea (Ibrahim and Pearce, 1983), and, most recently, hydrogen peroxide (Kerley et al., 1985). Because of chemical hazards, special equipment requirements, and modest improvement in feed value, the use of these intensive chemical treatments is limited.

Microbiological interventions to improve the use of lignocellulose materials have largely concentrated on the introduction of dietary agents that will favorably influence ruminal fermentation. Several drugs, most notably the ionophores, can improve the efficiency of weight gain by reducing methane loss and increasing propionate production (Chalupa, 1980). However, very few studies have been conducted on the genetics and genetic manipulation of ruminal bacteria (Forsberg et al., 1986; Hazelwood et al., 1986).

Research is needed to improve the bioavailability of lignocellulose. This would include plant breeding to improve the digestibility of pastures and forages and the vegetative portions of crop residues. Genetic manipulation of corn, for example, reduced the percentage of lignin by 40 percent. This resulted in a 30 percent increase of digestible dry matter from corn stover. When consumed, the forage caused a 1 pound per day increase in weight gain. Similar improvements might be possible with other forages (Council for Agricultural Science and Technology, 1986; Wedin et al., 1980). Additional research can help develop microbiological, chemical, and physical interventions that will improve the digestion of lignocellulose materials.

Improving Quality of Pastures and Forages

Genetic improvement in cash crops has resulted in significant increases in per acre yields. Similar yield improvement is possible for forages, although it has not been as widely exploited. For example, Bermuda grass yield potential has been doubled in the southeastern United States, but, in general, forages have not received the same level of research as cash crops. Improvement is also possible in nutrient availability, palatability, and the reduction of antimetabolite concentrations in plants. Reducing the amount of plant cell wall constituents, including hemicellulose, lignin, and silica, can improve nutrient availability.

Palatability is important because livestock must eat the forage to benefit from its nutritional content. The factors affecting palatability are little understood, and research could produce significant results. A variety of anti-metabolites and plant toxins are present in forage crops. Researchers are making progress in identifying and eliminating these chemicals. Plant breeders have lowered the indole alkaloid concentrations in a line of reed canary grass by one-third. Lambs fed the low-alkaloid canary grass gained weight at twice the daily rate of those fed the commercial variety (Marten et al., 1981). Similar results might be possible in reducing tannins in sorghum and lespedeza; coumarin in sweet clovers; cyanogenic glycosides in sorghum, Sudan grass, and white clover; and saponins in alfalfa.

Tall fescue is grown on 35 million acres in the United States. Poor animal performance and health problems have been reported in animals grazing some fescue pastures (Blaser et al., 1980; Goodman, 1952; Robbins, 1983; Studemann et al., 1973). The presence of the endophyte fungus *Acremonium coenophialum* has been implicated in poor animal performance in cattle grazing fescue (Hoveland et al., 1983). Improved performance has been reported in cattle grazing endophyte-free fescue. A number of fungus-free varieties have been developed; however, these varieties are more prone to insect damage due to loss of an insect toxin provided by the endophyte.

Fairly convenient and accurate methods of determining forage nutritive quality are available. Improvement in methods and availability, however, would enhance forage quality research.

Muscle-to-Fat Ratio

One goal of agricultural research and farming is to improve the nutritional quality of food. Growing evidence of the link between consumption of fat and heart disease has stimulated efforts to produce meat with lower fat content. Efficiency of conversion of feed to meat is improved when reduced fat content of meat is the goal because much more energy is required to produce fat than lean meat. Livestock fed on pastures are leaner than those fed grains, but leaner meat sometimes lacks the taste and texture desired by consumers (National Research Council, 1988). Perhaps more importantly, the price and grading structures for beef traditionally rewarded farmers for producing animals with higher fat content. Recent changes in grading standards have opened new markets for leaner products, but full adoption of research results is still hindered by economic incentives. Animal breeding, feeding intact rather than castrated males, and genetic engineering are producing leaner livestock. Consumer demand for leaner meat is growing. Genetic engineering has great potential to accelerate this progress. Hormones are also creating opportunities for decreasing fat content in meat. For example, porcine growth hormone increases growth rate, feed efficiency, and ratio of muscle to fat.

Animal Health Systems

Disease prevention through management has become an increasingly important research objective throughout the last decade. Nonetheless, technologies for disease treatment rather than management systems for disease prevention dominate current animal health systems. The subtherapeutic feeding of antibiotics and antibiotic treatment of diseased animals remain the mainstays of current animal health practices. Many alternative systems exist, however, and are widely practiced today (see the BreDahl, Kutztown, Thompson, and Coleman case studies). Some major commercial producers maintain animal health with reduced or no prophylactic feeding of antibiotics. They are able to achieve this by modified production systems, including reduced animal confinement, improved ventilation and waste management systems, and, in certain cases, the use of alternative technologies.

Antibiotics

The antibiotic era ushered in with the discovery of penicillin has permitted treatment and control of animal disease not previously possible. The application of inexpensive antibiotics to animal feed controlled many of the disease problems that were exacerbated by the strict confinement and intensive management of animals. Conversely, the use of antibiotics has facilitated the trend toward confinement housing and greater concentration of animals in production facilities (Council for Agricultural Science and Technology, 1981). About 9.9 million pounds of antibiotics are fed to livestock each year (Institute of Medicine, 1989). The widespread use of antibiotics in feed and to treat disease has reinforced a trend not to manage for disease prevention and to accept the costs of antibiotic feeding and use as a routine production expense. Feeds that contain antibiotics are widely used because, on the typical farm, they help animals use feed more efficiently. Research has demonstrated that the control of subclinical infection by feeding subtherapeutic levels of antibiotics results in increased production and growth (Freedeen and Harmon, 1983). It appears that antibiotic feeding also works by decreasing the total antigenic challenge to the animal's immune system. Immune systems that are stimulated for defense appear to channel nutrients to the need of the immune response and away from growth and production (Klasing et al., 1987).

Antibiotics will always be needed, to some extent, for the clinical treatment of disease. The extensive use of antibiotics in feed and for therapy, however, has the attendant risk of promoting the selection of resistant bacterial strains that may be passed on in food, become pathogenic in humans, and resist antibiotic treatment. Research to develop new antibiotics for use only in agriculture may not resolve the concerns about resistant strains, because resistance in bacteria tends to occur for groups of bacteria. Moreover, the research, development, and federal approval process of new antibiotics is slow and costly, and drug companies may become unwilling to take the financial risk for new antibiotic product development (Liss and Batchelor, 1987). It may be unwise, therefore, for animal agriculture and human medicine to assume that new antibiotics will always be developed to resolve resistance problems.

Alternatives to Antibiotic Use

The occasional adulteration of milk with antibiotics has been a problem for the dairy industry since the introduction of penicillin. To this day, penicillin and other antibiotic residues in milk, primarily from treatment for mastitis, remain a major concern for regulators and the dairy industry.

Mastitis, a common bacterial infection of the lactating dairy cow, remains a predominant reason for the use of antibiotic therapy. Considerable research into mastitis has revealed that the disease is controllable and, most importantly, largely preventable. National data indicate that approximately 50 percent of the dairy cattle in the country have mastitis (National Mastitis Council, 1987). But other studies have identified large numbers of dairy herds that have very low levels of mastitis, as evidenced by the low numbers of white blood cells in the commercial milk produced (Bennett, 1987).

Mastitis, like many other diseases of domestic animals, is a result of management practices. Dairy systems can be instituted and managed to reduce infection rates enough to greatly reduce intramammary use of therapeutic antibiotics. The opportunity to achieve this level of disease control is independent of the size of the dairy herd (Bennett, 1987).

The procedures and management for effective mastitis control have been widely published. A recent publication providing this information for producers and the allied industry is available nationwide (National Mastitis Council, 1987).

In general, effective disease control is achieved by:

- Pre- and postmilking disinfection of the udder;
- Proper milking machine function and use;
- Identification, diagnosis, and segregation of infected animals;
- Comprehensive prophylactic therapy of nonlactating animals;
- Maintenance of clean and dry cattle housing; and
- Culling chronically infected cattle.

Other common dairy disease problems that may result in antimicrobial residues in milk are largely preventable, and similar control strategies have been developed (Amstutz, 1980). The specifics of a control program must be tailored to each situation, and professional assistance is critical to the success of any program. Therefore, these systems often require a greater degree of management, knowledge, and information than is readily available to producers.

Swine Production

The trend in swine production is toward larger confinement operations (see Chapter 1). It is now common to find more than 1,000 sows per production unit. Without careful planning, design, and operation of these units, the risk of animal disease increases with the numbers of animals in the operation. Respiratory problems in pigs produced for market is a major

Partial confinement facilities provide shelter and open air space. Hogs raised in these systems generally have less incidence of disease and require fewer antibiotics than those raised in confinement facilities. *Credit:* Grant Heilman.

problem in swine production (National Research Council, 1988). A multitude of factors, including close proximity confinement, sanitation, and breed susceptibility, influence disease risk and the consequences of infection.

The practice of feeding livestock a wide variety of antibiotics at low or subtherapeutic levels has become commonplace as producers have adopted confinement production systems. Animals respond to antibiotic feeding by increasing feed intake and using less feed per unit of weight gained, which improves the growth rate. The mode of action of subtherapeutic antibiotics has not been fully explained, however. Their bacteriostatic and bacteriocidal activities are probably the primary causes of improved feed efficiency and growth rate (Zimmerman, 1986). However, additional theories have been proposed. Antibiotics may spare certain nutrients by reducing bacterial destruction of certain vitamins and amino acids or by favoring bacteria that synthesize these essential nutrients. Some studies have reported that antibiotics reduce the thickness of the intestinal wall and suggested that affected animals may absorb nutrients more efficiently. Antibiotics may also inhibit the growth of toxin-producing bacteria within the intestinal tract. For example, antibiotics may depress bacterial urease production, which would result in lower ammonia levels in the intestine and blood (ammonia is a powerful toxin). Antibiotics may also kill or inhibit pathogenic organisms in the intestine, thus reducing the incidence of subclinical or clinical disease states.

The significant benefits of antibiotic feeding have revealed the extent of disease problems in modern swine production. Because of the complex etiology and the pervasiveness of disease in swine, the subtherapeutic feeding of antibiotics will likely remain a simple and effective method of reducing disease loss in lieu of changes in production practices with greater emphasis on other methods of disease prevention.

Alternative management systems and techniques, however, can greatly reduce reliance on subtherapeutic feeding of antibiotics (Kliebenstein et al., 1981). Reduced confinement and the increased use of outdoor shelters and pastures are components of alternative livestock production systems that allow lowering or elimination of subtherapeutic feeding of antibiotics (see the BreDahl, Kutztown, Thompson, and Coleman case studies). Veterinary and medicine costs stemming from swine confinement production systems have been shown to be at least double those of a comparably productive pasture and hutch system. Kliebenstein et al. (1981) found that the total costs of producing 100 pounds of pork were $40.18 for the pasture system compared to $42.97 for the individual confinement unit system. In another example, veal calves raised in stall and pen confinement facilities have been shown to need five times the amount of antibiotics as hutch and yard calves (Friend et al., 1985). Using pastures and forages may improve other aspects of production, such as waste management and nutrition. Preventive disease management, however, is not as simple as redesigning production systems and facilities. Similar disease conditions can develop in these situations as well.

Hormonal Therapy

A wide variety of physiological hormones modulate the natural disease defense mechanisms of animals. Research is needed to further identify

Confinement systems have allowed the average size of hog operations to grow. This in turn has affected management practices and permitted a single individual to raise more hogs. Calves, hogs, and poultry in confinement livestock systems have greater risk of respiratory diseases, requiring use of antibiotics. *Credit:* Grant Heilman.

these hormones and determine what potential they represent for infectious disease therapy.

One of the natural reproductive hormones, prostaglandin $F_{2\alpha}$, has been shown to be useful as nonantibiotic therapy of the postpartum bovine uterus (Momont and Sequin, 1985). Uterine infection is a major cause of infertility in animals. Prostaglandin injections in the animal changes the natural reproductive hormone levels to the extent that the uterus has enhanced disease resistance.

Protection

Immunization to prevent diseases has been successfully used for decades to control many diseases associated with major herd and flock health problems or epidemics. For other important diseases, vaccinations have either not been effective or available. This is the case with acute coliform mastitis in dairy cattle.

Standard antibiotic treatments are not effective against coliform mastitis. Thus, veterinarians and producers attempt treatment with different and more costly antibiotics. Gentamicin, a drug reported by the Food and Drug Administration (FDA) to occur in meat from cull dairy cows, is being used for intramammary therapy (Paige and Kent, 1987). But advances in molecular immunology are beginning to reveal new opportunities for the development of effective vaccines. One vaccine developed with these techniques has been shown to modulate acute coliform mastitis in dairy cattle (Gonzalez et al., 1988).

Often, disease problems of newborn animals are devastating and result in major economic losses. There is a need for research to develop management systems, breeding programs, and immunological interventions that will enhance immunological protection against diseases in the newborn (Anderson et al., 1980).

Parasitism

Intensive use of chemical agents in the treatment and prevention of parasitic infestation has resulted in the appearance of ecto- and endoparasites resistant to many insecticides and deworming agents. Research is needed to determine the influence of parasitism and quantify its economic effects. Integrated control programs using management, breeding, nutrition, and chemical or biological agent interventions should be developed (Anderson et al., 1980). Parasitism of animals on pasture can be a major animal health problem and is often cited as a justification for the confinement housing and feeding of livestock.

Anthelmintics are administered on a regular or seasonal basis to treat subclinical parasitism before it induces a major production loss. The FDA establishes specific label instructions for use and withdrawal periods before slaughter. The prescribed withdrawal periods and use limitations should help producers avoid high residue levels in animal products. Research on the integrated management of parasites associated with animals grazing on irrigated and nonirrigated pasture is needed in order to break the cycles of parasitic infection and reinfection.

Genetic Resistance to Disease

The genetic selection of animals has placed emphasis on productivity and efficiency and has potentially reduced natural disease resistance. Studies in dairy cattle have shown that as milk production increased, resistance to mammary disease decreased. It is generally thought that animals free of disease will produce more milk. It is not known, however, whether antibiotic treatment and management or genetic resistance is the disease control strategy that will produce the most milk (Alrawi et al., 1979).

Use of genetic resistance for tolerance to disease and parasitism has been largely confined to tropical environments, where adaptation has been critical to animals' survival. Research is needed to develop improved breeds and strains of livestock that are more resistant to disease or tolerant of ecto- and endoparasites in various climates (Anderson et al., 1980).

Stress and Disease

The stress response in animals is not well understood. There is a consensus, however, that distressed animals become less able to adjust effectively to additional change (Selye, 1950). Various stressors, such as close confinement, transportation, and temperature, have been shown to affect disease resistance. Although little scientific information is available that quantifies animal stress, observational research on the effects of stressors on animal health and behavior has fueled the debate on animal welfare and the significance of stress (Mickley and Fox, 1987). It is generally agreed that reducing the stress associated with confinement, transportation, and temperature decreases incidence of disease in certain cases. It has yet to be determined whether or not stress resistance can be genetically altered. There is evidence, however, that swine bred for high muscle content are more susceptible to stress, and that the stress effects of transportation reduce meat quality. Research is needed to define more fully and quantify stress in food animals. Subsequently, research may be able to determine if low stress systems offer disease protection and economic advantages.

Technology and Advanced Diagnostics

The performance of animals in disease-free and diseased states is difficult to measure precisely. It is very difficult to accurately diagnose diseases, particularly those that are subclinical.

Immunological, biochemical, and electronic advances, however, are providing new tools for the rapid and accurate determination of disease and the measurement of disease effects. Immunodiagnostics employing biotechnological advances such as monoclonal antibodies and enzyme-linked antibodies have aided in disease control in virtually every food animal species. The electronic enumeration of white blood cells in cows' milk can provide every dairy producer with monthly information about the udder health status of every cow in the herd. Major advances in udder health nationwide may be possible because of this technology.

A comprehensive or systems approach to the maintenance of health in food-producing animals offers perhaps some of the greatest opportunities for increasing the safety, quality, and profitability of food animal products. The academic intradisciplinary approach to research has precluded investigations into the larger systems approach to animal disease problems. By examining the interaction of nutrition, management, animal disease resistance, and the biology of disease agents, it may become possible to identify the critical control points governing the susceptibility and spread of disease and select those that will provide the most cost-effective means of disease prevention.

PEST CONTROL IN CROPS

Control of pests (including insects, nematodes, mites, weeds, and pathogens) has been a major research activity in agriculture for decades. Crop management practices, rotations, genetic improvements through classical plant breeding, and synthetic organic chemicals are widely used to control pests in modern commercial agriculture. Steady progress has been made in these areas, and much of what has been accomplished is relevant to alternative agriculture. Plant breeding that has produced many economically significant pest-resistant varieties of major cultivars is particularly relevant. Nevertheless, most of agriculture relies on synthetic chemical pesticides, even though in many cases effective alternatives are now available.

Use of synthetic chemicals for pest control began in the 1940s, when the discovery of organic compounds such as dichloro diphenyl trichloroethane (DDT), benzene hexachloride (BHC), and (2,4-Dichlorophenoxy) acetic acid (2,4-D) heralded a revolution in pest management. Pesticides made it feasible to control many pests and pathogens for which no effective control measure was previously available. Consequently, pesticides contributed significantly to yield increases in the 1940s through the 1960s. Pesticide use, especially of insecticides, grew rapidly. Many farmers began to apply insecticides on a regular schedule, often with little attention to actual infestations. As a result, insecticide-resistant strains developed. Pesticides, particularly insecticides and fungicides, wiped out certain beneficial species, which often led to the emergence of their prey as an even more serious pest problem. Many of the original organochlorine insecticides had detrimental environmental effects and presented unacceptable risks to human health. Most of these compounds have been removed from agricultural use and replaced with less persistent organophosphate, carbamate, and synthetic pyrethroid insecticides.

Herbicide use has more than doubled since 1970, currently accounting for more than 65 percent of all pesticides applied. Today, more than 95 percent of all corn and soybean acres are treated with herbicides. As farmers continue to cut input costs in response to economic conditions, researchers are seeking more effective, economical, and ecologically sound ways to control pests. Nonetheless, pesticide expenditures are about 20 percent of total input costs, although this figure varies substantially by crop and region (see Chapters 1 and 4). Farmers and society could benefit from safer and more economical methods of controlling pests. Progress in this direction is under way as government, university, and industry researchers discover biological and genetic alternatives to the use of pesticides and devise a variety of

cultural and biological control strategies aimed at reducing and even eliminating pesticide use. These efforts remain small, however, compared with
the time and resources devoted to chemical pest control research. Nonetheless, there are promising results to report.

IPM Development

Led by entomologists, researchers began to recognize the problems associated with dependence on extensive insecticide use in the 1950s. They
developed concepts leading to what is now commonly referred to as integrated pest management (IPM). A central principle of IPM is the economic
threshold concept, which holds that the mere presence of a pest population
does not necessarily indicate an economically damaging situation where
benefits will exceed the cost of control. In principle, IPM is an ecologically
based pest control strategy that relies on natural mortality factors such as
natural enemies, weather, and crop management and seeks control tactics
that disrupt these factors as little as possible.

Traditionally, the term IPM has been associated with insect control, in
large part because formal IPM systems began with efforts to reduce insecticide use and avert a growing insect resistance problem, principally in cotton. Today, however, pest management systems that integrate a number of
tactics for control of plant diseases, weeds, and other pests are widely
practiced in a number of crops. For the purposes of this report the committee uses the term IPM to include the integrated control of insects, diseases,
and weeds. Ideally, the term IPM refers to control of all agricultural pests
through the use of an integrated approach.

IPM involves all aspects of crop production, including cultural practices
such as cultivation, fertilization, postharvest management of fields, scouting for pests, tillage practices, the use of genetically improved pest-resistant
varieties, rotations with other crops, and the use of biological controls (see
the Florida, Pavich, Ferrari, and Kitamura case studies). However, most
current IPM programs do not use all of these techniques. Current insect
IPM programs generally focus on the use of improved crop varieties, scouting for pests, better timing of pesticide applications, and the use of more
specific, less biologically active pesticides. The need for protection of naturally occurring biological control agents, such as predators and parasites, in
the crop ecosystem is widely recognized but often overlooked in practice.
In many situations, their populations cannot be preserved. Cultural practices, such as increasing a crop's ability to resist pests through nutrient
management techniques that promote crop health, appear to offer great
promise. But these practices are not well articulated or understood. They
are therefore underused and in need of quantitative research (see the Pavich
case study).

The common thread for all IPM programs is the concept of an economic
threshold below which pest populations or damage is tolerated. The determination of an economic threshold is difficult, however, because it is not a
constant. It varies depending on an individual farmer's pest problems,

stage of crop growth, and economic expectations. As economic thresholds differ, so will IPM programs. Decreased pesticide use is often associated with IPM, but it is not a mandatory component of the system. Most IPM programs decrease pesticide use, but many increase the number of pesticide applications in light of better knowledge of pest populations (Allen et al., 1987).

Significant federal support for IPM extension, research, and field studies began in 1972. The National Science Foundation (NSF), EPA, and USDA jointly funded this work known as the Huffaker Project through the Cooperative State Research Service (CSRS).

Reauthorized in 1979 as the Consortium for Integrated Pest Management (CIPM), this project supported interdisciplinary research to develop IPM systems in alfalfa, apples, cotton, and soybeans. In 1984, the project was reorganized and decentralized. It is now administered in four regions—the Northeast, North Central, South, and West. The regions set research priorities to match regional needs, and as a result, IPM research projects now deal with a wider range of pests on more crops. The funding for IPM research through CSRS has been about $3 million per year since 1972.

The Extension Service (ES) has operated an IPM program since 1973. This program is designed to implement IPM systems and to develop an independent capacity for IPM among grower organizations, consultants, and other private parties. Allen et al. (1987) estimate that private IPM consultants generate $400 million per year in economic activity. Since 1979, funding for IPM activities through the ES has been steady but relatively small at about 2 percent of the total extension budget, or $7 million per year.

Although effective, highly profitable, and relatively safe, IPM has been widely adopted only for some crops. Table 3-6 shows the implementation of IPM in 12 major crops. Implementing IPM programs for some crops in some regions, such as vegetables and ornamental flowers in the southeastern states, is particularly difficult (see the Florida fresh-market vegetable case study). The sophistication of IPM programs also varies greatly by crop, region, and individual grower. Insects and disease are an ever-present threat in hot areas with a long growing season. Some plant diseases are more consistently a problem in hot areas with high humidity, such as the southeastern United States. Fruit and vegetable crops have a high unit value, and the loss of even a small share of the crop could be costly. Under the current federally regulated grower-operated grading system, blemish-free produce often receives a higher price. This incentive encourages uniformly scheduled pesticide applications and works strongly against reductions in pesticide use. Extremely low tolerance levels for insects or their fragments in high-value canned and processed foods also encourage prophylactic pesticide applications. Faced with this stituation, a grower has a strong incentive to eliminate any chance of damage by applying pesticides on a schedule rather than depending on monitoring and IPM. In the context of these grading standards, IPM in high-value crops may involve a very low economic threshhold that results in a greater number of pesticide applications based on increased monitoring of pest populations.

TABLE 3-6   U.S. IPM Use in 12 Major Crops, 1986

Crop	Acres Planted (1,000s)	Acres Under IPM[a] (1,000s)	Percentage of Total Acres Under IPM
Alfalfa	26,748	1,273	4.7
Apples[b]	461	299	65.0
Citrus[b]	1,057	700	70.0
Corn	76,674	15,000	19.5
Cotton	10,044	4,846	48.2
Peanuts	1,572	690	43.8
Potatoes	1,215	196	16.1
Rice	2,401	935	38.9
Sorghum	15,321	3,966	25.8
Soybeans	61,480	8,897	14.4
Tomatoes	378	312[b]	82.5
Wheat	72,033	10,687	14.8

NOTE: IPM is defined broadly to include all acres where basic scouting and economic threshold techniques are reportedly used.

[a]Includes acres under IPM management by the Cooperative Extension Service, grower organizations, producer industries, or consultants.
[b]Data based in part on conversations with IPM entomologists in major growing regions.

SOURCES: U.S. Department of Agriculture. 1987. Agricultural Statistics. Washington, D.C.; U.S. Department of Agriculture. 1987. National IPM Program. Cooperative Extension Service. Washington, D.C.; U.S. Department of Agriculture. Forthcoming. Fruit Situation and Outlook Report. Economic Research Service. Washington, D.C.

Alternative Insect and Mite Control

When intervention is necessary, chemical insecticides and acaricides will continue to be important ways to protect many commercially produced crops. Research shows, however, that significant progress has been made with other approaches. Incorporating pest control into the overall management of a farm by modifying cultural practices or rotating crops, for example, is essential to effective alternative pest control strategies. The most limiting factor in the adoption of these strategies is the failure to introduce them as a part of an overall farm management system. When used in the environment of conventional agriculture, the effectiveness of many alternatives is diminished or lost.

Cultural controls for insects include modifying the pest habitat through use of crop rotations, increasing ecosystem diversity, adjusting the time of planting and harvest, precise management of water and fertilizer, modified cultivation and tillage practices, and improved sanitation. Cultural controls have demonstrated their effectiveness in many situations against such pests as pink bollworms on cotton in Texas. There, a short-season, early-harvest crop is followed by immediate shredding and plow-down on a uniform basis (often state-mandated) throughout the area. Rotating corn with soybeans is another common cultural practice that so far has virtually eliminated damage by corn rootworms.

Before resistant varieties of wheat were developed, wheat rust annually caused millions of dollars of damage in the United States. Fungicides are sometimes still needed to control isolated infestations. Breeders continue to develop high-yielding resistant varieties to help farmers cut costs and remain competitive. *Credit:* Agricultural Chemicals Division, Mobay Corporation.

Plant breeders have developed many widely grown cultivars with insect and disease resistance. Plant breeding, which is the science of manipulating a plant's genetic composition, has been a very effective control method for insects, mites, and particularly plant diseases. Resistant cultivars have dramatically reduced fungicide use in wheat and peanuts and the vulnerability of wheat, corn, and alfalfa to certain pests. The potential benefits from future research remain substantial. Research and development on wheat resistant to Hessian fly and wheat stem sawfly, alfalfa resistant to spotted alfalfa aphid, and corn resistant to European corn borer cost less than $10 million. The annual savings to farmers are estimated at hundreds of millions of dollars (National Research Council, 1987b). Recent developments in genetic technology promise to enhance this approach to crop improvement (Goodman et al., 1987).

Natural biological controls, such as antagonists, predators, and self-defense mechanisms, suppress most pests. Biological control of pests by natural enemies (parasites, predators, and insect pathogens) is partially or entirely effective on most potential pests. Additionally, this sort of control

Successful biological control of pests involves attracting and keeping beneficial insects in crop fields. Here, a spined soldier bug feeds on a Mexican bean beetle larva, one of the most damaging soybean insect pests. U.S. Department of Agriculture scientists have isolated the pheromone to attract the spined soldier bug. This development may aid the biological control of the Mexican bean beetle. *Credit:* Agricultural Research Service, U.S. Department of Agriculture.

is long-lasting if it is not disrupted by farming practices such as insecticide use, certain crop rotations, or unusual climatic conditions. Ironically, the cases in which insecticides reduce the beneficial insect population and new insect problems emerge best illustrate the importance of natural enemies (Ridgway and Vinson, 1977; Settle et al., 1986).

Importation and release of exotic natural enemies is another effective

Pheromones in the trap in this staked tomato field in Florida attract insects, which are counted to determine their populations. Pesticide applications, as described in the Florida case study, can then be made in response to the actual magnitude of the pest infestation, rather than on a routine schedule. *Credit:* Will Sargent.

biological control tactic (Osteen et al., 1981). The first example of classic biological control by an introduced exotic species occurred about a century ago, when the Australian vedalia beetle was introduced into California to control the cottonycushion scale insect of citrus. Since then, the introduction of exotic parasites, predators, and pathogens in the United States has controlled almost 70 insect pests (National Research Council, 1987a).

Augmentation of indigenous natural enemies is an important biological control technique. It is usually accomplished by mass release of natural enemies in the target field. But this means of pest management has been very costly in some cases. Further, its effectiveness is sometimes difficult to evaluate. In several instances, however, it has been very cost-effective. One example is the use of natural enemies in controlling the alfalfa weevil (Osteen et al., 1981). The release of insecticide-resistant natural enemies can be very effective in certain permanent crops if monitored and managed carefully. While this approach is not universally successful, it has shown great utility in controlling spider mite pest species in deciduous fruit and nut orchards (Hoy, 1985; Hoyt and Burts, 1974).

Insect diseases are highly selective in their action. Although these diseases are somewhat limited to use in special situations, some scientists are convinced of their potential, having observed epidemics caused by viral disease under natural conditions. The use of viruses to control insect pests has been inconsistent and generally disappointing in trials, however. Research is needed to improve virus formulation, production, and application in field situations.

Bacteria have also been used successfully. For example, *Bacillus thuringiensis* (Bt) controls certain lepidopterous larvae. Bt is also sometimes used for larvae "cleanup" just before the harvest of vegetable crops. *Bacillus thuringiensis israelensis*, a subspecies, has demonstrated promise for use against immature mosquitoes living in aquatic environments. Genetic engineering technology creates other possibilities for Bt. Researchers have transferred the toxin gene from Bt into tobacco to control certain leaf-feeding caterpillars. It may also be transferred to corn seed to control European corn borer. Some entomologists, however, have serious reservations about this process because of potential resistance selection in species feeding on the plant containing the toxin.

There are many types of control agents for insects that are relatively nontoxic to other organisms. Future research may prove their safety and usefulness in applied situations (National Research Council, 1987a). Recent advances in the understanding of insect ecology, biology, physiology, and biochemistry are providing new opportunities for insect control. The rearing and releasing of sterilized insect pests have successfully controlled the screwworm, which afflicts livestock. The technique has proved successful for eradicating new infestations of fruit fly species on the U.S. mainland. Prospects are good for similar results with other species. A similar strategy is to breed genetically altered pest species that can mate with pests to produce offspring unable to reproduce, feed, or perform other functions necessary for survival. Increased knowledge of such insect hormones as brain hormones, molting hormones, and the juvenile hormone has made it possible to synthesize them. Introduction of devices that emit synthetic hormones offers the potential to disrupt normal functions such as breeding, growth, and molting, thus controlling the pest population.

Insect pheromones are used commercially to monitor, detect, and predict insect populations and to control several insect pest species on a variety of crops. They have great promise for more widespread use when registration procedures under the Federal Insecticide, Fungicide and Rodenticide Act (FIFRA) are completed. University, USDA, and industry scientists have identified sexual attractant or aggregation pheromones for more than 436 insect species. More than 50 companies sell about 250 of these pheromones worldwide, mostly for population monitoring and detection traps, for which they are in great demand.

Pheromones dispensed over fields to confuse males and prevent mating successfully control insect pests in a number of crops, including grape leafroller, oriental fruit moth on peaches, pink bollworm on cotton, and tomato pinworm. Commercial formulations are economically competitive with insecticides, even taking application costs into account. Additionally, pheromones affect only the target species because of their extreme specificity.

During season-long pheromone disruption with no insecticides, treated grape fields in New York had crop damage below 1 percent. Similar fields treated with the insecticide carbaryl reported crop damage of 18 and 2.5 percent (Booth, 1988). Pink bollworm disruptant has been sold commercially in the Southwest for almost 10 years and has been applied on as many as 100,000 acres per year. New formulations have held infestations to 1 percent or less throughout the season with one application; consequently, insecticide applications have been reduced by nearly 90 percent (Baker et al., in press). Oriental fruit moth infestations on peaches are routinely held to a fraction of a percent by a single season-long application of disruptant; subsequent insecticide use is eliminated (Rice and Kirsch, in press). Continued development of new controlled-release technologies and cheaper synthetic pheromones will further improve the competitiveness of pheromones in the marketplace. Small companies specializing in pheromones will continue to be the innovators in developing new pheromone products—unless fees imposed by the government for entry into the registration process reach prohibitive levels and force them out of the market.

Alternative Plant Pathogen Control

The integration of a variety of methods has historically controlled plant diseases. The selection and development of varieties with specific or general (multigenic) resistance have helped to avoid overdependence on chemicals. Genetic resistance is the single most important defense against plant disease and *the* proven alternative to chemical control when available.

For a variety of reasons, most plant diseases cannot be directly controlled. For example, many of the fungi that infect plant roots have not been fully investigated, nor has their importance in affecting yield been quantified. In most cases, farmers cope with diseases by using good farm management practices and planting resistant varieties. When combined with the existing natural level of biological control, management and resistant varieties keep the majority of diseases in check. Nonetheless, disease can still cause significant yield loss. Moreover, germplasm for desirable resistance has not been identified for many of the world's important crops. More research is needed to better characterize available germplasm for genetic resistance to disease and plant transformation.

The development and durability of resistant varieties have been a challenge to plant pathologists and plant breeders. Genetic strategies to improve the durability of resistance include use of multilines and cultivar mixtures as well as multigenic or horizontal resistance. Modern genetic technology will speed the development of resistant crops. It should be possible to identify genes that confer resistance to a specific pathogen. These genes would then be introduced to the appropriate plant, without incorporating other genes that may confer detrimental characteristics. This gene transfer has been achieved to produce resistance to several plant viruses in tobacco plants (National Research Council, 1987a). Advances in understanding the genetic and molecular bases of disease in plants promise major improvements in plant disease control using genetic rather than chemical methods (Goodman, 1988).

Cultural practices such as crop rotations, alteration of soil pH, sanitation, and adjustment of the timing of planting and harvest to avoid peak periods of the pathogens complement genetic resistance in many situations. For example, raising soil pH with lime from 6.5 to 7.5 reduces the severity of fusarium wilts on tomato and potato crops in Florida (Jones and Woltz, 1981). Lowering soil pH with sulfur to 5.0 controls potato scab caused by *Streptomyces scabies* (Oswald and Wright, 1950). Forms of nitrogen also can play a significant role in disease severity. For example, ammonium nitrogen suppresses the disease take-all in wheat but nitrate favors it (Huber et al., 1968).

Tillage practices can have effects on pathogen populations and resultant diseases. Ecofallow is a form of conservation tillage that can reduce stalk rot of sorghum but permits increases in other diseases (Cook and Baker, 1983).

Harvesting and processing practices can also influence the inception of disease. The hydrostatic pressure from tank-washing potatoes causes water infiltration of pathogens into the lenticels of the tubers, predisposing them to attack by bacterial soft rot (Bartz and Kelman, 1985). Potatoes are then generally treated with a fungicide. There is enormous need and potential to control diseases by nonchemical methods (Cook and Baker, 1983). But there remains a lack of understanding of the underlying mechanisms that affect disease incidence and severity.

Synthetic chemical control of plant pathogens has become an increasingly important pest control tactic as agriculture has shifted toward intense cultivation of monocultures (Delp, 1983). Practices previously used to control pathogens, such as crop rotations, are not compatible with current crop specialization (Tweedy, 1983). Because commercial cultivars are genetically related, the loss of resistance to pathogens could cause serious problems if fungicides were not available. In California, the use of methyl bromide and chloropicrin soil fumigation resulted in huge increases in yield and quality in several crops (Wilhelm and Paulus, 1980). This combination is widely credited with saving the strawberry industry from high production costs and foreign competition.

Although the total amount of fungicides used in the United States is much less than the amount of herbicides or insecticides, the potential chronic health risk to humans is significant. Ninety percent (by weight) of fungicides applied are known to cause tumors in laboratory animals. Fungicide residues in food are responsible for the largest share of potential dietary oncogenic risk from pesticides. Developing fungicides not toxic to nontarget organisms, including humans, is difficult; very few new fungicides have reached the market. Only four nononcogenic fungicides have been introduced in the past 15 years that have captured greater than 5 percent of any food crop market (National Research Council, 1987b). In recent years, there has been a movement toward the development of highly specific systemic fungicides, but this has accelerated evolutionary selection of fungicide-resistant plant pathogens. Research to understand the mechanisms of resistance could aid the development of chemicals with new modes of action and better-targeted effects.

The introduction or application of biological control agents has not been very successful with plant pathogens because of the great complexity in microbial communities. Although many of the management practices that indirectly control diseases strike a balance between beneficial and deleterious microorganisms, there is insufficient knowledge to effectively develop and use biological control agents commercially (Schroth and Hancock, 1985). Little is known concerning the ecology, classification, and physiology of biological control organisms or the underlying mechanisms affecting the interactions among beneficial microorganisms, pathogens, and plants.

The potential to use microorganisms against microorganisms has stirred the interest of many investigators. A number of companies have pioneered efforts to develop biological control agents for plant pathogens. Several products have already reached the market. An avirulent, antibiotic-producing strain of *Agrobacterium* is available to control crown gall tumors of ornamental plants and orchard trees caused by *Agrobacterium tumefaciens* (National Research Council, 1987a). Plans are underway to market a rootcolonizing *Pseudomonas* bacterium as a control for *Rhizoctonia* and *Pythium* fungi in cotton.

Another interesting disease control possibility is to stimulate a plant's own defense system with chemicals or by inoculation with an avirulent

form of a pathogen. The citrus tristeza virus from Africa entered Brazil in
the 1920s and nearly decimated the citrus industry. In the 1950s, researchers
found a mild strain of the virus that protected trees from the severe strain.
Commercial inoculation with the mild virus began in the late 1960s and has
been very successful so far (National Research Council, 1987a). It remains
unlikely, however, that disease control in continuous crop monocultures in
certain regions, such as fruit and vegetable production in the East and
Southeast, will be possible without use of synthetic chemical fungicides
and fumigants. Disease pressures in areas with high temperatures and
humidity and long growing seasons are so severe that only dramatic changes
in production systems will enable widespread adoption of alternative disease control measures.

Alternative Nematode Control

Nematode control is particularly difficult. Strategies include genetic resistance, chemical control, and cultural methods such as rotations (see the
BreDahl, Kutztown, Thompson, and Kitamura case studies). Genetic resistance is successful in only a few cases. Chemical control, which is feasible
only in certain situations, relies on broad-spectrum, highly toxic, and often
volatile materials. It is expensive and hazardous. The decline of basic cultural practices such as rotations, particularly in the Midwest, has led to an
increase in nematodes in soybeans. Rotating corn with soybeans will control
most nematode problems. Current research for nematode control is focusing
on the development of effective cultural practices such as those traditionally
practiced before the advent of broad-spectrum nematocides.

Genetic research to develop nematode-resistant cultivars has been successful in sugar beets and tomatoes (Goodman et al., 1987). More research
is necessary to determine how various nematodes damage different crops
and how to modify practices if a combination of nematode species is present. Similarly, the accuracy and efficiency of techniques for estimating nematode populations needs to be improved.

The biological antagonism level of the soil must be determined if management decisions are to be based on an understanding of the relationship
between yield and population density of nematodes. A given number of
nematodes will affect the same crop differently in soils of differing biota.
More basic studies in biological control and interactions in the rhizosphere
are required. Improved assay techniques for assessing the biological antagonism coefficients of various soils must be developed.

One promising biological control agent is the pathogenic bacterium *Pas-
teuria penetrans*, which is effective against several economically important
nematodes. It is expensive to produce on a large scale, however. A less
expensive, but also less effective, biological control option is the use of
plants such as *Crotalaria spectabilis* that prevent the nematode from repro-
ducing. Coastal Bermuda grass *(Cynodon dactylon)* incorporated before
planting lespedeza, tobacco, or vegetable transplants protects against root-

knot nematodes (*Meloidogyne* spp.) (Burton and Johnson, 1987). Coastal Bermuda grass will also reestablish itself after the annual crop is harvested. These plants could be even more effective if they could be genetically engineered to produce nematode attractants or pheromones.

The selection and development of varieties for resistance and tolerance to nematode stress will continue. This may involve incorporation of appropriate genetic material into varieties already selected for production, economic, and marketing qualities. It is still important to develop biological and chemical nematocides that are systemic, easily associated with the root system, target organism specific, or a combination of these factors. These pesticides will allow flexibility in management decisions and compensation for management errors that have promoted or amplified nematode stress problems in a particular production system.

Alternative Weed Control

Farmers in the United States depend greatly on herbicides to control weeds. Nearly two-thirds of U.S. pesticide purchases are for herbicides. But a variety of other means, such as crop rotations, mechanical cultivations, competition with other plants, and biological control through natural enemies can control weeds (see Spray, BreDahl, Sabot Hill, Kutztown, Thompson, Pavich, and Lundberg case studies). In fact, growers are often unaware of the forces naturally controlling weeds. The purslane sawfly and the leafmining weevil, for example, help control purslane in California. These insects would be even more effective if their populations were not reduced by insecticide use. The moth *Bactra verutana* suppresses the weed *Cyperus rotundus* that infests cotton in Mississippi. More than 70 plant-feeding insects and plant pathogens have been introduced to control weeds in the United States; 14 weed species are now controlled in this way (National Research Council, 1987a; Osteen et al., 1981). Few weeds are controlled biologically in agriculture, however, although future opportunities are numerous. For example, many of the hundreds of species of carabid beetles are seed eaters and could play a role in weed control (Andres and Clement, 1984).

Cultural practices are currently the most effective alternative to herbicides. Cultivating, rotary hoeing, increasing the density of the crop plant to crowd out weeds, intercropping, timing of planting to give the crop a competitive advantage, and transplanting seedling crop plants to give them a head start on weeds are currently practiced and effective measures. Transplanting tomatoes to a high density has successfully controlled the growth of shade-intolerant redroot pigweed. Clover planted as an understory or living mulch reduces weed growth in corn. Several combinations of cover crops and tillage practices are effective in controlling weeds in corn and soybeans.

Weed-tolerant crops and crops that produce substances toxic to weeds are potentially promising approaches that have received little research attention. Naturally occurring phytotoxic allelopathic chemicals, however, may not always be safer than some of the more undesirable synthetic herbicides. Introducing weed diseases is also a possibility. The rust *Puccinia chondrillina* controls the rush skeleton weed, which is a problem in wheat and pasture areas. *Alternaria macrospora* can inhibit the growth of spurred anoda, a damaging weed in cotton production that is resistant to several cotton herbicides.

The development of herbicide-resistant crops may offer opportunities to substitute safer herbicides for more dangerous herbicides. For example, efforts are being made to develop crops resistant to the herbicide glyphosate, a compound with very low mammalian toxicity. Like other broad-spectrum herbicides, glyphosate has limited use in crop production because it destroys crops as well as weeds and therefore must be used before crop germination or with special application methods and equipment. In response to this problem, researchers have isolated glyphosate-resistant genes and successfully transferred them to poplar trees, tobacco, and tomatoes (Della-Croppa et al., 1987; Stalker et al., 1985). If the plants tolerate glyphosate, the herbicide could then be used as a postemergent treatment. In certain cases, this strategy could reduce weed control costs, improve weed control quality, and reduce human health hazards.

SUMMARY

Alternative farming encompasses a range of farming practices, including the use of crop rotations, IPM, biological and cultural pest control, use of organic materials to enhance soil quality, different tillage methods, and animal rearing techniques that involve less reliance on antibiotics and confinement. The unifying premises of alternative systems are to enhance and use biological interactions rather than reduce and suppress them and to exercise prudence in the use of external inputs.

Research has not fully addressed the integration of study results essential to the adoption of a number of alternative farming methods as unified systems. Although some components of alternative systems have been examined, they have been generally studied in isolation. Lack of systems research is a key obstacle to the adoption of a number of alternative farming practices. On the whole, land-grant universities and the USDA have not adequately integrated the results of this research into production systems.

Nonetheless, a significant amount of scientific evidence exists that supports the effectiveness of a range of alternative practices. There is a large body of information about the value of legumes in fixing nitrogen, improving soil quality, reducing erosion, and increasing yields of subsequent crops. IPM programs are effective, profitable, and increasingly adopted. Although biological and natural controls are underused, they have been demonstrated to be effective and warrant increased research support. Genetic engineering techniques should enhance this aspect of IPM. The integration of livestock into farming systems provides additional means for nutrient cycling. Improving forage digestibility needs further research, however.

The scientific basis for some of these practices and their interaction in agricultural systems is not always understood, but they work. Many farmers have adopted them and are using them profitably. The economics of these and other alternative farming practices and systems are discussed in the following chapter.

REFERENCES

Allen, W. A., E. G. Rajotte, R. F. Kazmierczak, Jr., M. T. Lambur, and G. W. Norton. 1987. The National Evaluation of Extension's Integrated Pest Management (IPM) Programs. VCES Publication 491–010. Blacksburg, Va.: Virginia Cooperative Extension Service.

Alrawi, A. A., R. C. Laben, and E. J. Pollack. 1979. Genetic analysis of California mastitis test records. II. Score for resistance to elevated tests. Journal of Dairy Science 62:1105–1131.

Amstutz, H. E., ed. 1980. Bovine Medicine and Surgery, 2d ed. Santa Barbara, Calif.: American Veterinary Publications.

Anderson, D. P., W. R. Pritchard, J. J. Stockton, W. G. Bickert, L. Bohl, W. B. Buck, J. Callis, R. Cypess, J. Egan, D. Gustafson, D. Halvorson, B. Hawkins, A. Holt, A. D. Leman, S. W. Martin, T. D. Njaka, B. I. Osburn, G. Purchase, W. W. Thatcher, H. F. Troutt, J. Williams, and R. G. Zimbelman. 1980. Animal health. Pp. 129–151 in Animal Agriculture Research to Meet Human Needs in the 21st Century, W. G. Pond, R. A. Merkel, L. D. McGilliard, and J. Rhodes, eds. Boulder, Colo.: Westview Press.

Andraski, B. J., D. H. Mueller, and T. C. Daniel. 1985. Phosphorus losses in runoff as affected by tillage. Soil Science Society of America Journal 49:1523–1527.

Andres, L. A., and S. L. Clement. 1984. Opportunities for reducing chemical inputs for weed control. Pp. 129–140 in Organic Farming: Current Technology and Its Role in a Sustainable Agriculture, Special Publication No. 46, D. F. Bezdicek and J. F. Power, eds. Madison, Wis.: American Society of Agronomy, Crop Science Society of America, Soil Science Society of America.

Baker, K. F., and R. J. Cook. 1982. Biological Control of Plant Pathogens. St. Paul, Minn.: American Phytopathological Society.

Baker, T. C., R. T. Staten, and H. M. Flint. In press. Use of pink bollworm pheromone in the southwestern United States. In Behavior-Modifying Chemicals for Insect Management: Applications of Pheromones and Other Attractants, R. L. Ridgway, R. M. Silverstein, and M. N. Inscoe, eds. New York: Marcel Dekker.

Baldani, V. L. D., J. I. Baldani, and J. Dobereiner. 1987. Inoculation of field-grown wheat (Triticum aestivum) with Azospirillum in Brazil. Biology and Fertility of Soils 4:37–40.

Balfour, E. B. 1976. The Living Soil and the Haughley Experiment. New York: Universe Books.

Barnes, D., G. Heichel, and C. Sheaffer. 1986. Nitro alfalfa may foster new cropping system. News, November 20. St. Paul, Minn.: Minnesota Extension Service.

Bartz, J. A., and A. Kelman. 1985. Infiltration of lenticels of potato tubers by Erwinia carotovora pv. carotovora under hydrostatic pressure in relation to bacterial soft rot. Plant Disease 69:69–74.

Bennett, R. H. 1987. Milk quality management is mastitis management. Pp. 133–150 in Proceedings of the 26th Annual Meeting of the National Mastitis Council, Inc. Arlington, Va.: National Mastitis Council.

Blaser, R. E., R. C. Hammes, Jr., J. P. Fontenot, and H. T. Bryant. 1980. Forage-animal systems for economic calf production. Pp. 667–671 in Proceedings of the XIII International Grassland Congress. Berlin: Akademie-Verlag.

Booth, W. 1988. Revenge of the "nozzleheads." Science 23:135–137.

Burton, G. W., and A. W. Johnson. 1987. Coastal Bermuda grass rotations for control of root-know nematodes. Journal of Nematology 19:138–140.

Chalupa, W. 1980. Chemical control of rumen microbial metabolism. Pp. 325–347 in Digestive Physiology and Metabolism of Ruminants, Y. Ruckebusch and P. Thivend, eds. Lancaster, England: MTP Press Limited.

Coleman, D. C., C. P. P. Reid, and C. V. Cole. 1983. Biological strategies of nutrient cycling in soil systems. Advances in Ecological Research 13:1–55.

Coleman, D. C., C. V. Cole, and E. T. Elliott. 1984a. Decomposition, organic matter turnover, and nutrient dynamics in agroecosystems. Pp. 83–104 in Agricultural Ecosystems: Unifying Concepts, R. Lowrance, B. R. Stinner, and G. J. House, eds. New York: Wiley/Interscience.

Coleman, D. C., R. E. Ingham, J. F. McClellan, and J. A. Trofymow. 1984b. Soil nutrient transformation in the rhizosphere via animal-microbial interactions. Pp. 35–38 in Invertebrate-Microbial Interactions, J. M. Anderson, A. D. M. Rayner, and D. W. H. Walton, eds. Cambridge, England: Cambridge University Press.

Cook, R. J. 1984. Root health: Importance and relationship to farming practices. Pp. 111–127 in Organic Farming: Current Technology and Its Role in a Sustainable Agriculture, Special Publication No. 46, D. F. Bezdicek and J. F. Power, eds. Madison, Wis.: American Society of Agronomy, Crop Science Society of America, Soil Science Society of America.

Cook, R. J. 1986. Wheat management systems in the Pacific Northwest. Plant Disease 70(9):894–898.

Cook, R. J., and K. F. Baker. 1983. The Nature and Practice of Biological Control of Plant Pathogens. St. Paul, Minn.: American Phytopathological Society.

Council for Agricultural Science and Technology. 1980. Organic and Conventional Farming Compared. Report No. 84. Ames, Iowa.

Council for Agricultural Science and Technology. 1981. Antibiotics in Animal Feeds. Report No. 88. Ames, Iowa.

Council for Agricultural Science and Technology. 1986. Forages: Resources for the Future. Report No. 108. Ames, Iowa.

Crowder, B. M., D. J. Epp, H. B. Pionke, C. E. Young, J. G. Beierlein, and E. J. Partenheimer. 1984. The Effects on Farm Income on Constraining Soil and Plant Nutrient Losses: An Application of the CREAMS Simulation Model. Research Bulletin 850. University Park, Pa.: Agricultural Experiment Station, Pennsylvania State University.

Dabney, S. M., G. A. Breitenbeck, B. J. Hoff, J. L. Griffin, and M. R. Milam. 1987. Management of subterranean clover as a source of nitrogen for a subsequent rice crop. Pp. 54–55 in The Role of Legumes in Conservation Tillage Systems, J. F. Power, ed. Ankeny, Iowa: Soil Conservation Society of America.

Della-Croppa, G., S. C. Bauer, M. L. Taylor, D. E. Rochester, B. K. Klein, D. M. Shah, R. T. Fraley, and G. M. Kishore. 1987. Targeting a herbicide-resistant enzyme from *Escherichia coli* to chloroplasts of higher plants. Bio/Technology 5:579–584.

Delp, C. 1983. Changing emphasis in disease management. Pp. 416–421 in Challenging Problems in Plant Health, T. Kommendahl and P. H. Williams, eds. St. Paul, Minn.: American Phytopathological Society.

Eggert, F. P., and C. L. Kahrmann. 1984. Response of three vegetable crops to organic and inorganic nutrient sources. Pp. 97–109 in Organic Farming: Current Technology and Its Role in a Sustainable Agriculture, Special Publication No. 46, D. F. Bezdicek and J. F. Power, eds. Madison, Wis.: American Society of Agronomy, Crop Science Society of America, Soil Science Society of America.

Elliott, L. F., R. I. Papendick, and D. F. Bezdicek. 1987. Cropping practices using legumes with conservation tillage and soil benefits. Pp. 81–90 in The Role of Legumes in Conservation Tillage Systems, J. F. Power, ed. Ankeny, Iowa: Soil Conservation Society of America.

Forsberg, C. W., B. Crosby, and D. Y. Thomas. 1986. Potential for manipulation of the rumen fermentation through the use of recombinant DNA techniques. Journal of Animal Science 63:310–325.

Foth, H. D. 1978. Fundamentals of Soil Science, 6th ed. New York: Wiley.

Freedeen, H. T., and B. G. Harmon. 1983. The swine industry: Changes and challenges. Journal of Animal Science 57(Suppl. 2):110–118.

Friend, T. H., G. R. Dellmeier, and E. E. Gbur. 1985. Comparison of Four Methods of Calf Confinement. 1. Physiology. Technical article 18960. College Station, Tex.: Texas Agricultural Experiment Station.

Gebhardt, M. R., T. C. Daniel, C. E. Schweizer, and R. R. Allmaras. 1985. Conservation tillage. Science 230:625–630.

Gerdeman, J. W. 1976. Vesicular-arbuscular mycorrhizae. Pp. 576–591 in The Development and Function of Roots, J. G. Torrey and D. T. Clarkson, eds. London: Academic Press.

Glenn, S., and J. S. Angle. 1987. Atrazine and simazine in runoff from conventional and no-till corn watersheds. Agriculture, Ecosystems and Environment 18(4):273–280.

Goldstein, W. A., and D. L. Young. 1987. An agronomic and economic comparison of a conventional and a low-input cropping system in the Palouse. American Journal of Alternative Agriculture 2(2):51–56.

Gonzalez, R. N., J. S. Cullor, D. E. Jasper, T. B. Farver, R. B. Bushnell, and M. Oliver. 1988. A mastitis vaccine? Don't laugh: New research and field tests have scientists very optimistic. Dairyman (May):20.

Goodman, A. A. 1952. Fescue foot in cattle in Colorado. Journal of the American Veterinary Medical Association 121:289–290.

Goodman, R. M. 1988. An agenda for phytopathology. Phytopathology 78:32–35.

Goodman, R. M., H. Hauptli, A. Crossway, and V. C. Knauf. 1987. Gene transfer in crop improvement. Science 236:48–54.

Granatstein, D. 1988. Reshaping the Bottom Line: On-farm Strategies for a Sustainable Agriculture. Stillwater, Minn.: The Land Stewardship Project.

Hall, J. K., N. L. Hartwig, and L. D. Hoffman. 1984. Cyanazine losses in runoff from no-tillage corn in "living" and dead mulches vs. unmulched, conventional tillage. Journal of Environmental Quality 13(1):105–110.

Harwood, R. R. 1983. International overview of regenerative agriculture. Pp. 24–35 in Proceedings of the Workshop on Resource-Efficient Farming Methods for Tanzania, J. M. R. Semoka, F. M. Shao, R. R. Harwood, and W. C. Liebhardt, eds. Emmaus, Pa.: Rodale Press.

Hazelwood, G. P., S. P. Mann, C. G. Orpin, and M. P. M. Romaniec. 1986. Prospects for the genetic manipulation of rumen microorganisms. In Recent Advances in Anaerobic Bacteriology, Anaerobic Discussion Group, Proceedings of the Fourth Annual Meeting, July 1985, S. P. Vorriello, ed. Boston: Martinus Nijhoff.

Heady, E. O. 1948. The economics of rotations with farm and production policy applications. Journal of Farm Economics 30(4):645–664.

Heady, E. O., and H. R. Jensen. 1951. The Economics of Crop Rotations and Land Use: A Fundamental Study in Efficiency with Emphasis on Economic Balance of Forage and Grain Crops. Research Bulletin 383, August. Ames, Iowa: Agricultural Experiment Station, Iowa State University.

Heichel, G. H. 1987. Legumes as a source of nitrogen in conservation tillage systems. Pp. 29–35 in The Role of Legumes in Conservation Tillage Systems, J. F. Power, ed. Ankeny, Iowa: Soil Conservation Society of America.

Hendrix, P. F., R. W. Parmelee, D. A. Crossley, Jr., D. C. Coleman, E. P. Odum, and P. M. Groffman. 1986. Detritus food webs in conventional and no-tillage agroecosystems. Bioscience 36(6):374–380.

Holben, F. J. 1956. History of the Jordan plots: Practices, changes, and people. Pp. 5–11 in Jordan Soil Fertility Plots, Agricultural Experiment Station Bulletin 613, F. J. Holben, ed. University Park, Pa.: Pennsylvania State University.

Holland, E. A., and D. C. Coleman. 1987. Litter placement effects on microbial and organic matter dynamics in an agroecosystem. Ecology 68:425–433.

House, G. J., B. R. Stinner, D. A. Crossley, Jr., E. P. Odum, and G. W. Langdale. 1984. Nitrogen cycling in conventional and no-tillage agroecosystems on the southern Piedmont: An ecosystem analysis. Journal of Soil and Water Conservation 39:194–200.

Hoveland, C. S., S. P. Schmidt, C. C. King, Jr., J. W. Odom, E. M. Clark, J. A. McGuire, L. A. Smith, H. W. Grimes, and J. L. Holliman. 1983. Steer performance and association of Acremonium coenophialum fungal endophyte on tall fescue pasture. Agronomy Journal 75:821.

Howard, A. 1943. An Agricultural Testament. New York: Oxford University Press.

Hoy, M. 1985. Recent advances in genetics and genetic improvements in Phytosiidae. Annual Review of Entomology 30:345–370.

Hoyt, S. C., and E. C. Burts. 1974. Integrated control of fruit pests. Annual Review of Entomology 19:231–252.

Huber, D. M., C. G. Painter, H. C. McKay, and D. L. Peterson. 1968. Effect of nitrogen fertilization on take-all of winter wheat. Phytopathology 58:1470–1472.

Ibrahim, M. N. M., and G. R. Pearce. 1983. Effects of chemical pre-treatments on the composition and in vitro digestibility of crop by-products. Agricultural Wastes 5:135.

Ingham, R. E., J. A. Trofymow, E. R. Ingham, and D. C. Coleman. 1985. Interactions of bacteria, fungi, and their nematode grazers: Effects on nutrient cycling and plant growth. Ecological Monographs 55:119–140.

Institute of Medicine. 1989. Human Health Risks with the Subtherapeutic Use of Penicillin or Tetracyclines in Animal Feed. Washington, D.C.: National Academy Press.

Jones, J. P., and S. S. Woltz. 1981. Fusarium-incited diseases of tomato and potato and their control. Pp. 157–168 in Fusarium: Diseases, Biology, and Taxonomy, P. E. Nelson, T. A. Toussoun, and R. J. Cook, eds. University Park, Pa.: Pennsylvania University Press.

Kerley, M. S., G. C. Fakey, Jr., and L. L. Berger. 1985. Alkaline hydrogen peroxide treatment unlocks energy in agricultural by-products. Science 230:820–822.

Kilkenny, M. R. 1984. An Economic Assessment of Biological Nitrogen Fixation in a Farming System of Southeast Minnesota. M.S. thesis, University of Minnesota, St. Paul.

Klasing, K. C., D. E. Laurin, R. C. Peng, and D. M. Fry. 1987. Immunologically mediated growth depression in chicks: Influence of feed intake, corticosterone and interleukin I. Journal of Nutrition 117:1629–1637.

Kliebenstein, J. B., C. L. Kirtley, and M. L. Killingsworth. 1981. A comparison of swine production costs for pasture, individual, and confinement farrow-to-finish production facilities. Special Report 273. Columbia, Mo.: Agricultural Experiment Station, University of Missouri.

Klopfenstein, T. J. 1978. Chemical treatment of crop residues. Journal of Animal Science 46:841.

Klopfenstein, T. J., and F. G. Owen. 1981. Value and potential use of crop residues and by-products in dairy rations. Journal of Dairy Science 64(6):1250.

Koerner, P. T., and J. F. Power. 1987. Hairy vetch winter cover for continuous corn in Nebraska. Pp. 57–58 in The Role of Legumes in Conservation Tillage Systems, J. F. Power, ed. Ankeny, Iowa: Soil Conservation Society of America.

Liss, R. H., and F. R. Batchelor. 1987. Economic Evaluations of Antibiotic Use and Resistance: A Perspective. Report of Task Force 6. Reviews of Infectious Diseases 9(Suppl. 3):297–312.

Marten, G. C., R. M. Jordan, and A. W. Hovin. 1981. Improved lamb performance associated with breeding for alkaloid reduction in reed canary grass. Crop Science 21:295.

Meisenbach, T. 1983. Alternative Farming Task Force Report. Lincoln, Nebr.: University of Nebraska.

Mickley, L. D., and M. W. Fox. 1987. The case against intensive farming of food animals. Pp. 257–272 in Advances in Animal Welfare Science 1986/1987, M. W. Fox and L. D. Mickley, eds. Boston: Martinus Nijhoff.

Mitchell, M., and J. Nakas, eds. 1986. Microbial and Faunal Interactions in Natural and Agro-Ecosystems. Amsterdam: Martinus Nijhoff.

Momont, H. W., and B. E. Sequin. 1985. Prostaglandin therapy and the postpartum cow, No. 17. The Bovine Proceedings (April):89–93.

National Mastitis Council. 1987. Current Concepts of Bovine Mastitis, 3d ed. Arlington, Va.: National Mastitis Council.

National Research Council. 1987a. Biological Control in Managed Ecosystems. Pp. 55–68 in Research Briefings 1987. Washington, D.C.: National Academy Press.

National Research Council. 1987b. Regulating Pesticides in Food: The Delaney Paradox. Washington, D.C.: National Academy Press.

National Research Council. 1988. Designing Foods: Animal Product Options in the Marketplace. Washington, D.C.: National Academy Press.

Neely, C. L., K. A. McVay, and W. L. Hargrove. 1987. Nitrogen contribution of winter legumes to no-till corn and grain sorghum. Pp. 48–49 in The Role of Legumes in Conservation Tillage Systems, J. F. Power, ed. Ankeny, Iowa: Soil Conservation Society of America.

Olson, R. A., K. D. Frank, P. H. Grabouski, and G. W. Rehm. 1981. Economic and Agronomic Impacts of Varied Philosophies of Soil Testing. No. 6695 Journal Series. Lincoln, Nebr.: Agricultural Experiment Station, University of Nebraska.

Oltjen, R. R., D. E. Ullrey, C. B. Ammerman, D. R. Ames, C. A. Baile, T. H. Blosser, G. L. Cromwell, H. A. Fitzhugh, D. E. Goll, Z. Helsel, R. E. Hungate, L. S. Jensen, N. A. Jorgensen, L. J. Koong, D. Meisinger, F. N. Owens, C. F. Parker, W. G. Pond, R. L. Preston, and H. S. Teague. 1980. Animal Nutrition and Digestive Physiology. Pp. 69–91 in Animal Agriculture Research to Meet Human Needs in the 21st Century, W. G. Pond, R. A. Merkel, L. D. McGilliard, and J. Rhodes, eds. Boulder, Colo.: Westview Press.

Osteen, C. D., E. B. Bradley, and L. J. Moffitt. 1981. The Economics of Agricultural Pest Control: An Annotated Bibliography, 1960–1980. Bibliographies and Literature of Agriculture No. 14. Economics and Statistics Service. Washington, D.C.: U.S. Department of Agriculture.

Oswald, J. W., and D. N. Wright. 1950. Potato scab control. California Agriculture 4(4):11–12.

Paige, J. C., and R. Kent. 1987. Tissue residue briefs. FDA Veterinarian 2(6):10–11.

Phillips, R. E., R. L. Blevins, G. W. Thomas, W. W. Frye, and S. H. Phillips. 1980. No-tillage agriculture. Science 208:1108–1113.

Pimentel, D., G. Berardi, and S. Fast. 1984. Energy efficiencies of farming wheat, corn, and potatoes organically. Pp. 151–161 in Organic Farming: Current Technology and Its Role in a Sustainable Agriculture, Special Publication No. 46, D. F. Bezdicek and J. F. Power, eds. Madison, Wis.: American Society of Agronomy, Crop Science Society of America, Soil Science Society of America.

Power, J. F. 1987. Legumes: Their potential role in agricultural production. American Journal of Alternative Agriculture 2(2):69–73.

Randall, G. W., and P. L. Kelly. 1987. Soil test comparison study. Pp. 145–148 in A Report on Field Research in Soils. Miscellaneous Publication No. 2(Revised)–1987. St. Paul, Minn.: University of Minnesota Agricultural Experiment Station.

Rice, R. E., and P. Kirsch. In press. Mating disruption of the oriental fruit moth in the United States. In Behavior-Modifying Chemicals for Insect Management: Applications of Pheromones and Other Substances, R. L. Ridgway, R. M. Silverstein, and M. N. Inscoe, eds. New York: Marcel Dekker.

Ridgway, R. L., and S. B. Vinson, eds. 1977. Biological Control by Augmentation of Natural Enemies. New York: Plenum Press.

Robbins, J. D. 1983. The tall fescue toxicosis problem. Pp. 1–4 in Proceedings of a Tall Fescue Toxicosis Workshop. Athens, Ga.: Cooperative Extension Service, University of Georgia College of Agriculture.

Sauer, T. J., and T. C. Daniel. 1987. Effect of tillage system on runoff losses of surface-applied pesticides. Soil Science Society of America Journal 51:410–415.

Schroth, M. N., and J. G. Hancock. 1985. Soil antagonists in IPM systems. Pp. 415–431 in Biological Control in Agricultural IPM Systems, M. A. Hoy and D. C. Herzog, eds. Orlando, Fla.: Academic Press.

Selye, H. 1950. The Physiology and Pathology of Exposure to Stress: A Treatise Based on the Concepts of the General Adaptation Syndrome and the Diseases of Adaptation. Montreal: ACTA Publications.

Settle, W. H., L. T. Wilson, D. L. Flaherty, and G. M. English-Loeb. 1986. The variegated leaf hopper: An increasing pest of grapes. California Agriculture 40(7&8):30–32.

Shrader, W. D., and R. D. Voss. 1980. Soil fertility: Crop rotation vs. monoculture. Crops and Soils Magazine 7:15–18.

Smith, J. S. C. 1988. Diversity of United States hybrid maize germplasm: Isozymic and chromatographic evidence. Crop Science 28:63–69.

Springman, R. E., T. C. Daniel, E. E. Schulte, and L. G. Bundy. 1986. Soil fertility guidelines for conservation tillage corn. University of Wisconsin Extension Bulletin A3369.

Stalker, D., W. R. Hiatt, and L. Comai. 1985. A single amino acid substitution in the enzyme 5-enolpyruvylshikimate-3-phosphate synthase confers resistance to the herbicide glyphosate. Journal of Biological Chemistry 26:4724–4728.

Studemann, J. A., S. R. Wilkinson, W. A. Jackson, and J. J. Jones, Jr. 1973. The association of fat necrosis in beef cattle with heavily fertilized fescue pastures. Pp. 9–22 in Proceedings of the Fescue Toxicity Conference. Columbia, Mo.: Cooperative Extension Service, University of Missouri.

Tweedy, B. G. 1983. The future of chemicals for controlling plant diseases. Pp. 405–415 in Challenging Problems in Plant Health, T. Kommendahl and P. H. Williams, eds. St. Paul, Minn.: American Phytopathological Society.

U.S. Department of Agriculture. 1978. Improving Soils with Organic Wastes. Report to the Congress in Response to Section 1461 of the Food and Agriculture Act of 1977 (P.L. 95–113). Washington, D.C.

U.S. Department of Agriculture. 1980. Report and Recommendations on Organic Farming. Washington, D.C.

U.S. Department of Agriculture. 1987. Fertilizer Use and Price Statistics, 1960–85. Statistical Bulletin No. 750. Economic Research Service. Washington, D.C.

University of Kentucky. 1978. Liming Acid Soils. Leaflet AGR-19. Lexington, Ky.

Van Dyne, D. L., and C. B. Gilbertson. 1978. Estimating U.S. Livestock and Poultry Manure and Nutrient Production. Economics, Statistics, and Cooperatives Service. Washington, D.C.: U.S. Department of Agriculture.

Varco, J. J., W. W. Frye, M. S. Smith, and J. H. Grove. 1987. Legume nitrogen transformation and recovery by corn as influenced by tillage. P. 40 in The Role of Legumes in Conservation Tillage Systems, J. F. Power, ed. Ankeny, Iowa: Soil Conservation Society of America.

Voss, R. D., and W. D. Shrader. 1984. Rotation effects and legume sources of nitrogen for corn. Pp. 61–68 in Organic Farming: Current Technology and Its Role in a Sustainable Agriculture, Special Publication No. 46, D. F. Bezdicek and J. F. Power, eds. Madison, Wis.: American Society of Agronomy, Crop Science Society of America, Soil Science Society of America.

Webb, J. R. 1982. Rock phosphate-superphosphate experiment. Pp. 7–9 in Northern Research Center Annual Progress Report. ORC81-14, Vol. 22. Clarion-Webster Research Center. Ames, Iowa: Iowa State University.

Webb, J. R. 1984. Rock phosphate-superphosphate experiment. Pp. 5–7 in Northwest Research Center Annual Report. ORC83-29. Ames, Iowa: Iowa State University.

Wedin, W. F., H. J. Hodgson, J. E. Oldfied, K. J. Frey, C. W. Deyoe, R. S. Emergy, L. Hahn, V. W. Hays, C. H. Herbel, J. S. Hillman, T. J. Klopfenstein, W. E. Larson, V. L. Lechtenberg, G. C. Marten, P. W. Moe, D. Polin, J. M. Sweeten, W. C. Templeton, P. J. VanSoest, R. L. Vetter, and W. J. Waldrip. 1980. Feed production. Pp. 153–191 in Animal Agriculture Research to Meet Human Needs in the 21st Century, W. G. Pond, R. A. Merkel, L. D. McGilliard, and J. Rhodes, eds. Boulder, Colo.: Westview Press.

Wilhelm, S., and A. O. Paulus. 1980. How soil fumigation benefits the California strawberry industry. Plant Disease 64:264–270.

Williams, W. A., and J. H. Dawson. 1980. Vetch is an economical source of nitrogen in rice. California Agriculture 34(8&9):15–16.

Young, C. E., B. M. Crowder, J. S. Shortle, and J. R. Alwang. 1985. Nutrient management on dairy farms in southeastern Pennsylvania. Journal of Soil and Water Conservation 40(5):443–445.

Zachariassen, J. A., and J. F. Power. 1987. Soil temperature and the growth, nitrogen uptake, dinitrogen fixation, and water use by legumes. Pp. 24–26 in The Role of Legumes in Conservation Tillage Systems, J. F. Power, ed. Ankeny, Iowa: Soil Conservation Society of America.

Zimmerman, D. R. 1986. Role of subtherapeutic levels of antimicrobials in pig production. Journal of Animal Science 62(Suppl. 3):6–17.

4

Economic Evaluation of
Alternative Farming Systems

INTEREST IN ALTERNATIVE FARMING SYSTEMS is often motivated by a desire to reduce health and environmental hazards and a commitment to natural resource stewardship. But the most important criterion for many farmers considering a change in farming practices is the likely economic outcome.

Wide adoption of alternative farming methods requires that they be at least as profitable as conventional methods or have significant nonmonetary advantages, such as preservation of rapidly deteriorating soil or water resources. Economic performance can be improved in several ways:

- Lowering per unit expenditures on production inputs;
- Increasing output per unit of input;
- Producing more profitable crops and livestock;
- Reducing capital expenditures on machinery, irrigation equipment, and buildings;
- Reducing natural crop and animal losses;
- Reducing income loss through commodity price fluctuations; and
- Making fuller use of available land, labor, and other resources.

Several economic analyses of alternative farming systems were conducted in the 1970s. A review found most of these studies were methodologically flawed, however, and used prices, technologies, and policies that are of limited relevance today (Lockeretz et al., 1984). In particular, energy and land values have fallen, real interest rates have risen, inflation has slowed, cash market commodity prices have generally declined beginning in 1982, and a wide range of government policies have exerted greater influence on farmer decision making. These factors are dynamic and constantly influence agricultural producers and policies.

Nonetheless, a growing body of contemporary data supports the economic viability of alternative farming practices and systems. The committee reviewed and interpreted available literature on the economics of alternative methods and systems, focusing on the general areas of pest control, diversification, nutrient sources, and the effect of government and market price structures on the adoption of alternative practices. Economic findings from the case study farms are presented in this chapter.

ECONOMIC ASSESSMENTS OF ALTERNATIVE METHODS

Understanding the overall economic implications of alternative farming systems requires research at several levels, including individual components of crop and livestock enterprises, whole-farm studies, and national and international analyses.

Traditionally, most evaluations of the economic impact of adopting alternative farming practices have focused principally on the cost and returns of adopting a specific farming method. For example, many studies at the farm level have estimated the economic benefits of integrated pest management (IPM), crop rotations, and manure management options. Such studies generally assume no other changes in the farm operation, input or output, or prices. These studies fall into a broader literature on farm management that employs partial budget analysis techniques.

Fewer studies have considered the impact of alternative farming systems on the economic performance of the whole farm. At the aggregate level, the committee could identify no useful studies of the potential effects of widespread adoption of alternative agricultural systems.

Most aggregate studies are flawed in their methods and assumptions regarding the effectiveness of alternative systems and the impact of commodity policy on farm management. The common approach has been to compare conventional farming practices with the economic performance of a similar farm, assuming total withdrawal of certain categories of farm inputs. These studies usually assume or project substantial reductions in per acre yields in many crops and then project the effect of these reductions in the context of strong export demand and limited commodity supplies. These assumptions and conditions often result in projected food production shortfalls that do not accurately reflect the constant change of markets or the production capabilities of many available alternative systems. The committee could identify no aggregate studies that compare the costs and benefits of conventional agriculture with successful alternative systems. Such analyses are needed but will be complex, involving a wide range of factors.

Economic Studies of Farming Practices

Economic analyses of single enterprises or their components usually employ partial budgeting techniques that estimate the change in production costs, profits, and risks accompanying a specific change in farming practice

(Boehlje and Eidman, 1984). Results are often expressed as a change in the net return over cash production costs per acre or per unit of output. Methodologically, partial budget studies focus on short-term net returns, including labor, and generally do not take into account off-farm impact or long-term changes in the productivity of the natural resource base. They also assume no change in farm size, enterprise combinations, prices of commodities or inputs, or other variables.

Despite these limitations, this method is practical and easy to understand. Partial budget study findings can be augmented by drawing on additional analyses from specialists in biology, ecology, and physical science. In recent years, biological, physical, and social scientists have made much progress in their collaborative research efforts in developing new methodologies for estimating the economic consequences of farming systems and practices.

Partial budgeting is reported to be the most widely used method of estimating changes in income of an individual farm as a result of adopting IPM (Allen et al., 1987; Osteen et al., 1981). The landmark research on the economics of crop rotations by Heady (1948) and Heady and Jensen (1951) was based essentially on the partial budgeting approach, because the only aspect of the farm operation assumed to vary was the crop rotation. Contemporary research that includes a greater consideration of biological and economic factors is presented later in this discussion (Goldstein and Young, 1987; Helmers et al., 1986). The review by Allen et al. (1987) of the agricultural, economic, and social effects of IPM is another example of the multidisciplinary approach to partial budgeting analysis.

Whole-Farm Analysis of Alternative Methods

Frequently, a farming method that appears profitable when analyzed at a component level may prove less attractive from the perspective of the whole farm, particularly in relation to other possible practices or combinations of practices.

Analysis at the whole-farm level recognizes that a farmer's decision to adopt one or more farming practices is not made in isolation from the rest of the farm enterprise. Perhaps the most important factor in adopting any management system or combination of crops is the net return to the farm family. The successful commercial farmer must assess the compatibility of proposed alternative practices with other practices already in place, taking into account a farm's physical and biological resources and anticipated changes in crop yields, livestock productivity, production costs, farm programs and policy, and labor and machinery requirements. These and other factors will strongly affect the farm operator's cash flow and the farm's profitability and long-term economic viability.

Whole-farm studies typically use one of two approaches: linear budget (risk programming) or overall farm surveys. Both approaches attempt to examine the effect of different farming practices or production systems at the farm level, taking into account all components of the farming system and operation, such as land-use patterns, pest control practices, and nutrient management.

Microeconomic programming or planning models analyze farm decision making based on particular resource and financial assumptions as well as estimated relationships between management choices and crop or animal production levels. The usefulness and validity of these models depend on the availability of reliable experimental or empirical data on input and output relationships in specific agricultural systems. When such data are present, whole-farm planning models can analyze the economic consequences of a wide range of alternative production systems. A principal objective of the committee's research recommendations is the development of such a knowledge and information base (see the Executive Summary).

Partial- and whole-farm analyses can take a short- or long-term perspective. For short-term analyses, some resources and technologies are assumed fixed, and management decisions are made among existing alternatives. Long-term studies are more complex and difficult because many more variables are changeable, including technologies and policies. A critical need identified by the committee is expanded multidisciplinary research on long-term technological trends and policy changes and how these trends and changes are likely to influence the relative costs and benefits of various farming systems. For example, the committee suspects that biotechnology will greatly increase technological options in support of alternative agricultural systems, and that society's environmental and public health goals will tend to support producers successfully adopting these technologies. The committee cannot go further in quantifying these trends, however, because the necessary knowledge base and analytical framework do not exist.

Farm surveys are based on empirical measures of the performance of agricultural production systems. It is often difficult to draw cause and effect inferences from surveys, however. For example, farm operators' technological choices and management abilities greatly influence profitability. But it is difficult to separate the contribution of technology from that associated with managerial skill. Nonetheless, the performance of agricultural systems as captured in well-designed surveys implicitly reflects the interaction of these factors. Experimental data on alternative agricultural systems are clearly lacking, and relatively few well-designed surveys have been undertaken. The literature is beginning to grow, however, and a number of solid studies have reached conclusions indicating the prospective economic benefits of alternative production systems.

The Transition to Alternatives

Most economic studies of alternative production at the whole-farm level take a static approach, ignoring the year-to-year difficulties associated with the transition from one system to another. Moreover, the assumptions used generally ignore uncertainty stemming from the weather, crop yields, management skills, prices of inputs and products, government policies, and other variables. As a result, these studies must be interpreted cautiously.

Several whole-farm studies have examined the financial impact of changing from conventional to alternative farming practices (Hall, 1977; Osteen et al., 1981; Reichelderfer and Bender, 1979). These studies recognize that a farm's economic performance can change significantly during a multiyear evolution from conventional to alternative practices (Dabbert and Madden, 1986).

Many factors can influence the economic performance of farms during the transition to alternative practices. The use of certain kinds of pesticides and fertilizers may have disrupted natural predators and other biota. Reestablishing these populations and the balance among them can occur quickly or require several years (Koepf et al., 1976; U.S. Department of Agriculture, 1980). Although crop rotations will generally increase yields, decrease pesticide costs, and, in the case of legumes, decrease fertilizer costs, the full benefits of crop rotations may take several years to materialize. Depending on the prices of farm commodities and inputs, adoption of a rotation sometimes reduces net farm income, particularly during the initial years of a transition (Dabbert and Madden, 1986). For example, including a forage legume in a rotation may not sufficiently decrease production costs and increase the yields of cash grain crops to compensate for the reductions in their acreage—especially when cash grain prices are supported far above market levels (Duffy, 1987; Goldstein and Young, 1987). Farmers may also need a few years of experience to acquire the additional knowledge and management skills necessary for more diversified operations. The economic impact of a farmer's decision to change from conventional to alternative farming methods on all or part of a farm operation will vary depending on factors such as climate, soil type, crops and livestock produced, cropping history of the farm, the farmer's skills, and many other considerations.

Because of these factors, most farmers adopt alternatives gradually. Although the transition may be difficult, successful alternative systems tend to reduce variability of net returns (Helmers et al., 1986). The consistency of yield and return to the farm family is a potential benefit of alternative agriculture that deserves further study.

Comparative Regional Cost of Production

Production cost per unit of output is one of the most important short-term measures of the economic performance of an agricultural operation, production system, or sector. Comparing per unit production costs for a given crop by region is a good indicator of regional absolute advantage—or the inherent suitability of an area or farm for the profitable production of a given crop.

Another common measure—production costs per acre—is widely used in comparative analyses. This measure, however, differs significantly from per unit production costs. Per acre costs do not take into account the actual

TABLE 4-1 Regional per Bushel and per Acre Production Cost Estimates and Yields, 1986

Crop	Corn Belt-Great Lakes	Southeast
Corn		
Total variable costs (dollars)		
Per bushel	0.93	1.81
Per acre	118.68	120.15
Yield/acre (bushels)	126	66
Soybeans		
Total variable costs (dollars)		
Per bushel	1.31	3.15
Per acre	49.93	67.89
Yield/acre (bushels)	37	21

NOTE: The Corn Belt and Great Lakes region includes Minnesota, Wisconsin, Michigan, Iowa, Missouri, Indiana, Illinois, and Ohio. The Southeast region includes Kentucky, Tennessee, Alabama, Georgia, South Carolina, North Carolina, and Virginia.

SOURCE: U.S. Department of Agriculture. 1987. Economic Indicators of the Farm Sector—Costs of Production, 1986. ECIFS 6-1. Economic Research Service. Washington, D.C.

yields harvested; they reflect the level of inputs applied on a per acre basis. Consequently, per acre production costs do not as accurately reflect the productivity of a cropping system or an area for a particular crop. Likewise, high per acre costs for fertilizer and pesticides do not necessarily indicate high per unit costs or low productivity. For example, farmers in highly productive corn-growing regions generally use more fertilizer and other inputs per acre because they can afford it based on the high yields they will achieve, not because the area is unsuited to corn production. This is particularly true when market or government support prices are high.

In contrast to the limitations of per acre costs, per bushel costs are good indicators of an area's suitability for production of a given crop. The example in Table 4-1 shows this and indicates the superiority of per unit production cost figures in defining the productivity of regions. The Corn Belt-Great Lakes region is highly suited to corn and soybean production in terms of rainfall, soils, and temperature, particularly in contrast to the Southeast region. Per acre production cost estimates, however, do not reveal this advantage as clearly as per unit production costs do; the total per acre variable costs are similar for these regions. The same costs expressed on a per bushel basis, however, show that it requires far less cash expenditure to produce a bushel of corn or soybeans in the Corn Belt-Great Lakes region than in the Southeast. Table 4-2 shows total variable costs and fertilizer and pesticide cost estimates per unit of production for various regions producing corn, soybeans, and hard red winter wheat.

Per unit production costs reflect what actually happens during a given growing season. Many things, such as too much or too little rain, cold

TABLE 4-2 Regional per Acre Yield and Selected per Bushel Production Cost Estimates for Corn, Soybeans, and Hard Red Winter Wheat, 1986

Crop	Delta	Corn Belt-Great Lakes	Northeast	Northern Plains	Southeast	Southwest	Central Plains	Southern Plains
Corn								
Yield/acre (bushels)	—	126	89	116	66	117	—	—
Total variable costs/bushel (dollars)	—	1.53	2.22	1.59	2.34	1.92	—	—
Fertilizer and pesticide costs/bushel (dollars)	—	0.54	0.67	0.39	1.07	0.54	—	—
Soybeans								
Yield/acre (bushels)	19	38	—	34	22	—	—	—
Total variable costs/bushel (dollars)	2.80	1.31	—	1.21	3.15	—	—	—
Fertilizer and pesticide costs/bushel (dollars)	1.31	0.64	—	0.49	1.69	—	—	—
Hard red winter wheat[a]								
Yield/acre (bushels)	—	—	—	27	—	72	32	21
Total variable costs/bushel (dollars)	—	—	—	1.60	—	2.37	1.18	2.22
Fertilizer and pesticide costs/bushel (dollars)	—	—	—	0.69	—	0.74	0.31	0.77

NOTE: A dash indicates that data were not reported.

For corn, the Corn Belt-Great Lakes region includes Illinois, Indiana, Iowa, Michigan, Minnesota, Missouri, Ohio, and Wisconsin; the Northeast, Maryland, New York, and Pennsylvania; the Northern Plains, Colorado, Kansas, Nebraska, and South Dakota; the Southeast, Alabama, Georgia, Kentucky, North Carolina, South Carolina, Tennessee, and Virginia; and the Southwest, California and Texas. For soybeans, the Delta region includes Arkansas, Louisiana, and Mississippi; the Corn Belt-Great Lakes region includes Illinois, Indiana, Iowa, Michigan, Minnesota, Missouri, Ohio, and Wisconsin; the Northern Plains, Kansas, Nebraska, and South Dakota; and the Southeast, Alabama, Georgia, Kentucky, North Carolina, South Carolina, Tennessee, and Virginia. For hard red winter wheat, the Northern Plains region includes Idaho, Montana, and Wyoming; the Southwest, Arizona and California; the Central Plains, Colorado, Kansas, Nebraska, and South Dakota; and the Southern Plains, New Mexico, Oklahoma, and Texas.

[a]Averages for 1984–1986.

SOURCE: U.S. Department of Agriculture. 1987. Economic Indicators of the Farm Sector—Costs of Production, 1986. ECIFS 6-1. Economic Research Service. Washington, D.C.

spells, pests, hail, or soil fertility problems, can affect productivity in farming. These factors, as well as diverse soil types, climates, and levels of pest infestation, often account for large regional differences in per unit costs for a given crop, despite fairly similar per acre production costs.

Increased efficiency and lower per unit production costs are essential for agricultural producers to remain competitive in domestic and international markets. Alternative systems can often help achieve these goals. To better understand the role and viability of specific alternative agriculture systems, however, far greater knowledge of regional differences in production costs, their variability, and their causes is needed. Such understanding will help

- Explain how and why some farmers within regions and in different regions of the country can produce a given crop at markedly lower per unit costs than their neighbors or producers in other regions;
- Identify the production cost advantages and disadvantages stemming from soil, water, weather, pests, and other natural factors;
- Target technologies, management approaches, and policy decisions that most effectively reduce these costs and make the most of regional advantages; and
- Better understand how commodity, conservation, regulatory, and other policies influence on-farm management decisions and production costs.

Methods for Comparing Production Costs

A variety of farm accounting systems and methods can be used to calculate per acre and per unit production costs. Most farmers use some system of recordkeeping to track expenditures and determine profits and losses at the end of each season.

Most states and the U.S. Department of Agriculture (USDA) collect and analyze farm budget data. A variety of private organizations have developed recordkeeping systems that farmers can use for estimating cash flow, working with lenders, tracking returns to certain investments, identifying areas where profits could be increased, and preparing income tax statements. Some lenders require these records. Many of these recordkeeping systems are very sophisticated and have been used to study the distribution of per acre and per unit production costs for major commodities. The quality of individual farmers' recordkeeping, however, has a great effect on the quality of the data reported. The committee has reviewed several farm budget and cost of production studies, including Southwestern Minnesota Farm Business Management Association data, Southwest Kansas Farm Management Association data, and data compiled and published by the USDA.

Reports by these and other organizations use a variety of different measures, assumptions, and formats in collecting, analyzing, and reporting data. They are not random samples and do not generally employ sampling techniques. As a result, care must be exercised in drawing inferences from data and findings associated with different data sets. To the extent possible, the USDA tries to use consistent definitions and accurate methods in its published reports on state average production costs. The level of aggregation reported, however, masks much of the variability within states in the costs incurred on individual farms.

Additional insights can be extracted from analyses of comparative production costs on particular groups of farms within a given region. A common analytical approach is to separate a sample of farms producing a given crop into groups based on a given indicator or particular farm characteristic. The results of one such analysis of dryland wheat farms in southwest Kansas are shown in Table 4-3.

The sample of 3,000 farms was divided into quartiles by income. The first column in the table reports average yields, costs, and acreage for the 750 farms—or 25 percent—reporting highest income; the second column reports the same information for the 750 farms reporting the lowest income.

These data show

- Low-income farms incur per unit production costs nearly twice those of high-income farms ($3.66 versus $1.87 per bushel).
- The yields on low-income farms are about 9 percent less than on the high-income farms even though the per bushel production costs are almost double.
- All variable costs per acre were greater on the low-income farms. The per acre differential was greatest for machinery hire ($7.57), fertilizer ($7.53), machinery repair ($6.02), and herbicides and insecticides ($5.28).

Insights into the potential benefits of certain alternative production systems arise from identifying the cost factors that tend to distinguish high-income low-cost producers from less profitable but otherwise similar farms.

Some important factors contributing to higher per acre costs in Kansas wheat production and corn and soybeans grown in southwest Minnesota are summarized in Table 4-4. The difference in fertilizer and pesticide per acre and per bushel production cost for high-cost and low-cost corn and soybean farms in Minnesota are presented in Table 4-5. Per bushel fertilizer and pesticide costs were 144 percent greater for high cost soybean farms in 1986 and 60 percent greater on corn farms in 1987. Variable costs associated with machinery and repairs are also consistently high on low-income farms, in part because these farms are smaller on average and machinery costs are spread over fewer acres. These data are consistent with national average cost of production data for major crops (Table 4-6).

Alternative Agriculture and Production Costs

Alternative production systems are designed to enhance beneficial biological interactions and improve economic performance through better nutrient management and pest control. When successfully adopted, most alternative systems greatly influence fertilizer and pest management costs (see all case

TABLE 4-3   Cost of Production for Dryland Wheat in Southwest Kansas, 1986

Costs	25 Percent of Farms with Highest Income (per acre)	25 Percent of Farms with Lowest Income (per acre)
Crop production costs		
Hired labor	$ 4.02	$ 4.35
Repairs	8.90	14.92
Seed crop insurance	2.17	3.18
Fertilizer-lime	2.62	10.15
Machine hire	8.09	15.66
Storage-marketing	1.68	3.97
Fees-conservation-auto expenses	1.06	3.14
Gas-fuel-oil	6.91	9.99
Personal property tax	0.27	0.46
General insurance	0.45	1.13
Utilities	1.31	2.27
Herbicide-insecticide	1.19	6.47
Interest on operating costs (12%)	3.48	6.81
Interest on machinery investment (12%)	3.62	5.31
Total operating costs	$ 45.77	$ 87.81
Depreciation		
Motor vehicles	$ 13.01	$ 12.69
Machinery	3.98	9.13
Buildings	1.50	5.01
Total depreciation	$ 18.49	$ 26.83
Total production costs	$ 64.26	$114.64
Total production costs/bushel	$ 1.87	$ 3.66
Management, labor, and land costs		
Management charge[a]	$ 4.17	$ 3.81
Operation, unpaid labor[b]	10.12	20.44
Land charge[c]	27.82	25.40
Total management, labor, land costs	$ 42.11	$ 49.65
Total management, labor, land costs/ bushel	$ 1.23	$ 1.58
Total costs	$106.37	$164.29
Total costs/bushel	$ 3.10	$ 5.24
Wheat acres	1,482	734
Wheat yield/acre (bushels)	34.35	31.36

[a]5 percent of yield per acre times $2.43 per bushel.
[b]$15,000 per operator divided by wheat acres.
[c]33.33 percent of yield per acre times $2.43 per bushel.

SOURCE: B. L. Flinchbaugh, Kansas State University, correspondence, 1988.

TABLE 4-4 Major Inputs Resulting in Higher per Acre Costs: High-Cost
Farms Versus Low-Cost Farms, Selected Studies

Year Location/Crop	Input	Difference in Variable Costs Between High- and Low-Cost Farms	
		Dollars/ Acre	Percentage of Total Difference
1985 Kansas/wheat	Repairs	4.46	20.9
	Machine hire	2.65	12.4
	Fertilizers	1.59	7.5
	Pesticides	0.99	4.6
1986 Kansas/wheat	Machine hire	7.57	18.0
	Fertilizers	7.53	17.9
	Repairs	6.02	14.3
	Pesticides	5.28	12.6
1986 Minnesota/soybeans	Pesticides	5.09	24.1
	Repairs	3.01	14.2
	Fertilizers	0.24	1.1
1987 Minnesota/corn	Repairs	19.37	36.3
	Fertilizers	8.00	15.0
	Pesticides	4.61	8.6

SOURCES: Kansas Cooperative Extension Service. 1987. The Annual Report—Management
Information—Kansas Farm Management Associations. Manhattan, Kans.: Kansas State University;
Olson, K. D., E. J. Weness, D. E. Talley, P. A. Fales, and R. R. Loppnow. 1987. 1986 Annual
Report, Revised. Southwestern Minnesota Farm Business Management Association. Economic
Report ER87-4. St. Paul, Minn.: University of Minnesota; Olson, K. D., E. J. Weness, D. E. Talley,
P. A. Fales, and R. R. Loppnow. 1988. 1987 Annual Report: Southwestern Minnesota Farm
Business Management Association. Economic Report ER88-4. St. Paul, Minn.: University of
Minnesota.

studies). Regional cost of production studies based on farm recordkeeping
systems (Goldstein and Young, 1987; Kansas State University, 1987; Olson
et al., 1981, 1986, 1987) and the committee's limited case studies indicate
that the most profitable alternative and conventional farms are often those
that successfully cut back on fertilizer, pesticide, and machinery expenses
while sustaining high levels of crop production.

The extent and causes of variability in production costs warrant careful
study in assessing agricultural commodity, conservation, and regulatory
policies. High target prices, deficiency payments, and disaster provisions
that compensate farmers for crop losses are principal causes of inefficient
input use. Current farm programs base payments on historical per acre
yield levels, multiplied by a per bushel deficiency payment rate. The per
bushel deficiency payment is the difference between the government-set
target price and loan rate or the market price, whichever difference is less.
When deficiency payments are large, during periods of protracted low crop
prices, farmers have greater incentive to apply fertilizers and pesticides in
greater amounts to produce the most bushels per acre and collect the

TABLE 4-5   Fertilizer, Pesticide, and Total Variable Costs for Minnesota Corn and Soybeans (in dollars)

Year Location/Crop	Total Variable Costs		Fertilizer and Pesticide Costs	
	Per Acre	Per Bushel	Per Acre	Per Bushel
1986 Minnesota/soybeans				
Average high-cost farm	70.24	2.62	17.67	0.66
Average low-cost farm	49.10	1.07	12.34	0.27
Difference	21.14	1.55	5.33	0.39
1987 Minnesota/corn				
Average high-cost farm	153.88	1.24	55.64	0.45
Average low-cost farm	100.58	0.65	43.03	0.28
Difference	53.30	0.59	12.61	0.17

SOURCES: Olson, K. D., E. J. Weness, D. E. Talley, P. A. Fales, and R. R. Loppnow. 1987. 1986 Annual Report, Revised. Southwestern Minnesota Farm Business Management Association. Economic Report ER87-4. St. Paul, Minn.: University of Minnesota; Olson, K. D., E. J. Weness, D. E. Talley, P. A. Fales, and R. R. Loppnow. 1988. 1987 Annual Report: Southwestern Minnesota Farm Business Management Association. Economic Report ER88-4. St. Paul, Minn.: University of Minnesota.

TABLE 4-6   National Average Cost of Production for Selected Inputs, 1986

Input	Corn	Grain Sorghum	Wheat	Rice	Soybeans	Cotton
	Dollar Cost per Acre					
Custom operation	6.70	3.49	5.38	49.06	3.77	14.19
Seed	16.82	3.92	5.97	24.14	8.54	8.16
Fertilizers	45.51	17.88	14.30	31.20	6.41	20.14
Pesticides	19.21	9.27	3.25	5.73	18.93	50.32
Fuel	9.52	10.73	6.06	26.67	4.80	18.71
Subtotal	97.76	45.29	34.96	136.80	42.45	111.52
Total variable costs	118.74	59.25	44.36	242.85	52.04	193.19
Fixed cash costs (interest, insurance, and overhead)	70.83	32.27	30.77	71.72	51.88	65.33
Total variable and fixed costs	189.57	91.52	75.13	314.57	103.92	258.52
	Percentage of Total Variable Costs					
Pesticides and fertilizers	55	46	40	15	49	37
	Percentage of Total Variable and Fixed Costs					
Pesticides and fertilizers	34	30	23	12	25	27

SOURCE: U.S. Department of Agriculture. 1987. Economic Indicators of the Farm Sector—Costs of Production, 1986. ECIFS 6-1. Economic Research Service. Washington, D.C.

highest payments. Between 80 and 95 percent of major commodity producers currently participate in federal commodity programs. (For a further discussion of loan rates and target prices, see "The Power of Policy" section in Chapter 1.)

Farm programs strongly influence input use, planting decisions, and the use of marginal lands in generally productive areas. The effect of these programs on input use differs greatly by region. Two basic types of inefficiency can arise from federal commodity programs:

1. Excess input use to achieve higher yields and maximize government program payments, and

2. Use of inputs to expand crop production onto marginal lands, or to support the production of crops in regions poorly suited to a particular crop.

In areas well suited to the production of a crop (for example, corn in the Corn Belt), high government payments encourage excessive and inefficient input use (Olson et al., 1981; Randall and Kelly, 1987). In areas where production of a crop is inherently more difficult (for example, corn in the Mississippi Delta), government payments often subsidize the production of crops that would not otherwise be profitable (see Tables 4-1 and 4-2).

There are sound correlations, although complicated and often poorly documented, between the economic and environmental performance of farming systems. Efficient systems generally are associated with fewer environmental problems because cropping patterns, fertility, and pest control practices are matched to the strengths and limitations of the resource base and follow sound biological and agronomic principles. Both types of inefficiencies identified above can lead to environmental problems, such as water pollution, soil erosion, or loss of wildlife habitat. For example, when Corn Belt farmers overapply nitrogen fertilizer, the result can be nitrates in surface water and groundwater. When western farmers produce cotton or other crops with irrigation in the face of inherent environmental, resource, and climatic limitations, the result can be salinization of soils and water and depletion of aquifers.

The adoption of policies that reduce or remove incentives for these inefficiencies would reduce environmental damage and enhance the competitiveness and profitability of U.S. agriculture. By reducing program crop acreage in high per unit cost regions or on high per unit cost farms and expanding production in low per unit cost regions or on particularly efficient farms, policy reform has the potential to

- Reduce average per unit production costs;
- Improve efficiency of input use;
- Reduce environmental consequences of inefficient input use;
- Lower federal program costs; and
- Cause farmers to abandon certain high-cost crops in certain areas.

Production cost analyses can yield important insights into the economic and environmental performance of farming systems. Studies based on actual farm records for operations within a given region appear particularly promising. More in-depth assessments designed to distinguish features of low-cost farms, in contrast to high-cost farms, could guide agricultural researchers and extension specialists toward the most important technical and managerial factors underlying profitability.

Production system and technology changes designed to attain environmental and public health goals must also help to reduce costs. The unique and highly variable interrelationships among resources, management, policy, and economics must be much better understood and quantified so that reliable and realistic estimates can be provided to policymakers regarding the tradeoffs, costs, and consequences inherent in policy choices.

The case studies and the committee's review of available cost of production studies support a number of important conclusions that warrant further exploration and analysis.

- Within a given region for a specific crop, average production costs per unit of output on the most efficient farms is typically 25 percent less, and often more than 50 percent less, than average costs on less efficient farms. There is a great range in the economic performance of seemingly similar or neighboring farms.
- Average production costs per unit of output also vary markedly among regions, although not as dramatically as among individual farms.
- High-income and low-cost farms are often larger. The causes and effects of this, however, deserve study.
- Certain variable production expenses—machinery expenses, pesticides, fertilizers, and interest charges (excluding land)—account disproportionately for differences in per unit production costs.

ALTERNATIVE PEST MANAGEMENT STRATEGIES

Successful alternative pest management strategies include a range of methods. Examples include traditional IPM insect control systems, systems based primarily on cultural practices such as rotations and short growing seasons, and biological control systems that use no synthetic chemical pesticides.

IPM

IPM is a pest control strategy based on the determination of an economic threshold that indicates when a pest population is approaching the level at which control measures are necessary to prevent a decline in net returns—that is, when the predicted value of the impending crop damage exceeds the cost of controlling the pest. In this context, IPM rests on a set of ecological principles that attempt to capitalize on natural pest mortality

This tractor pulls an alfalfa cutter in fields planted with a corn, alfalfa, and wheat rotation. *Credit:* Larry Lefever from Grant Heilman.

Irrigation makes production of high-cost, high-value crops, such as lettuce, possible in the arid central valley of California. *Credit:* Richard Steven Street.

Salinated soil stunts irrigated cotton growth in a California field. *Credit:* U.S. Department of Agriculture.

Alternative cropping systems research farm. *Credit:* Rodale Press, Inc.

Most growers burn rice fields after harvest to remove crop stubble and prepare fields for the next rice crop. The Lundberg Farm case study describes an alternative method for incorporating rice stubble into the soil, which eliminates the need for burning. *Credit:* Philip Wallick.

More than 28 million acres of highly erodible land have been removed from cultivation and enrolled in the Conservation Reserve Program. This CRP land is in New Mexico. *Credit:* Soil Conservation Service, U.S. Department of Agriculture.

Soil erosion and sedimentation in Clinton County, Missouri. *Credit:* Chevron Chemical Company.

Dairy farm in Elizabethtown, Pennsylvania. *Credit:* Isaac Geib from Grant Heilman.

Center pivot irrigation in southeastern Nebraska. *Credit:* Reinke Manufacturing Company, Inc.

These Emperor grapes were grown with no synthetic chemical pesticides or fertilizers. *Credit:* Richard Steven Street.

This farm in the sandhills of Nebraska was irrigated and planted with continuous corn. Wind has eroded the soil in the white wedge-shaped section. The sandhills, which need cover throughout the year, are far more suited to grazing or small grain production than cultivated row crops. *Credit:* A. D. Flowerday.

This waist-deep field of switch grass was established on cultivated cropland removed from production under the Conservation Reserve Program of the Food Security Act of 1985. *Credit:* Soil Conservation Service, U.S. Department of Agriculture.

Nitrate occurs naturally in groundwater. Levels over 3 micrograms/liter are thought to be caused by human activity. (Data were not available from all counties.) *Credit:* U.S. Geological Survey.

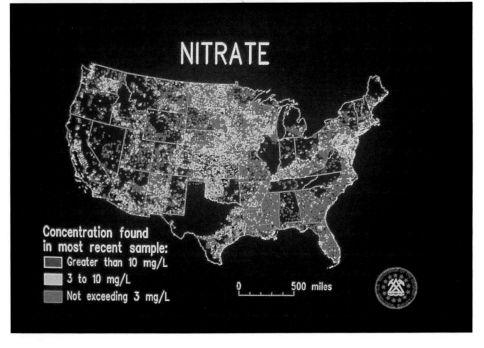

Contour strip cropping of corn, alfalfa, and wheat in Lancaster County, Pennsylvania. *Credit:* Larry Lefever from Grant Heilman.

factors; pest-predator relationships; genetic resistance; and the timing and selection of a variety of cultural practices, such as tillage, pruning, plant density, and residue management. In practice, however, IPM is generally based on scouting fields to determine pest or disease populations or infestation levels, more precise timing and application of pesticides derived from scouting, better knowledge of consequences of various levels of pest and predator populations, rotations, and more precise timing of planting. To further advance IPM systems, more research is needed into the development of economic thresholds and agroecosystem pest biology (Frisbie and Adkisson, 1985).

Individual farmers generally determine economic thresholds based on individual objectives and experience. Farmers producing the same crop are often willing to accept different levels of pest damage prior to the implementation of control measures. Most economic thresholds determine the most profitable short-term pest control strategy, taking account of the prices of the crop or livestock and the cost of the proposed control action. Other factors in developing thresholds and IPM methods include population trends of the target pest and its natural enemies, anticipated damage under various scenarios (including taking no action), potential increase in value of the crop, possible costs of adopting rotations, cost of pest scouting, probable effectiveness of chemical or nonchemical control techniques, and changes in pesticide application costs.

For insect control, most IPM systems rely on precise application of specific pesticides. In some instances, the number of pesticide applications per acre actually increases with scouting because a threatening pest population is discovered that would otherwise be missed, and more selective materials are used in the hope of minimizing the disruption of ecological balance in other pest species. The total volume of pesticides applied usually declines, however, because of more precise timing and selective applications of pesticides (Allen et al., 1987).

For disease control, the use of rotations, planting dates, weather monitoring, and resistant varieties are the most common components of IPM programs. Few formal IPM programs for weeds have been developed because many alternative weed control strategies already exist and are currently practiced by farmers.

In addition to focusing on increasing net farm income, the development of economic thresholds must take into account specific biological characteristics of the target pest and the long-term implications of current pest control actions. For example, it may be profitable in a given season to tolerate a certain number of weeds in a field; but from a long-term standpoint, increased weed pressure from weed seeds in the soil may become a serious problem (Coble, 1985). Long-term adverse implications can also arise from recurrent use of the same or related pesticides. The emergence of pesticide-resistant pest populations is a particularly worrisome phenomenon (National Research Council, 1986b). Neglecting pest resistance may lead to higher volume or more frequent applications of pesticides, with

TABLE 4-7   Percentage Differences in Yields, Crop Values, Pesticide
Applications, and Pest Control Costs for IPM Users Compared With Nonusers

Crop/State or Region	Yield/Acre	Dollar Value/ Unit of Production	Pesticide Applications	Pest Control Costs/Acre (including scouting)
Alfalfa seed/Northwest[a]	+17	+3	+107	NS
Almonds/California	+118[b]	NA	NA	NA
Apples/Massachusetts	+12	−8	−4	−23
Apples/New York	+21	+3	+15	−6
Corn/Indiana	+10	+4.5	+41	+45
Cotton/Mississippi[c]	+20	NS	NA	+32
Cotton/Texas	+30	+5	NA	+40
Peanuts/Georgia	+11	NS	+10	−11
Soybeans/Virginia	+9	+4	+38	+23
Stored grain/Kentucky	NR	NS	NA	−14
Tobacco/North Carolina	+0.5	NS	−17	NS

NOTE: This study surveyed 3,500 growers. NA = Not available. NR = Not relevant. NS = Change
not significant, less than 1 percent.

[a]Northwest includes Washington, Idaho, Oregon, Montana, and Nevada.
[b]Three-year average.
[c]Compared with low IPM users.

SOURCE: Adapted from Allen, W. A., E. G. Rajotte, R. F. Kazmeirczak, M. T. Lambur, and G. W.
Norton. 1987. The National Evaluation of Extension's Integrated Pest Management (IPM)
Programs. VCES Publication 491-010. Blacksburg, Va.: Virginia Cooperative Extension Service.

serious economic consequences for the individual farmer, the industry, and
even the nation (Hueth and Regev, 1974). Human health risks, environmental impacts, and water-quality degradation are other serious concerns.

Several studies have estimated the farm level and aggregate monetary
benefits and costs associated with development and adoption of IPM programs (Osteen et al., 1981). A 1987 evaluation of the Extension Service's
IPM programs for insects on nine crops in 10 individual states and 5 states
in the Northwest is the most comprehensive review to date (Table 4-7). The
results show higher average per acre yield in every case for IPM users over
nonusers growing the same crop in the same state. In 13 of the 15 cases,
growers using IPM received the same or higher prices for their crop. Only
one case reported a lower per unit price; the other did not report these
data. In every case the net return per acre was higher for IPM users versus
nonusers, primarily as a result of increased yields and prices received by
growers using IPM systems. In some states, growers using IPM systems
were the only group able, on average, to earn a profit from sale of those
commodities studied (Allen et al., 1987). The study's survey of 3,500 growers using IPM showed a $54 million increase in net return for these growers,
compared with growers not using IPM (Allen et al., 1987).

In most cases reported by Allen et al. (1987), IPM resulted in more pesticide applications per growing season. In nine cases growers made more

applications, in two cases less, and in four cases growers did not report the data. In four of the cases with significant changes in pest control costs (including scouting), costs increased; in four cases control costs decreased; in six cases costs remained about the same; and in one case they were not reported. Although pest control costs often increase or remain constant with IPM, growers nearly always profit from higher yields and improved fruit quality (Allen et al., 1987). As a percentage of total cash operating costs, pest control costs for IPM users were quite variable, ranging from 2 to 22 percent for the nine crops studied. Many other studies have shown that IPM can result in fewer pesticide applications and lower pest control costs (Kovach and Tette, 1988; Office of Technology Assessment, 1979; Shields et al., 1984; Smith and Barfield, 1982).

Although IPM users sometimes make more applications, they generally achieve a higher degree of control using a smaller volume of pesticide. This observation is reinforced by a review of 42 IPM studies. In every case, pesticide use or the cost of production or both decreased with IPM. Twenty-four cases reported increased net returns, 2 cases reported no change, and 16 cases made no report (Allen et al., 1987).

This review of the Extension Service's IPM program estimated that the adoption of IPM for the nine commodities in 15 states would result in an estimated $578 million in additional returns for producers of these crops. This result was obtained using an enterprise budgeting approach, based on data obtained from farmers in telephone interviews regarding pesticide costs and changes in yields. The study assumed that prices of pesticides and farm commodities would not change as a result of adoption of IPM. Effects on consumers or producers of possible reductions in commodity prices that could accompany increases in crop yield were not examined, nor were the secondary economic and environmental benefits likely to follow from widespread adoption of IPM. The same study also estimated the gross revenues of the private pest management consultant industry at about $400 million per year. The economic benefits associated with the expansion of this industry were not included in the total estimate of economic benefits related to wider use of IPM.

Several studies have gone beyond the farm level to project regional economic implications of IPM adoption. Frisbie and Adkisson (1985) reported a variety of regional and statewide economic assessments of IPM systems. Studies on cotton in Texas reported regional economic benefits ranging from $63 million to $192 million, which translated into estimated overall statewide economic benefits of $92 million to $305 million (Frisbie and Adkisson, 1985). Massachusetts apple growers using IPM systems reported increased net returns of $98.00 per acre; a separate study of northeastern apple producers reported increases of $25.00 per acre. Increased net returns of $28.00 per acre with a 75 percent reduction in pesticide use for alfalfa IPM in the North Central region were observed (Frisbie and Adkisson, 1985).

Because IPM programs are grounded in economic threshold principles, they nearly always result in increased returns for growers. They generally achieve this through better knowledge of pest and predator populations; more accurate, precisely timed, and better measured pesticide applications that reduce overall pest control costs; and use of improved crop varieties. Together, this results in higher yields and improved crop quality. In certain cases, notably cotton and alfalfa, IPM programs have resulted in dramatic decreases in applications and the volume of pesticides used. This is because IPM programs incorporate newer pesticides that are effective at far lower rates, and previously used pesticides are used with greater precision (see the Florida fresh-market vegetable case study). In other cases, applications remain constant or even increase while the volume of pesticide applied declines. Most current insect IPM systems make only modest use of cultural and biological pest control methods. They focus primarily on scouting and optimizing pesticide use. (Prominent exceptions are the control of corn rootworm through rotation and control of pink bollworm with short-season cotton.) In contrast, most IPM systems to control plant disease take advantage of rotations and cultural practices. One common example is the control of wheat root pathogens through rotations. The possible economic and environmental benefits of biological and cultural pest control techniques may become more compelling as biological and genetic pest control methods are improved.

The case study farms that use IPM—the Kitamuras' processing tomato operation in California and the Florida fresh-market vegetable operations—demonstrate the effectiveness of IPM. They also demonstrate the differences in pest control needs in different regions of the country. Disturbance of endemic insect and disease populations because of pesticide use are partially responsible for these differences. Each case study farm using IPM reports substantial reductions in pest control costs. Insect and disease pressures, however, are generally greater in Florida than in most regions of California. Growers in Florida, therefore, report greater savings per acre even though they continue to apply far more pesticides than either the Kitamura Farm or the majority of California processing tomato growers using IPM.

The Kitamura Farm uses cultural practices including irrigation, rotations, and sanitation; monitoring; and pesticides as a part of an IPM program that has decreased pest control costs by more than $45.00 per acre. On 160 acres of processing tomatoes the Kitamuras report a savings of $7,318. Their savings, however, are well above the average $7.70 savings per acre for California processing tomato growers using IPM (Antle and Park, 1986). One reason for this is their virtual elimination of insecticide applications, compared with an average reduction from 1.7 to 1.5 sprays per season for other growers. The Kitamuras' success is an indication of the degree to which management, innovation, and increased knowledge can influence the success of alternative systems.

Improved control and quality with comparable yields ensure the profitability of IPM in processing tomatoes. California processing tomato growers

Farmers use plastic mulch on staked tomatoes in Florida to seal in soil fumigants, control weeds, conserve soil moisture, and reduce nitrogen fertilizer leaching. Laser-guided machinery levels fields to a uniform slope of 0.15 percent. This helps to control irrigation and improves water efficiency. Buried pipes at the end of the rows supply water. Wide rows permit pesticide applications from elevated tractors. *Credit:* Will Sargent.

using IPM had virtually none of their tomatoes rejected for insect damage; non-IPM growers had a rejection rate of 5.6 percent. The yield on the Kitamura Farm is equal to or just below the historic county average. In 1986, however, their yield was 35.5 tons per acre, 6.3 tons per acre above the county average, with essentially no insect or mold damage.

In the hot dry climate of California's Central Valley, where there is virtually no rainfall between April and October, the Kitamuras are able to nearly eliminate fungicide and insecticide applications. In contrast, the hot humid climate of south Florida has an average year-round daily temperature of 74° and an average rainfall of 54 inches. These factors make it an excellent environment for insects, fungi, and other pests. Even Florida vegetable growers using IPM continue to need substantial quantities of pesticides. Nonetheless, they achieve far greater per acre savings from the adoption of IPM than do California growers. In the committee's case study, Florida producers who used IPM had direct pest control costs ranging from $200 to $300 per acre compared with $450 to $700 per acre for nonusers of IPM.

Scouting and better timing of pesticide applications that reduce the need for subsequent applications are primarily responsible for savings. A Univer-

A vacuum sucks insects off strawberry plants. This newly developed machine may enable farmers to forgo several insecticide applications for certain pests on certain crops. In field trials on strawberries, populations of lygus bugs were effectively controlled. Spider mite populations decline because lady bugs, which prey on the mites, avoid the vacuum and return later to feast on the mites. *Credit: Richard Steven Street.*

sity of Florida study of 40 tomato farms using IPM programs indicated that growers reduced insecticide inputs by about 21 percent (K. Pohronezny, University of Florida, interview, 1986). While generally effective at reducing insecticide applications, IPM in Florida has not yet been effective at reducing soil fumigant applications or the use of bactericides. Without rotations, soil fumigation will remain necessary. Reductions in fungicide use are generally far less than decreases in insecticide use under IPM. Even a marked increase in research and development investments in the development of IPM programs for Florida commercial vegetable production will not eliminate the need for several insecticide, fungicide, or nematocide treatments per season for many years to come. Yet, such a goal is clearly within reach in other regions.

Alternative Weed Control Practices

Various combinations of cultivation, tillage practices, cover crops, and rotations are widely used to control weeds and reduce or eliminate herbicide use. Five of the committee's case studies involve farms using various combinations of these methods for weed control in cash grain and livestock operations. These include the Spray, BreDahl, Sabot Hill, Kutztown, and Thompson farms.

A common feature of farms using alternative production methods is the integration of individual practices, such as weed control, into the overall management of the farm. For example, controlling weeds by rotations, cover crops, and cultivation complements fertilization, erosion control, and animal forage and feed requirements. This method of weed control also reduces pest problems caused by certain insects, plant pathogens, and nematodes. Farm size can limit the practicality of some of these methods on some crops, however. To switch from herbicides to these methods, sole operators of large row-crop farms (over 2,000 acres) will have to hire farm help with new knowledge, change cropping patterns, and acquire new management skills.

The Spray brothers plant 400 acres in a corn-soybeans-small grain-red clover hay rotation. They have not used herbicides for 15 years. They attribute this success to a program of rotations, tillage, cultivation, rotary hoeing, and the timing of planting. Many farms with successful alternative production systems, such as the BreDahl, Thompson, and the Kutztown farms, use a similar combination of techniques to control weeds.

The Sabot Hill Farm, on the other hand, has adopted a unique approach to weed control. The operators were faced with a $26,000 annual bill to control Johnsongrass with herbicides on 500 acres of corn and soybeans. They decided to alter their enterprise from cash grain farming to forage production, incorporating the Johnsongrass into a mixed forage hay crop. The objective of the operation was changed to maximize food and feed output with a minimum of purchased inputs. In the process, a problem weed requiring costly annual control was converted into a crop. Corn and soybeans are now grown on only 325 acres, and weed control expenses have been cut to $6,000. The hay crop is fed to their own livestock and sold to area farmers.

The Kutztown Farm controls weeds almost exclusively through rotations and cultivation. Corn is usually rotary-hoed once and then cultivated several times. No specific weed control measures are needed on any of the small grain crop acres. However, wet seasons and muddy conditions can interfere with cultivation, resulting in severe weed problems. This is a characteristic difficulty of weed control based on traditional cultivation techniques. Animal manure that is used as fertilizer can also contain weed seeds and may increase weed problems, a problem that has occurred on the Kutztown Farm. It illustrates the need to continually assess, refine, and take advantage of the interactions of farm management practices—in this case, sustaining fertility through manure applications relative to the ease of weed control.

The Thompson Farm has successfully addressed the shortcomings of cultivation-based weed control strategies with a combination of modified ridge-tillage planting, rotations, and cultivation. Conventional or modified ridge-tillage planting generally makes cultivation for weed control more

A cultivator kills weeds between rows during the first cultivation of ridge-tilled corn on the Thompson farm. To be effective, cultivation must be done at the proper time in the grow- ing season and with care to avoid damage to growing plants. *Credit:* Dick Thompson, the Thompson Farm.

effective and provides far more control than cultivation without ridge tilling (see the Thompson Farm case study). The Thompsons also include small grains and a hay crop in the rotation with corn and soybeans. This rotation appears to provide additional weed control through competition, allelopathy, or a combination of the two.

The Thompsons' per acre costs of production for corn are $96.20 less than the costs for the production practices reported by Iowa State University (ISU); the Thompsons' costs for soybeans are $44.45 less per acre (see the Thompson Farm case study). Eighty dollars of the savings for corn production are from reduced fertilizer use and the elimination of herbicides. The Thompsons' average per acre yields are 140 bushels of corn and 50 bushels of soybeans—well above the county average of 124 bushels of corn and 40 bushels of soybeans. The Thompsons also derive a slight savings in fuel costs from fewer (seven or eight) trips across the fields. ISU reported nine trips using conventional methods.

While per acre yields on the Thompson Farm are high and the per unit costs of production are low, the overall effect on net farm income cannot be simply derived from these figures. Rotations provide much of the weed control, plant nutrients, and input cost savings on the Thompson Farm. However, the Thompson Farm rotation reduces total acreage planted with cash grain crops over their 5- and 6-year rotations. Acreage planted with

Rotary hoeing is another weed control practice. The tightly spaced blades of the hoe spin and cut weed seedlings just below the soil surface. This tillage tool is used very early in the season, before crops grow more than a few inches. The hoe moves over the crop rows, dislodging weeds that have just germinated and are shallowly rooted. The deeper-rooted crop seedlings recover quickly. *Credit:* Dick Thompson, the Thompson Farm.

forage or hay crops during part of the rotation generally earns less gross income than cash crops. Accordingly, the economic performance of the Thompson Farm or other farms using rotations must be evaluated over the full life of the rotation. This rule is true when comparing whole-farm costs, returns, and profits per acre or per unit harvested.

To evaluate the effectiveness of ridge tillage for a conventional corn and soybean enterprise, the Thompsons have analyzed three systems of weed control for the common corn and soybean rotation: ridge tillage without herbicides, ridge tillage with preplant application of the herbicide metolachlor, and conventional tillage without herbicides. The performance of conventional tillage without herbicides has clearly demonstrated why farmers have so widely adopted herbicides. On these plots, the weather has often interfered with cultivation; weed infestation has increased over time and yields have declined.

In contrast, the two ridge-tillage systems have resulted in similar yields. Significantly, however, broadleaf weed infestation has increased in the herbicide-treated area. But in the absence of herbicides, there has been no increase in weeds (see the Thompson Farm case study). Ridge tillage is also a form of reduced tillage that provides other benefits, such as reduced erosion and a warmer seedbed for more rapid germination. It is clear that the economic and environmental advantages of controlling weeds through ridge tillage planting may be of significant value to many midwestern corn and soybean farmers.

Quantifying the Benefits of Pesticides

Under the Federal Insecticide, Fungicide and Rodenticide Act (FIFRA), the U.S. Environmental Protection Agency (EPA) must weigh the benefits and risks of pesticides. The agency is required to determine whether the risks presented by each use of a pesticide are reasonable in light of associated benefits. Accordingly, pesticide regulatory assessments are an important source of information on the economics of chemical pest control strategies.

Since the reform of FIFRA in 1972, the EPA has issued a number of formal policy statements describing acceptable methods for assessing human health risks from exposure to pesticides. The scientific basis for risk assessment has evolved steadily over the past 17 years. There is, moreover, a great deal of research under way in the public and private sectors to identify the toxicologic potential of pesticides. In contrast, considerably less effort is directed toward estimating pesticide benefits. Neither the EPA nor the USDA has developed a set of formal guidelines for calculating the benefits of pesticides under regulatory review. Nor are data on pesticide efficacy, an ingredient in any benefit assessment, routinely gathered and reported to the EPA. Benefits calculations for pesticides often employ different methods and assumptions.

Pesticides that meet the EPA's risk criteria may be subject to a special EPA regulatory review. Formal benefits assessments are conducted only during this review and do not generally contain detailed economic analyses of alternative nonchemical or IPM strategies (U.S. Congress, 1988; U.S. Environmental Protection Agency, 1982, 1985, 1986). The effect of this practice is to assume that the economic value of nonchemical or integrated control strategies is near zero. Consequently, benefits assessments tend to overstate the economic benefits of the individual pesticide under review as well as the impact of pesticide cancellation (U.S. Congress, 1988).

The thoroughness and quality of benefits assessments under FIFRA are an important public policy issue in the context of the economic and regulatory incentives or disincentives for adoption of alternative agriculture. Alternative production systems generally reduce reliance on pesticides and hence reduce the benefits associated with their use. As alternative production systems and nonchemical control options are developed, refined, and more fully incorporated into pesticide risk and benefit assessments, the balance between acceptable risks and benefits is likely to change. In turn, more pesticides may be subject to regulatory restrictions aimed at reducing risks. Such actions will most likely create further economic incentives for farmers successfully using nonchemical pest control methods. It may be necessary, however, to retain some uses of more hazardous compounds to control occasional outbreaks of certain pests. For this purpose a prescriptive use category for pesticides used in IPM programs could be developed.

While IPM and other alternative systems often require fewer pesticides on a per acre basis, pesticides will remain routine and occasionally invaluable production inputs in most crops for the foreseeable future. Progress toward wider adoption of alternatives, however, will continue to raise methodological issues for pesticide risk-benefit balancing. Alternative systems typically reduce reliance on pesticides through a complex combination of practices, including land-use decisions, rotations, cultural practices, selection of genetically resistant cultivars, and IPM programs. Benefits assessment techniques must be developed to take all these factors into account.

The EPA and the USDA should jointly develop and formally adopt a set of improved procedures for assessing the economic value of pesticides in the context of risk-benefit decision making already required by federal law. The benefits of a pesticide should be characterized as the difference between the total value of harvested commodities and the total value of the same crop using the next best alternative, which may involve an alternative cropping sytem that requires little or no pesticide use. Consideration of the costs of health and environmental risks of pesticides should be included in these analyses.

The Economics of Biological Methods of Pest Control

Most insects, pathogens, and other pests are kept from reaching damaging levels by natural enemies (see Chapter 3). Manipulations of a crop or populations of its natural enemies are important biological methods of pest control. Scientists have identified successful and cost-effective biological methods of control for many crops, typically by breeding resistant varieties, augmenting natural enemies, or introducing new predators or parasites. Nonetheless, a wide range of crop pests remain virtually impossible to control without the use of pesticides, particularly in certain regions and when farmers do not use crop rotations. Some remain largely uncontrollable even with pesticides. As with many alternative practices, a broader range of biological control options and techniques are possible within diversified agricultural ecosystems (see Chapter 3). There is a need for research on the specific effects of diversification in crop systems on pest populations and biological methods of control.

Several studies have examined the economic impacts and cost-benefit ratios associated with the development and dissemination of natural biological controls (Osteen et al., 1981; Reichelderfer, 1981). Reichelderfer (1981) lists six factors important to the success of biological control strategies.

1. The target pest consistently occurs, causes light or moderate damage, and is the major pest species of a high-value crop.

2. The biological control agent is effective and relatively risk free. Its effect on the pest population is not highly variable.

3. The price of the biological control agent, if it is marketed, is low. The research costs to develop the agent are justified by the economic impact of the target pest.

4. The biological control use or enhancement costs or both are low. Low costs can directly result from ease of use or be a function of economies of scale realized from the applicability of the method over large areas of use.

5. By net benefit criteria, the biological option compares favorably with available nonbiological control alternatives. This can be the result of its lower cost or its greater effect on yield or both.

6. Institutional arrangements exist or can be made easily to facilitate regional implementation, if necessary.

Many biological pest control techniques can be used in IPM systems. These include the use of pest predators or parasites, selection of pest-resistant plant cultivars, use of insect pheromones, release of sterile males, immunization of host plants, and use of bacterial insecticides. Some successful efforts using biological techniques are listed in Table 4-8. These include insect control by other insects, plant disease control by viruses, reproductive suppression by release of sterile males, and disease and insect control by the breeding of resistant strains. More than 100 host-specific insects have been introduced for the control of weeds. Plant-feeding insects or pathogens now partly or completely control at least 14 weed species. Currently, however, the viability of such techniques in agriculture is extremely limited because plant-feeding insects often damage crops.

Even though the process of developing an effective biological technique is sometimes expensive, the ratio of monetary benefits to costs can be very high (Batra, 1981). A study of the effects of introduction of six parasitic species for biological control of alfalfa weevils in the 11 northeastern states found that 73 percent of the alfalfa acreage in the region no longer requires the use of insecticides for protection against alfalfa weevils. This acreage was expected to increase as the six species become more prevalent in the region. More than a dozen additional natural enemies have been released; the incidence of reports of severe weevil infestation has declined steadily since 1962 (Day, 1981). This is one of the most successful examples of biological control of pests in crops.

A study of the potential economic impact of the introduction of the parasite *Bathyplectis curculionis* for biological control of alfalfa weevils in the eastern half of the United States concluded that $44 million per year could be saved in reduced crop loss and expenditures on insecticides. Insecticide use could be reduced by 1,100 tons. Total production of alfalfa would increase by only 1 percent as a result of biological control, thus avoiding any serious impacts on the market price of alfalfa. The greatest reductions in yield loss and insecticide application were estimated in the southern states, where insect pests are more severe because of warm winters (Zavaleta and Ruesink, 1980).

Classical plant breeding to develop new varieties is the most successful

TABLE 4-8 Selected Examples of Biological Controls

Component	Strategy		
	Regulation of the Pest Population	Exclusionary Systems of Protection	Self-Defense
Pest agent used against itself	Pheromone gossypol to control pink bollworm in cotton in Egypt, South America, and the United States; Sterile males to control screw-worm in the United States; Mosquitoes genetically incapable of vectoring the malaria agent used to displace capable types[a]	Avirulent strain K-84 of Agrobacterium for control of crown gall on fruit trees and ornamental plants in several countries; Ice-minus strains of Pseudomonas syringae to exclude ice-nucleation-active strains from leaves of frost-sensitive plants[a]	Mild strains of citrus tristeza virus to protect citrus against virulent strains of the virus in Australia and Brazil; Resistance to tobacco mosaic virus (TMV) in tobacco plants genetically engineered to express the coat-protein gene of TMV[a]
National enemies and antagonists (classic biological control agents)	Wasps for control of the alfalfa weevil in the United States; Predatory snail for control of snail vector of schistosomiasis agent in Puerto Rico; Puccinia rust for control of skeleton weed in Australia and the United States; Bacillus thuringiensis for control of certain caterpillars—used worldwide	Phlebia gigantea applied to pine stumps to exclude the pine root-rot fungus Heterobasidion annosum; Nonpathogenic Lactobacillus strains used to exclude Escherichia coli from the intestinal lining and protect piglets against neonatal scours[a]; Toxin gene from B. thuringiensis expressed in Pseudomonas on corn roots for protection against certain soil insects[a]	"Immunization" (induced resistance) of cucumbers and other plant species against Colletotrichum (anthracnose) by inoculating their leaves with tobacco necrosis virus[a]; Toxin gene from B. thuringiensis expressed in tobacco leaves for control of certain leaf-feeding caterpillars[a]
Host plant or animal	Crotalaria grown as a trap plant; root-knot nematode infects this plant but does not reproduce—minor use in the United States	Dense sowings of cereal-grain crops to preempt the establishment of weeds—used worldwide	Genetic resistance to southern corn leaf blight in corn in the United States; Genetic resistance to Hessian fly in wheat in the United States

[a]Experimental stage only.

SOURCE: National Research Council. 1987. Biological Control in Managed Ecosystems. Pp. 55-68 in Research Briefings 1987. Washington, D.C.: National Academy Press.

biological method of pest control. Genetic engineering promises to accelerate breeding for pest resistance. Disease and insect resistance have been
bred into many major grain and field crops, often with significant economic
payoffs. Federal, state, and private agencies spent $9.3 million developing
wheat resistant to the Hessian fly and wheat stem sawfly, alfalfa resistant
to the spotted alfalfa aphid, and corn resistant to the European corn borer.
Benefits to farmers in increased yields are estimated at several hundred
million dollars annually, not including savings from reduced pest control
expenditures (National Research Council, 1987c).

Many of the committee's case study farms use some type of biological
control as a strategy in highly successful pest management programs. Generally, these control methods are just one feature of an integrated production system. The Ferrari and Pavich case studies describe a high degree of
classical or natural biological control; the Thompson case study provides
insights into novel and highly successful strategies to manage weeds.

The Pavich Farm operation grows grapes on about 1,125 irrigated acres in
California and Arizona. A combination of complementary biological and
cultural practices control weeds, insects, mites, and diseases. Weeds are
controlled with a permanent cover and occasional mowing of perennial rye
grass and the occasional use of hand labor to pull weeds from among the
vines. The Paviches believe that this permanent ground cover also provides
the necessary habitat for many beneficial predators and parasites that feed
on potentially damaging pests. Soil fertility was maintained until recently
with a permanent leguminous cover crop and now is accomplished with
the application of compost. Some evidence suggests that the compost is
also helpful in controlling rootknot nematodes, although additional research is needed to understand more fully how compost affects nematode
populations.

The Paviches recently removed the legume cover crop from the vineyards
because too much nitrogen was being fixed. This caused excessive foliar
growth, which shaded the berries and provided a favorable pest habitat.
Trace elements are applied aerially as a foliar spray at least once a year or
on occasions when pest infestations are high, in the belief that they improve
plant health and the ability to fight pests. High levels of calcium relative to
nitrate are maintained in the soil, in the belief that this also reduces disease
and mite populations.

The Paviches cite expert vine dressing, proper soil nutrient balances, a
permanent ground cover supporting a population of beneficial insects, and
the advice of qualified field entomologists as practices that effectively control insects, mites, and diseases. The Paviches rarely use insecticides. In
1986, they applied none to their Arizona vineyard and made one application
of the broad-spectrum insecticide methomyl to 23 percent (142 acres) of
their California crop. The Arizona vineyard is relatively isolated, which
reduces pest pressure and the immigration of secondary pests, such as
spider mites, from other fields. The California operation, on the other hand,
is surrounded by vineyards that are regularly treated with pesticides for
infestations of leafhoppers and spider mites. Nonetheless, the Paviches

rarely spray for leafhopper and have only sprayed once on 40 acres in 15 years to control spider mites. The Paviches use no synthetic chemical fungicides to control grape diseases; however, sulfur is applied several times per season for this purpose.

The Paviches do not report pest control costs in a way directly comparable to the University of California (UC) accounting system applied to neighboring grape operations. The committee estimated production costs to be roughly equivalent at $2.14 per box for the UC system and $2.20 for the Paviches. The yield on the Pavich operation, however, is well above the UC average, at 653 boxes versus 522 boxes per acre. While the effectiveness of the Pavich system in profitably producing high-quality grapes with little or no pesticide treatments is evident, how and why the system works remains largely unknown. Research essential to understanding the interactions of the important cultural, pest control, and other practices on the Pavich Farm or similar farms has never been undertaken. As a result, agricultural research and extension personnel are not yet able to identify how and under what circumstances other producers might successfully apply some or all of the methods used on the Pavich Farms.

The Ferrari Farm produces conventional and certified organic tree fruits in California. The organic acres use a variety of biological control methods, including pheromones to control oriental fruit moths, the codling moth granulosis virus (CMGV), and the release of predacious mites. These methods effectively control the oriental fruit moth and mites. CMGV costs about the same as conventional miticide applications at $25.00 per acre. For the Ferraris, CMGV is effective in apples. It holds damage to between 1 and 3 percent, although more applications are necessary than with a conventional insecticide. CMGV control has not been effective in walnuts, however, although refinements in application technology are expected to improve control.

The breeding and use of pesticide-resistant predacious mites in almond groves has also been very successful. California almond growers have adopted the use of these mites, developed at the University of California at Berkeley, with great economic benefit. The per acre saving for all aspects of control using predacious mites versus chemical control is calculated at $34.00 per acre. Approximately 67 percent of all almond growers in California now use these beneficial mites, with an estimated economic benefit expressed as a net value of about $24 million. The ratio of agricultural benefits to research cost is estimated at about 30 to 1 (Hoy, 1985). Genetic engineering techniques promise to provide plant breeders with important new tools to breed genetically resistant crop cultivars and beneficial insects for use in biological systems.

In spite of these and other successes, biological control remains underresearched and underused relative to its potential—even as many economically important pests, notably soil-borne pests and insect pathogens, are not effectively controlled by chemical or other means in many regions and important crops (National Research Council, 1987a, 1987b).

One reason for this lack of support is that the availability of relatively inexpensive, effective pesticides has clearly dampened interest in biological control. Another constraint has been the sporadic nature of publicly funded research and education efforts toward the adaptation and implementation of biological control systems in the field. Biological control is dependent on public sector research and development for two principal reasons. First, some effective control techniques decrease the need for purchased inputs and, hence, undermine future commercial market potential. Ironically, the more effective and long term a biological system of pest control is, the more difficult it can be to interest a private company in making the necessary investments in bringing the product to the marketplace (Booth, 1988). Second, biological control research is often location and management system specific. Effective biological control systems must be carefully researched and tailored in light of seasonal weather patterns, crop conditions, and pest population trends and interactions.

Public funds for the development and delivery of biological pest control products or systems to growers are often lacking, as are funds to adequately assess conditions on individual farms. Private and public research and development expenditures for chemical control technologies in the United States have been estimated as at least five times greater than those spent for biological control (National Research Council, 1987b). As a result, scientific opportunities to research new biological control methods remain largely unexploited. In general, a relatively modest effort has been made to fully use those biological control systems that have been discovered.

Public and private sector collaboration is needed to improve delivery of successful experimental results to growers (National Research Council, 1987a). The Technology Transfer Act of 1986 is a constructive step toward this goal. It will encourage the research and marketing of biological pest control products at public research institutions by sharing profits more equitably from the commercialization of products flowing from public sector research. In the first year under this law, the Agricultural Research Service (ARS) received 28 licenses valued at $33 million.

ALTERNATIVE ANIMAL DISEASE PREVENTION STRATEGIES

There are two principal costs associated with animal disease. The first is that associated with decreased production. Death or losses in milk, body weight, or eggs sharply reduce or eliminate the profitability of the sick animal. Second, the cost of treatment is incurred. This cost can be compounded by ineffective treatments and the recurrence of disease. Because of growing market awareness of animal disease, animal welfare, and the potential for drug residues, there may be additional costs for market discriminations, incentives and disincentives, and regulatory fees. In addition, the new trace back provisions of the Food Safety Inspection Service monitoring may improve enforcement of food safety regulations.

Animal death and disease losses cost billions of dollars each year. Deaths of beef and dairy cattle are estimated to cost $4.6 billion. This figure has not changed significantly in 10 years. For the dairy industry alone, losses from the mammary disease mastitis are estimated at $180 per cow or $2.0 billion annually (National Mastitis Council, 1987). These losses arise from medical costs, decreased milk production, and death. Losses due to respiratory infections in cattle and swine are estimated at $800 million per year (National Research Council, 1986a). When these costs are combined with the losses that result from the condemnation of animals and the discounting of price for moribund, poor-quality animals, the cost of animal disease increases further. Production losses from most chronic and subclinical diseases of most food-producing animals, however, are virtually impossible to measure.

Veterinarians, universities, drug companies, and regulatory agencies generally address animal health by treating infected herds or animals primarily with prophylactic feeding of and therapeutic treatment with antibiotics. Disease management systems not reliant on antibiotics are not widely used, although they are being increasingly emphasized in research. Producers generally seek advice on how to treat a clinically diseased animal instead of how to manage for lower disease incidence. Veterinarians who are able to charge clients for their treatment services and public and private research reinforce this approach to animal health. Although recent research has focused on management systems to reduce disease, the current animal health system reflects the fact that until lately, universities tended to research causes and cures to satisfy the demands of producers. Regulatory agencies continue to approve drugs with little consideration of the costs and benefits of alternatives. Because intensive confinement facilities generally increase the risk of disease, the animal industry appears, at best, to be holding its own in terms of combating disease (Council for Agricultural Science and Technology, 1981). Yet, in almost every instance of clinical animal disease, it is more cost-effective to prevent the disease than to attempt to treat it.

As a result of research that links udder health to economic criteria, the dairy industry is currently evaluating the economic impact of mastitis prevention and control. For example, mastitis may cause an $18,000 loss in a typical 100-cow herd (National Mastitis Council, 1987). Two simple alternative approaches to mastitis control, postmilking teat disinfection and comprehensive nonlactating therapy, may provide excellent health and cost approximately $1,000 per year. An additional $1,000 per year should be included for labor and management. In a 2- to 3-year period, these methods can reduce mastitis losses by more than 95 percent. These rough approximations indicate that preventative mastitis control expenditures have a cost-benefit ratio that exceeds 8 to 1 (Barbano et al., 1987). The producer will receive additional benefits through premiums for milk quality. The processor will realize greater returns in cheese yields. The consumer will have a product with greater refrigerated shelf life and greatly reduced potential for antibiotic residues (Barbano et al., 1987).

Considerable systems research is needed in other diseases and species to determine the costs and benefits for subclinical and chronic disease prevention. Even though the marginal cost for disease prevention will exceed the marginal benefits at some point, most of the industry is far from that point. The industry can invest in disease prevention with confidence that it would be cost-effective. Major advances will have to be made in subclinical disease monitoring and modeling, however, before the more thorough cost-benefit analyses necessary to convince producers of the economic benefits of disease prevention will be available. Support for this work will have to come from public research institutions and the Congress. Veterinary health maintenance organizations may also provide an alternative economic and health maintenance philosophy needed to reorient producers from disease treatment and its attendant costs and risks to one of disease prevention.

Alternative Animal Production Systems

A number of studies have documented the profitability and productivity of alternative animal production systems. These systems are generally characterized by less confinement of animals, greater use of pastures, a lower incidence of disease, and, consequently, less use of antibiotics.

Controlled-environment systems that typically involve confinement of animals in stalls, pens, or cages are widely used in the poultry, pork, and veal industries. Intensive animal production tends to have performance characteristics similar to intensive crop production. Capital, technology, and chemicals are substituted for labor and management, resulting in systems that are productive and profitable under favorable economic conditions but more vulnerable to routine fluctuations in input and output prices. Intensive systems also present greater potential health and environmental hazards (Kliebenstein and Sleper, 1980). Confinement systems are further criticized on animal welfare grounds and because animals in confinement usually exhibit greater incidence of disease (Friend et al., 1985; Kliebenstein et al., 1981). The subsequent treatment with antibiotics creates additional costs and may contribute to antibiotic residues in animal food products.

The principal advantages of controlled-environment systems are that they permit a larger operation and greater control of the animal's environment and feeding so that more of the animal's feed intake is converted into body weight. In most species, these systems are the most efficient in converting feed to body weight (Table 4-9) (Lidvall et al., 1980). On the other hand, confinement systems require great capital investments to construct and generally involve higher maintenance and medical expenses. Before the Federal Tax Reform Act of 1986, capital-intensive animal confinement facilities or single-purpose agricultural structures enjoyed important tax advantages, including tax credits and rapid depreciation. These advantages helped to defer the expenses of the construction of these facilities, making them more affordable. The Federal Tax Reform Act of 1986 eliminated investment tax credits for single-purpose agricultural facilities. The act also lengthened depreciation for these facilities to 5 to 7 years. The depreciation period was

TABLE 4-9 Comparative Performance of Pasture With Hutch, Partial Confinement, and Total Confinement Swine Production Systems

Performance Indicator	Swine Production System		
	Pasture With Hutch	Partial Confinement	Total Confinement
Number of sows/system	29.5	63.6	105.8
Conception rate (percent)	82	81	78
Litters/sow/year	1.67	1.68	1.97
Average live farrow/litter	10.2	10.0	9.8
Average number weaned/litter	7.7	7.7	7.6
Percentage of live births	75	77	78
Wean weight (pounds)	33.5	30.0	18.6
Average daily gain during finish period (pounds)	1.40	1.29	1.46
Market weight (pounds)	230	220	216
Days to 230 pounds	205	215	210
Total feed required to produce 1 pound of pork (pounds)	4.16	4.21	3.87

SOURCE: Lidvall, E. R. 1985. A Comparison of Three Farrow-Finish Pork Production Systems. Knoxville, Tenn.: University of Tennessee.

extended to 10 years in 1988. Controlled-environment confinement facilities are also thought to require less labor than pasture or low-confinement systems. Several studies have shown, however, that as these facilities and equipment age, labor costs for repair and maintenance increase, and total labor costs for these systems can equal the costs of alternative systems (Killingsworth and Kliebenstein, 1984).

Alternative animal production systems have long existed. These modifications of traditional animal husbandry systems have been refined to take advantage of current knowledge of animal nutrition and health care. Several major analyses of confinement versus pasture and hutch systems for swine have shown that confinement and pasture systems produce relatively equivalent returns (Kliebenstein and Sleper, 1980; Lidvall et al., 1980). Nine years of data from an ongoing comparison of pasture with hutch, partial-confinement, and total-confinement hog production facilities in Tennessee are summarized in Table 4-9.

Confinement facilities for swine production generally result in greater feed efficiency and the greatest return per unit of labor. Herds are often larger and produce more litters per year, thus producing greater gross income. Pasture or low-confinement systems, in contrast, require less capital investment and provide the highest and most consistent returns per unit of input. They provide the highest returns when livestock prices are low or feed prices are high. This consistency of return is an important consideration in the long-term viability of these systems and their effect on net farm family income. Low-confinement systems usually provide the greater return per animal for all types of swine operations (feeder or farrow to finish). The animals generally exhibit less disease than those in total confinement facilities (Kliebenstein and Sleper, 1980; Lidvall et al., 1980).

Most poultry and egg production facilities in the United States are under controlled-environment and confinement conditions. The tight caging of birds allows more controlled feeding, climate, and production and decreases space and labor costs. The day and night cycle of modern egg production facilities is altered to as much as 22 hours of light per 24-hour period to increase production. Lighting is sometimes dimmed to reduce fighting aggravated by close caging. In contrast, alternative poultry or egg production systems generally do not cage the birds and usually permit uncontrolled access to feed. Although alternative production systems are often profitable, these systems are relatively few in number because of the drive for uniformity in the vertically integrated poultry and egg industry. Animal science research at land-grant institutions has reinforced this trend, with little funding directed toward the understanding of alternative production systems.

STUDIES OF DIVERSIFICATION STRATEGIES

The trend toward more specialized, high-yield agricultural production systems is well established and reflects many technological and socioeconomic factors that are firmly embedded in recent history and agricultural policy. Alternative farming systems, particularly for farms producing coarse grain and oilseed crops, small grains, and forages, generally depend on crop rotations and a number of other diversification strategies. These strategies are often contrary to the specialization and intensification characteristic of most agricultural operations. Federal commodity programs have accelerated specialized production by greatly reducing the risks of producing only one or two crops. Today's highly specialized farms would not be possible without federal program subsidies. Diversification, a basic alternative concept, also reduces risks by spreading risks among a number of crops and animals. The result is more consistent overall farm yield among a number of crops and less need for federal income support. However, the precise extent of diversification's effects deserves further study.

Integrated Crop and Livestock Systems

The most common diversification strategy remains the combination of crop and livestock enterprises. Many studies have documented the agronomic and economic benefits associated with the interaction of cropping and livestock enterprises on diversified farms (Heady and Jensen, 1951). Further evidence of the potential for positive interaction is contained in the Spray, BreDahl, Sabot Hill, Kutztown, and Thompson case studies.

Recent research has provided insight into whether livestock must be included as a farm enterprise in order to attain the economic benefits of crop-livestock interactions, particularly those related to soil fertility manage-

Diversification is a basic alternative strategy that can reduce production costs and help protect natural resources. Crop rotations in combination with contour strip cropping and conservation tillage can reduce erosion. Legumes on this Lancaster County, Pennsyl- vania, farm fix nitrogen for next season's wheat or corn crop. The triangular field hutches are used by hogs or calves. A nursery is shown in the upper right corner. *Credit:* Grant Heilman.

ment. Analysis of a diversified crop-livestock farm in Pennsylvania esti- mated that, at 1978 to 1982 prices (which were relatively high), profits would have increased with liquidation of the beef herd, the purchase of manure at going market prices, and the sale of crops directly on the market rather than through the feedlot (Domanico et al., 1986).

Similarly, the Rodale Research Center conversion experiment (4 years of data from a 15-acre field with 72 replicated plots) indicated that although livestock manure could speed the conversion from chemical-intensive to reduced-input methods, comparable yields could be obtained without ma- nure after an adjustment period using legumes (Brusko et al., 1985).

Crop-livestock operations are well suited to the adoption of many alter- native practices. Crop rotations using cover crops, such as leguminous hays, are readily suited to livestock operations. These rotations reduce fertilizer and pesticide needs and provide a valuable feed source. Many legumes are quality hay crops. In crop-livestock operations, hay crops with a market value ordinarily lower than cash grains have economic value as a feed source in addition to their value as a source of nitrogen. Keeping a portion of a farm's land in a cover crop may provide additional erosion control benefits and allow the planting of feed grains on more suitable land. Manure also becomes a valuable source of soil organic matter, nitrogen, and other nutrients such as potassium and phosphorus. Diversified crop-livestock operations also have greater protection from input (feed) and output (animal product) price fluctuations.

Crop Diversification Strategies

Crop diversification methods, which include rotations, polyculture, intercropping, and double cropping, have been found to be profitable in many situations. The primary advantages of diversification include reduction or elimination of certain diseases and weeds, reduced erosion, improved soil fertility and tilth, increased yields as a result of rotational effects, reduced need for nitrogen fertilizer (in cases using legumes in crop rotation), and reduction of financial risks resulting from changing crop prices.

Interplanting different crops in a given field has been known to suppress leaf diseases in cereal grain crops and powdery mildew in wheat (Wagstaff, 1987). Many instances of the beneficial effects of polyculture have been reported, with documented reductions in insect damage and increases in crop yields and net returns (Dover and Talbot, 1987; National Research Council, 1987b). The mixed grass and native weed ground cover used by the Paviches provide another example of the benefits of polyculture diversification (see the Pavich case study). Disadvantages can include increased machinery requirements and expense (for example, when forage crops are added); need for additional buildings, fences, and watering facilities when livestock or poultry are added; increased complexity of the farmer's management of production and marketing; and reduction of acreage planted with government-supported crops.

Crop rotations, in particular, are proven and used successfully in various regions of the country. Polycultures, intercropping, double cropping, and other techniques are far less common but used with various degrees of success in certain areas. These practices all have potential benefits for agriculture. It is important, however, that data bases are developed to support their adoption. Most of these practices involve trade-offs and some may aggravate certain pest problems.

Legume-Based Crop Rotations

Legume-based rotations are one of the most common and effective diversification strategies practiced in U.S. agriculture. The total nitrogen fixed by legumes currently grown in the United States has a potential value of $1.6 billion (Heichel, 1987) compared to the $4.3 billion spent on nitrogen fertilizers in 1987 (National Fertilizer Development Center, 1988). Leguminous nitrogen could be even more valuable if legume management and fertilizer application practices were improved. The inclusion of leguminous forages in rotations served as a primary source of nitrogen until the 1950s, when low-cost nitrogen became widely available. Studies in progress for more than 100 years in Great Britain, Illinois, and Missouri have demonstrated the capacity of legumes in rotations to sustain high levels of grain production over long periods without the use of nitrogen fertilizers (Power, 1987).

Results of a series of rotation experiments conducted in various midwestern states during the 1930s and 1940s provided early evidence of the economic advantages of legume-based crop rotations (Heady, 1948; Heady and Jensen, 1951). Analysis of data from experiments in Illinois, Iowa, and Ohio found a greater total volume of grain was produced per acre using certain rotations including clover or alfalfa compared with continuous corn production. The net return over variable cost was calculated for each of the rotations under a variety of pricing assumptions. In most instances, continuous corn was found to be less profitable than rotations with legumes and grains, even when the forage was assumed to have no monetary value. For example, analysis of data from an experiment in Ohio determined that a rotation of corn-corn-corn-wheat-alfalfa (C-C-C-W-A) provided a 12.6 percent higher net return over variable cost compared with continuous corn (C-C) from 1937 to 1943, even when the value of forage was assumed to be zero. These findings were based on the prices and technology prevailing in the 1930s and 1940s, when pesticides, low-cost fertilizers (particularly nitrogen), and modern cultivars were not available, and government programs had a much less dramatic effect on crop prices and farmers' land-use decisions.

There are important regional distinctions associated with the use of legumes in rotations. The level of precipitation in a given area strongly influences the usefulness of legumes. Alfalfa and other legumes can dry out the subsoil to a greater depth than corn. Consequently, in arid and semiarid regions or during drought conditions in subhumid and humid regions, introduction of a deep-rooted legume in a rotation may suppress subsequent yields of corn or other crops. Whenever soil moisture is not a limiting factor, however, legumes in rotations with cash grains will increase yields and can supply some or all of the nitrogen needed by corn or small grains. During the drought years from 1933 to 1940, an experiment in Iowa involving continuous corn production estimated an average net return of $10.81 per acre (assuming 1940–1944 average prices), a return more than double the income earned by a corn-oats-clover (C-O-Cl) rotation (Shrader and Voss, 1980). During the more favorable weather of 1941 to 1948, however, the rotation earned $17.19 per acre compared to $0.42 per acre for continuous corn (Heady and Jensen, 1951).

Benefits from rotations in addition to nitrogen fixation have been noted in Chapter 3 and are referred to as rotational effects (Heichel, 1987). The

TABLE 4-10   Effect of Previous Crop on Corn Yield

| Nitrogen Rate (pounds/acre) | Previous Crop | | | |
	Corn	Soybean	Wheat	Wheat-Alfalfa
	Corn Yield (bushels/acre)			
0	70	109	108	115
40	92	129	127	127
80	102	137	139	139
120	108	142	139	143
162	116	148	142	146
200	119	148	145	144

SOURCE: Adapted from Lager, D. K., and G. W. Randall. 1981. Corn production as influenced by previous crop and N rate. Agronomy Abstracts. American Society of Agronomy. Madison, Wisconsin. p. 182. In Power, J. F. 1987. Legumes: Their potential role in agricultural production. American Journal of Alternative Agriculture 2(2):69–73.

rotational effect is the increase in grain or other crop yields following the planting of the field with another crop. Much of this increase in yield is thought to stem from the well-documented pest control benefits of rotations (Baker and Cook, 1982). Data reported by Power (1987) show that yields of a grain crop grown in rotation are 10 to 20 percent greater than those of continuous grain, regardless of the amount of fertilizer applied (Table 4-10).

A more recent study in southeastern Minnesota examined the nitrogen contribution and other benefits of legumes in a crop rotation with corn or soybeans or both (Kilkenny, 1984). A linear programming model projected results for a 400-acre farm with a 60-cow dairy herd. The study concluded that if the monetary value of the nitrogen fixed were ignored and if nitrogen fertilizer is assumed to be free, the most profitable cropping system in terms of current net returns over cash operating costs is continuous corn on about two-thirds of the acreage, with about one-third of the acreage in a 3-year corn-oats/alfalfa-alfalfa (C-O/A-A) rotation. As the assumed price of nitrogen increases, the profit-maximizing crop rotations feature increasing proportions of legumes. With an assumed nitrogen price of $0.115 per pound (the 1980–1982 price), a corn-soybean rotation was found to be more profitable than continuous corn. If the price of nitrogen increased to $0.69 per pound, however, the most profitable rotation shifts from corn-soybeans toward continuous soybeans on more of the acreage in combination with the 3-year C-O/A-A rotation. Continuous soybeans, however would probably not be sustainable because of disease—notably, brownstem rot.

An 8-year experiment conducted recently by University of Nebraska scientists compared 13 cropping systems, including rotation, using only manure for fertilizer and no herbicides or other pesticides. The crops, which included corn, soybeans, grain sorghum, and oats with sweet clover, were grown in various rotations and in continuous cropping systems. The results confirmed the findings of studies done in the first half of this century (Heady, 1948; Heady and Jensen, 1951) using more primitive cultivars and no synthetic chemical pesticides: rotations can produce higher yields per acre than continuous monocropping systems. Different fertilization regimes, including manure only, were found to have little impact on yields and profitability. The continuous cropping systems were found to require higher pesticide expenditures and be subject to greater year-to-year variation in yields and profits per acre compared with the various rotations (Helmers et al., 1986).

For specialized operations growing government-supported crops, off-farm-purchased inputs are a significant part of total variable input costs and total operating costs (that is, variable costs plus fixed costs such as insurance, overhead, and interest). In 1986, the national average pesticide and fertilizer costs per acre were 55 percent of total variable input costs for corn, 46 percent for grain sorghum, 40 percent for wheat, 49 percent for soybeans, and 37 percent for cotton. These two inputs accounted for 34 percent of total operating costs for corn, 30 percent for grain sorghum, 23 percent for wheat, 25 percent for soybeans, and 27 percent for cotton. On farms using rotations, these costs may be markedly reduced or even eliminated (see the Spray, BreDahl, Sabot Hill, Kutztown, and Thompson case studies).

Legume-based rotations are not without costs, however. There are the direct costs of establishing a stand as well as the opportunity costs of foregoing the production of higher-value cash grain crops in certain years of the rotation. The costs and returns of a leguminous rotation, therefore, must be calculated over the length of the rotation.

The Effect of Government Programs on Legume-Based Rotations

Comparing the profitability of a legume-cash grain rotation not enrolled in the federal commodity program with systems receiving federal per acre income and price support payments is complicated. Several provisions of the farm programs, notably the base acres and cross-compliance requirements, can impose significant economic penalties in terms of lost federal payments to farmers incorporating certain rotations into their operation. As a result, the economic and ecological benefits of rotations are often foregone because of the financial incentives and rules of the federal commodity programs.

These economic disadvantages seem to have been overcome by an alternative rotation studied in the Palouse area of eastern Washington (Goldstein and Young, 1987; Young and Goldstein, 1987). This rotation, called the perpetuating alternative legume system (PALS), featured the biennial legume black medic, which has been observed to reseed itself for as long as 30 years following establishment. The PALS rotation is 3 years: spring peas plus medic in year one, medic in year two, and winter wheat in year three (P/M-M-W) (Table 4-11). The only synthetic chemical applied during this rotation is an insecticide applied to the peas. The rotation controls almost all the weeds in wheat; harrowing during seedbed preparation provides adequate control of the rest.

TABLE 4-11   Estimated Fertilizer and Pesticide Use for Conventional
Management and PALS[a]

| Crop | Fertilizer (pounds/acre) | | | Pesticides | | | |
	N	P	S	Herbicide	Rate (units/acre)	Insecticide or Fungicide	Rate (units/acre)
Conventional[b]							
Winter wheat	130	30	25	Difenzoquat methyl sulfate	3.0 pints	Benomyl	1.5 pounds
				Bromoxynil	1.5 pints		
Spring barley	80	0	0	Triallate	1.25 quarts		
				Bromoxynil	1.5 pints		
Winter wheat	130	30	25	Difenzoquat methyl sulfate	3.0 pints	Benomyl	1.5 pounds
				Bromoxynil	1.5 pints		
Spring peas	0	0	0	Triallate	1.25 quarts	Phosmet	1.5 pounds
				Dinoseb-amine	0.8 pounds		
PALS							
Peas + medic	0	0	0	Triallate	1.25 quarts	Phosmet	1.5 pounds
				Dinoseb-amine	0.8 pounds		
Medic	0	0	0	0	0	0	0
Winter wheat	0	0	0	0	0	0	0

[a]Perpetuating alternative legume system. A low-input system with a three-year pea plus
medic-medic-wheat rotation with pesticides used only on peas.

[b]Four-year wheat-barley-wheat-pea rotation with fertilizer and pesticide inputs each year.

SOURCE: Goldstein, W. A., and D. L. Young. 1987. An agronomic and economic comparison of a
conventional and a low-input cropping system in the Palouse. American Journal of Alternative
Agriculture 2(Spring):51–56.

A more common crop sequence in the Palouse is a 4-year rotation of
wheat-barley-wheat-peas (W-B-W-P). In this rotation, it is necessary to use
two herbicides as well as a systemic fungicide application for each crop. The
pesticides applied to the peas include the same insecticide used in the pea
crop year of the PALS rotation. Fertilizer applied to the conventional 4-year
rotation includes 130 pounds of nitrogen, 30 pounds of phosphorus, and
25 pounds of potassium per acre. The barley receives 80 pounds of nitrogen
per acre. No fertilizer is applied in the PALS rotation.

Input costs per year are dramatically higher in the conventional system,
at $129.40 per acre compared with $56.82 per acre for the PALS system. The
majority of this difference is comprised of fertilizer and pesticide costs that
are $57.52 per acre greater for the conventional system (Table 4-12).

In contrast to input costs, annual crop yields were similar during 2 trial
years at three sites. PALS wheat yields averaged 62.6 bushels per acre
compared with 60.3 bushels on the conventional plots. The largest differences occurred during the drought of 1985; yields for the PALS experimental plots averaged 83 percent more than those of the conventional plots. In
1984, when rainfall was close to normal, the PALS wheat yields were 3
percent less than the conventional yields.

TABLE 4-12   Costs of Conventional and Alternative Rotations per Acre of Rotation per Year

Inputs	Costs/Acre (dollars)	
	Conventional (W-B-W-P)[a]	PALS[b] (P/M-M-W)
Fertilizers and pesticides (application and product)	72.52	15.00
Field operation (tillage, planting, and harvest)	45.44	35.00
Overhead and crop insurance	11.44	6.82
Total	129.40	56.82
Average yield of winter wheat (bushels/acre)	60.3	62.6

[a]Four-year wheat-barley-wheat-pea rotation with fertilizer and pesticide inputs each year.
[b]Perpetuating alternative legume system. A low-input system with a three-year pea plus medic-medic-wheat rotation with pesticides used only on peas.

SOURCE: Goldstein, W. A., and D. L. Young. 1987. An agronomic and economic comparison of a conventional and a low-input cropping system in the Palouse. American Journal of Alternative Agriculture 2(Spring):51–56.

In three out of four scenarios, including market price and government program price assumptions, the PALS rotation was equal or more profitable on a per acre basis than the conventional rotation. The conventional system is significantly more profitable than the PALS rotation only under high-yielding (good weather) conditions with government price supports (Table 4-13). Profits are greater in this instance primarily because a greater percentage of the acreage (75 percent of the total) produces government-supported crops. Under low-yielding conditions, the productivity of the conventional rotation is reduced to such an extent that, even assuming government support prices, the net income of the two systems is roughly equivalent. Assuming market prices and no government program payments or requirements, the PALS rotation is always more profitable.

IMPACT OF GOVERNMENT POLICY

Crops eligible for price and income supports are planted on more than 70 percent of the cropland in the United States. These include feed grains, wheat, cotton, rice, soybeans, and sugar. From 80 to 95 percent of the acres producing these crops are currently enrolled in federal programs. Dairy farmers also enjoy income protection through a price support program, import quotas, and marketing orders for milk. The marketplace has more of an influence on prices of other commodities such as fruits, vegetables, livestock, poultry, and hay and forage crops. However, many factors influence the supply and demand for these commodities as well as practices used to produce a crop. Grading and cosmetic standards, for example, are

TABLE 4-13   Gross Returns, Variable Costs, and Net Returns (dollars/acre of rotation/year) Under Conventional and PALS Management, High and Low Yielding Conditions, and Market and Target Prices, 1986

	Conventional[a]		PALS[b]	
	High Yield	Low Yield	High Yield	Low Yield
1986 Market prices				
Gross returns	176.00	136.00	118.00	93.00
Variable costs	129.40	129.40	56.82	56.82
Net returns	46.60	6.60	61.18	36.18
1986 Government target prices				
Gross returns	274.20	210.60	170.80	132.60
Variable costs	129.40	129.40	56.82	56.82
Net returns	144.80	81.20	113.98	75.78

[a]Four-year wheat-barley-wheat-pea rotation with fertilizer and pesticide inputs each year.
[b]Perpetuating alternative legume system. A low-input system with a three-year pea plus medic-medic-wheat rotation with pesticides used only on peas.

SOURCE: Goldstein, W. A., and D. L. Young. 1987. An agronomic and economic comparison of a conventional and a low-input cropping system in the Palouse. American Journal of Alternative Agriculture 2(Spring):51–56.

applicable to various fruit, vegetable, and meat products. These standards are basically designed to control supply and price of individual crops. Acreage reduction programs influence the amount of land available to produce hay crops; water pricing policies affect costs of production on irrigated crops. Trade policies here and abroad affect the flow of farm commodities into and out of the U.S. market.

Government price and income support programs can have significant unintended effects. During the early to mid–1980s, the programs tended to price U.S. exports out of highly competitive world markets because federal support prices (the loan rates) were held rigidly high during a period of declining world market prices. The programs have also encouraged surplus production of certain commodities by reducing risks. They have provided economic incentives for farmers to continue to grow certain crops, even in periods of surpluses. Over the years, the programs also have contributed to soil erosion and surface water and groundwater pollution by encouraging the cultivation of marginal lands and subsidizing excessive and inefficient use of inputs. Further, producers pay no price for offsite environmental consequences of production. In many parts of the United States, producers now routinely strive for higher yields than those profitable in the absence of government programs designed to reduce risk. In other areas farmers grow crops with a high risk of failure from weather or pest conditions because government programs absorb all of the risk.

The price support payment that a farmer receives per acre is based on the farm's historical yields, an average of the yield on supported crop acreage in the previous 2 to 5 years (frozen at 90 percent of 1985 program payment yield in the Food Security Act of 1985), and the target price established by Congress and the USDA through legislation. Deficiency payments per bushel of established yield are the difference between the target price and market price or support price (loan rate), whichever difference is less. For many crops, the target price has been far above the market price for most of this decade (see Figures 1-30 and 1–33 in Chapter 1).

High target prices can promote higher levels of inputs, thereby contributing to surplus production. This is illustrated by the theoretical example presented in Figure 1-30 in which a farmer will produce 19,000 bushels at the market price and 24,000 bushels at the target price (this example does not take into account annual set-aside requirements). It costs the farmer more to produce the additional 5,000 bushels than they are worth on the market. The additional 5,000 bushels cost taxpayers $10,000 in government payments ($2.00 per bushel × 5,000 bushels).

Commodity programs also influence which crops are planted and the economic and environmental impacts associated with land-use decisions. The cross-compliance provision of the Food Security Act of 1985 is designed to control production of program commodities by limiting a farmer's ability to increase base acres. It also serves as an effective financial barrier to diversification into other program crops, especially if a farmer has no established base acres for those crops. Cross-compliance stipulates that in order to enroll land from one crop acreage base in the program, the farmer must not exceed his or her acreage base for *any* other program crop. The practical impact of this provision is profound, particularly if a farmer's acreage base for other crops is zero. For example, a farmer with corn base acreage and *no* other crop base acres would lose the right to participate in *all* programs if *any* land on his or her farm was planted with other program crops such as wheat or rye (oats are currently exempt) as part of a rotation. If a farm had base acreage for two or more crops when cross-compliance went into effect in 1986, the farm must stay enrolled in both programs each year to retain full eligibility for benefits from both programs.

High government support prices also influence planting decisions. Throughout the 1970s, soybean prices averaged more than twice the corn target prices. In recent years, soybean prices have strengthened markedly in contrast to corn. Yet, soybean stocks have fallen to their lowest level in a decade, even though prospects for increased demand in the United States and abroad are very good. The total acres planted with soybeans are declining because of high government support payments for other crops, most notably corn. Moreover, considerable acreage is now producing corn because farmers must continue to plant their corn base every year to preserve their current level of eligibility for future corn program payments. Even though commodity prices may rise somewhat as a result of the 1988 drought, the programs will remain an attractive option to most growers. This is because target prices and deficiency payments are likely to remain substantial. Farmers have become more efficient, and interest and rents

TABLE 4-14 Average Annual Target Prices as a Percentage of Total Economic Costs

Crop	1978–1981[a]	1982–1985[b]	1986–1990[c]
Corn	95	110	141
Cotton	82	107	111
Rice	106	125	153
Soybeans[d]	82	85	91
Wheat	94	108	123

NOTE: Total economic costs cover all fixed and variable production costs for an operator with full ownership of the land and other capital assets.

[a]Crop years covered by the Food Security Act of 1977.
[b]Crop years covered by the Food Security Act of 1981.
[c]Forecasts under current legislation for crop years covered by the Food Security Act of 1985. Minimum target prices for grains and cotton and the minimum soybean loan rate under the Agricultural Act of 1949, as amended, were assumed for 1988–1990.
[d]Soybean loan rate as a percentage of soybean total economic costs.

SOURCE: U.S. Department of Agriculture. 1988. Investigations of Changes in Farm Programs. 201-064/80069. Washington, D.C.

have declined, making deficiency payments even more valuable (Table 4-14).

The Effect of Rotations on Base Acres and Federal Deficiency Payments

For farms currently participating in commodity programs, the transition from continuous cropping to rotations will decrease gross farm income by reducing a farm's acreage base eligible for federal deficiency payments. The magnitude of this reduction depends on the size of the deficiency payment. Table 4-15 illustrates the reduction in deficiency payments due to the loss of corn base acres resulting from the adoption of a corn-oats-meadow-meadow (C-O-M-M) rotation. When complete, the change from continuous corn to a C-O-M-M rotation on 1,000 base acres would cost this farm about $90,000 per year in deficiency payments. Overall farm income, however, depends on a number of factors, including the market for new crops, incorporation of livestock into the operation, the possible increase in corn yield, and the type of rotation adopted. Nonetheless, the loss of current and future income from ineligibility for government programs presents a significant obstacle to the adoption of alternatives.

The previously discussed PALS studies of wheat farms in Washington and additional work on cash grain farms in Iowa further illustrate the strong economic influence of the target price and base acres provisions of the farm programs. Almost no pesticides or fertilizers were used in the PALS rotation. This reduced variable production costs per acre to about half that of the conventional rotation, or $56.82 versus $129.40 per acre (Goldstein and

TABLE 4-15   Reduction of Deficiency Payments and Corn Acreage Base Following Change From Continuous Corn to C-O-M-M[a] Rotation on 1,000-Acre Farm

Years Since Adopting C-O-M-M Rotation	Corn Base (acres)	Corn Planted (acres)	Set-Aside[b] (acres)	Corn Yield (bushels/acre)[c]	Deficiency Payments[d] (dollars)
0	1,000	800	200	147	142,296
4	550	250	110	173	52,332
8	250	250	50	173	52,332

[a]Represents a corn-oats-meadow-meadow rotation.

[b]Assumes 20 percent corn base set-aside.

[c]Based on Duffy, M. 1987. Impacts of the 1985 Food Security Act. Ames, Iowa: Department of Economics, Iowa State University.

[d]Corn production times 1987 deficiency payment ($1.21/bushel), ignoring the statutory $50,000 limit on payments.

Young, 1987) (see Table 4-12). Wheat yields were nearly identical. PALS reduced pea yields about 10 percent from the conventional rotation yields, however, because of competition with the medic.

The high support price for wheat greatly affects the comparative profitability of PALS and conventional rotations. When the revenue from sale of all crops in the rotation was based on government deficiency payments, favorable growing conditions, and subsequent high yields, the conventional rotation earned $144.80 per acre, compared with $113.98 per acre for the PALS rotation. These figures assumed 1986 target prices for wheat and barley that were 45 and 35 percent higher than market prices, respectively. But when market prices were used in calculating net returns, the positions were reversed. The PALS rotation returned an estimated $61.18 per acre over variable costs versus $46.60 for the conventional rotation (see Table 4-13).

The cause of the disparity in net returns is that the PALS rotation produced wheat, a price-supported crop, on only one-third of the acreage each year. PALS wheat yields averaged 62.6 bushels per acre, whereas conventionally produced wheat yields averaged 60.3 bushels per acre. The conventional rotation, however, produced program crops on 75 percent of the acreage each year (2 years of wheat, 1 year of barley in a 4-year rotation). But when less favorable growing conditions were assumed, the net returns of the conventional rotation declined dramatically, even assuming government price supports. Under government support and less favorable weather conditions, PALS earned only $5.42 less per acre than the conventional rotation.

An analysis of five rotations in Iowa reached similar conclusions. Without government payments, continuous corn was found to be the least profitable of the rotations at $56.00 per acre average net return over variable cost compared with $90.00 for a corn-soybeans-corn-oats (C-B-C-O) rotation and

TABLE 4-16   Returns per Acre by Nitrogen Fertilizer Application Rates, Rotation, and Government Program Participation[a]

| Rotation | Dollars/Acre | | | |
	No Program Participation	Basic Participation (20 percent set aside)	Full Participation (35 percent)[b]	N (pounds/acre)
C-C-C-C	56	222	221	240
C-C-C-O	61	187	186	180
C-B-C-O	90	177	175	120
C-C-O-M	64	151	150	120
C-O-M-M	67	113	112	40

NOTE: Crops in rotations are abbreviated by the following: C is corn; O, oats; B, soybeans; and M, meadow.

[a]Returns over variable costs only.
[b]35 percent includes 20 percent set aside and 15 percent paid land diversion.

SOURCE: Duffy, M. 1987. Impacts of the 1985 Food Security Act. Ames, Iowa: Department of Economics, Iowa State University.

$67.00 for a corn-oats-meadow-meadow (C-O-M-M) rotation (Duffy, 1987). But with government program payments and a 20 percent set-aside, continuous corn earned annually on average $222.00 per acre, compared with $177.00 and $113.00 for the C-B-C-O and C-O-M-M rotations, respectively. In recent years the feed grain program encouraged higher per acre corn yields, continuous corn production, and greater use of pesticides and nitrogen fertilizer. Duffy (1987) incorporated prevailing input assumptions into his study: for continuous corn, 240 pounds of nitrogen per acre was applied; for the C-B-C-O and C-O-M-M rotations, the application rates were 120 and 40 pounds, respectively. By encouraging high-yield, continuous corn production, the program has increased the corn surplus in spite of acreage set-aside requirements designed to reduce production, while exacerbating the potential for surface water and groundwater pollution (Table 4-16) (Duffy, 1987).

Impact of Research and Technology Transfer

Alternative farming systems are based on better management and information rather than the use of commercial products. Hence, there may be fewer opportunities and incentives for current input producers to develop and market inputs for alternative farming systems. Markets may be created, however, for companies offering management advice on better crop rotation strategies, efficient manure use, IPM, and other such practices and technologies. More resources should be allocated to collection of data about alternative farming systems regarding costs and the value and variability of resource requirements, yields, and other performance measures ordinarily incorporated into farm management budgets. A data base should be developed to integrate findings from the various biological and physical sciences, financial analyses, and estimates of the impact of farm practices on human health, water quality, and the environment.

SUMMARY

Research has begun to demonstrate the economic benefits of alternative farming systems and how current policies impose incentives and disincentives for the selection of various types of farming systems.

The committee's case studies provide examples of several profitable alternative operations. Additionally, several farm surveys provide general information about the overall financial performance of farmers using low-input methods, such as those who practice organic farming. But many questions remain unanswered. Farm surveys do not provide conclusive evidence regarding the advantages and disadvantages of different farming methods because many factors are randomized or not constant. Somewhat more systematic data are available regarding the economic performance of IPM programs. IPM has been highly successful in many instances. Farmers who use IPM usually reduce the amount of pesticides applied and increase their net returns compared with farmers who apply pesticides on a regular schedule.

Diversification strategies such as crop rotations can decrease input costs and increase crop yields. Experimental results must be interpreted with caution, however, when used to project the results of widespread adoption. Nonetheless, rotations have the potential to simultaneously increase farm income and reduce farm program expenses. Forage legumes in the crop rotation have the added advantage of supplying nitrogen. But when cash grain prices are supported far above the market level, many farmers would reduce their net farm incomes if they shifted from growing only price-supported crops, such as corn and soybeans, to legume-based rotations—unless commodity program rules are reformed.

Livestock are an essential component of some diversified alternative cropping systems. Many alternative farming systems, however, do not depend on livestock. Examples include perennial crop systems such as orchards and vineyards, and vegetable and other annual crop farms that use legumes as green manure crops or import organic residues from off the farm. Diversification can reduce risks and variability of net returns to farm families. For these reasons, it should be studied in more detail.

Very little is known about the aggregate impacts of possible widespread adoption of alternative farming methods. Future economic research on alternative farming methods should examine social and aggregate costs and benefits. This research should be integrated with that of other agricultural disciplines, the Extension Service, and the private sector to apply the results at the farm level.

REFERENCES

Allen, W. A., E. G. Rajotte, R. F. Kazmeirczak, Jr., M. T. Lambur, and G. W. Norton. 1987. The National Evaluation of Extension's Integrated Pest Management (IPM) Programs. VCES Publication 491-010. Blacksburg, Va.: Virginia Cooperative Extension Service.

Antle, J. M., and S. K. Park. 1986. The economics of IPM in processing tomatoes. California Agriculture 40(3&4):31–32.

Baker, K. F., and R. J. Cook. 1982. Biological Control of Plant Pathogens. St. Paul, Minn.: American Phytopathological Society.

Barbano, D. M., R. J. Verdi, A. I. Saeman, D. M. Galton, and R. R. Rasmussen. 1987. Impact of mastitis on dairy product yield and quality. In Proceedings of the 26th Annual Meeting of the National Mastitis Council. Arlington, Va.: National Mastitis Council.

Batra, S. W. 1981. Biological control of weeds: Principles and prospects. Pp. 45–59 in Biological Control in Crop Production. Beltsville Symposia in Agricultural Research, No. 5, G. C. Papavizas, B. Y. Endo, D. L. Klingman, L. V. Knutson, R. D. Lumsden, and J. L. Vaughn, eds. Totowa, N.J.: Allanheld, Osmun.

Boehlje, M. D., and V. R. Eidman. 1984. Farm Management. New York: Wiley.

Booth, W. 1988. Revenge of the "nozzleheads." Science 23:135–137.

Brusko, M., G. DeVault, F. Zahradnik, C. Cramer, and L. Ayers, eds. 1985. What the research reports haven't told you. Pp. 20–28 in Profitable Farming Now!, M. Brusko, G. DeVault, F. Zahradnik, C. Cramer, and L. Ayers, eds. Emmaus, Pa.: Regenerative Agriculture Association.

Coble, H. D. 1985. Development and implementation of economic thresholds for soybeans. Pp. 295–307 in CIPM Integrated Pest Management on Major Agricultural Systems, R. E. Frisbie and P. L. Adkisson, eds. College Station, Tex.: Texas A&M University.

Council for Agricultural Science and Technology. 1981. Antibiotics in Animal Feeds. Report No. 88. Ames, Iowa: Council for Agricultural Science and Technology.

Dabbert, S., and P. Madden. 1986. The transition to organic agriculture: A multi-year model of a Pennsylvania farm. American Journal of Alternative Agriculture 1(3):99–107.

Day, W. H. 1981. Biological control of alfalfa weevil in the northeastern United States. Pp. 361–374 in Biological Control in Crop Production. Beltsville Symposia in Agricultural Research, No. 5, G. C. Papavizas, B. Y. Endo, D. L. Klingman, L. V. Knutson, R. D. Lumsden, and J. L. Vaughn, eds. Totowa, N.J.: Allanheld, Osmun.

Domanico, J. L., P. Madden, and E. J. Partenheimer. 1986. Income effects of limiting soil erosion under organic, conventional and no-till systems in eastern Pennsylvania. American Journal of Alternative Agriculture 1(2):75–82.

Dover, M. J., and L. M. Talbot. 1987. To Feed the Earth: Agro-Ecology for Sustainable Development. Washington, D.C.: World Resources Institute.

Duffy, M. 1987. Impacts of the 1985 Food Security Act. Ames, Iowa: Department of Economics, Iowa State University.

Friend, T. H., G. R. Dellmeier, and E. E. Gbur. 1985. Comparison of Four Methods of Calf Confinement. 1. Physiology. Technical article 18960. College Station, Tex.: Texas Agricultural Experiment Station.

Frisbie, R. E., and P. L. Adkisson. 1985. Integrated Pest Management on Major Agricultural Systems. MP-1616. College Station, Tex.: Texas Agricultural Experiment Station.

Goldstein, W. A., and D. L. Young. 1987. An agronomic and economic comparison of a conventional and a low-input cropping system in the Palouse. American Journal of Alternative Agriculture 2(2):51–56.

Hall, D. C. 1977. The profitability of integrated pest management: Case studies for cotton and citrus in the San Joaquin Valley. Bulletin of the Entomological Society of America 23:267–274.

Heady, E. O. 1948. The economics of rotations with farm and production policy applications. Journal of Farm Economics 30(4):645–664.

Heady, E. O., and H. R. Jensen. 1951. The Economics of Crop Rotations and Land Use: A

Fundamental Study in Efficiency with Emphasis on Economic Balance of Forage and Grain Crops. Research Bulletin 383. Ames, Iowa: Agricultural Experiment Station, Iowa State University.

Heichel, G. H. 1987. Legumes as a source of nitrogen in conservation tillage systems. Pp. 29–35 in The Role of Legumes in Conservation Tillage, J. F. Power, ed. Ankeny, Iowa: Soil Conservation Society of America.

Helmers, G. A., M. R. Langemeier, and J. Atwood. 1986. An economic analysis of alternative cropping systems for east-central Nebraska. American Journal of Alternative Agriculture 1(4):153–158.

Hoy, M. 1985. Recent advances in genetics and genetic improvements in Phytosiidae. Annual Review of Entomology 30:345–370.

Hueth, D., and U. Regev. 1974. Optimal agricultural pest management with increasing pest resistance. American Journal of Agricultural Economics 56(3):543–552.

Kansas State University. Cooperative Extension Service. 1987. The Annual Report: 1987 Management Information, Kansas Farm Management Associations. Manhattan, Kans.: Kansas State University.

Kilkenny, M. R. 1984. An Economic Assessment of Biological Nitrogen Fixation in a Farming System of Southeast Minnesota. M.S. thesis, University of Minnesota, St. Paul.

Killingsworth, M. L., and J. B. Kliebenstein. 1984. Estimation of production cost relationships for swine producers using differing levels of confinement. Journal of the American Society of Farm Managers and Rural Appraisers 48(2):32–36.

Kliebenstein, J. B., and J. R. Sleper. 1980. An Economic Evaluation of Total Confinement, Partial Confinement, and Pasture Swine Production Systems. Research Bulletin 1034. Columbia, Mo.: University of Missouri-Columbia.

Kliebenstein, J. B., C. L. Kirtley, and M. L. Killingsworth. 1981. A comparison of swine production costs for pasture, individual, and confinement farrow-to-finish production facilities. Special Report 273. Columbia, Mo.: Agricultural Experiment Station, University of Missouri-Columbia.

Koepf, H. H., B. D. Peterson, and W. Schaumann. 1976. Bio-dynamic Agriculture: An Introduction. Spring Valley, N.Y.: Anthroposophic Press.

Kovach, J., and J. P. Tette. 1988. A survey of the use of IPM by New York apple producers. Agriculture, Ecosystems and Environment 20:101–108.

Lidvall, E. R., R. M. Ray, M. C. Dixon, and R. L. Wyatt. 1980. A Comparison of Three Farrow-Finish Pork Production Systems. Reprint from Tennessee Farm and Home Science No. 116.

Lockeretz, W., G. Shearer, D. H. Kohl, and R. W. Klepper. 1984. Comparison of organic and conventional farming in the Corn Belt. Pp. 37–48 in Organic Farming: Current Technology and Its Role in a Sustainable Agriculture, D. F. Bezdicek and J. F. Power, eds. Madison, Wis.: American Society of Agronomy, Crop Science Society of America, Soil Science Society of America.

National Fertilizer Development Center. Tennessee Valley Authority. 1988. Unpublished data.

National Mastitis Council. 1987. Current Concepts of Bovine Mastitis, 3d ed. Arlington, Va.: National Mastitis Council.

National Research Council. 1986a. Animal Health Research Programs of the Cooperative State Research Service—Strengths, Weaknesses, and Opportunities. Washington, D.C.: National Academy Press.

National Research Council. 1986b. Pesticide Resistance: Strategies and Tactics for Management. Washington, D.C.: National Academy Press.

National Research Council. 1987a. Agricultural Biotechnology: Strategies for National Competitiveness. Washington, D.C.: National Academy Press.

National Research Council. 1987b. Biological Control in Managed Ecosystems. Research Briefings 1987. Washington, D.C.: National Academy Press.

National Research Council. 1987c. Regulating Pesticides in Food: The Delaney Paradox. Washington, D.C.: National Academy Press.

Office of Technology Assessment. 1979. Pest Management Strategies. Working Papers. Vol. 2. Washington, D.C.: Office of Technology Assessment. 169 pp.

Olson, K. D., E. J. Weness, D. E. Talley, P. A. Fales, and R. R. Loppnow. 1986. 1985 Annual Report: Southwestern Minnesota Farm Business Management Association. Economic Report ER86–1. St. Paul, Minn.: University of Minnesota.

Olson, K. D., E. J. Weness, D. E. Talley, P. A. Fales, and R. R. Loppnow. 1987. 1986 Annual Report, Revised: Southwestern Minnesota Farm Business Management Association. Economic Report ER87–4. St. Paul, Minn.: University of Minnesota.

Olson, R. A., K. D. Frank, P. H. Grabouski, and G. W. Rehm. 1981. Economic and Agronomic Impacts of Varied Philosophies of Soil Testing. Nebraska Agricultural Experiment Station. No. 6695 Journal Series. Lincoln, Nebr.: Agricultural Experiment Station, University of Nebraska.

Osteen, C. D., E. B. Bradley, and L. J. Moffitt. 1981. The Economics of Agricultural Pest Control: An Annotated Bibliography, 1960–80. Bibliographies and Literature of Agriculture No. 14. Economics and Statistics Service. Washington, D.C.: U.S. Department of Agriculture.

Power, J. F. 1987. Legumes: Their potential role in agricultural production. American Journal of Alternative Agriculture 2(2):69–73.

Randall, G. W., and P. L. Kelly. 1987. Soil test comparison study. Pp. 145–148 in A Report on Field Research in Soils. Miscellaneous Publication No. 2 (Revised)–1987. St. Paul, Minn.: University of Minnesota Agricultural Experiment Station.

Reichelderfer, K. H. 1981. Economic feasibility of biological control of crop pests. Pp. 403–417 in Biological Control in Crop Production, Beltsville Symposia in Agricultural Research, No. 5, G. C. Papavizas, B. Y. Endo, D. L. Klingman, L. V. Knutson, R. D. Lumsden, and J. L. Vaughn, eds. Totowa, N.J.: Allanheld, Osmun.

Reichelderfer, K. H., and F. E. Bender. 1979. Application of a simulative approach to evaluating alternative methods for the control of agricultural pests. American Journal of Agricultural Economics 61(2):258–267.

Shields, E. J., J. R. Hyngstrom, D. Curwen, W. R. Stevenson, J. A. Wyman, and L. K. Binning. 1984. Pest management for potatoes in Wisconsin—A pilot program. American Potato Journal 61:508–517.

Shrader, W. D., and R. D. Voss. 1980. Soil fertility: Crop rotation vs. monoculture. Crops and Soils Magazine 7:15–18.

Smith, J. W., and C. S. Barfield. 1982. Management of preharvest insects. Pp. 250–325 in Peanut Science and Technology, H. E. Pattee and C. T. Young, eds. Yoakum, Tex.: American Peanut Research and Education Society.

U.S. Congress, House. Committee on Government Operations, Subcommittee on the Environment, Energy, and Natural Resources. 1988. Hearing on Environmental and Economic Benefits of Low Input Farming, April 28, Washington, D.C.

U.S. Department of Agriculture. 1980. Study Team on Organic Farming. Report and Recommendations on Organic Farming. Washington, D.C.

U.S. Environmental Protection Agency. 1982. Ethylene Bisdithiocarbamates Decision Document: Final Resolution of Rebuttal Presumption Against Registration. Washington, D.C.

U.S. Environmental Protection Agency. 1985. Captan: Special Review Position Document 2/3. Washington, D.C.

U.S. Environmental Protection Agency. 1986. Alachlor: Special Review Technical Support Document. Washington, D.C.

Wagstaff, H. 1987. Husbandry methods and farm systems in industrialized countries which use lower levels of external inputs: A review. Agriculture, Ecosystems and Environment 19:1–27.

Young, D. L., and W. A. Goldstein. 1987. How government farm programs discourage sustainable cropping systems: A U.S. case study. Paper No. 15. Pp. 443–460 in How Systems Work, The Proceedings of the Farming System Research Symposium. Fayetteville, Ark.: University of Arkansas.

Zavaleta, L. R., and W. G. Ruesink. 1980. Expected benefits from nonchemical methods of alfalfa weevil control. American Journal of Agricultural Economics 62(4):801–805.

PART TWO

The Case Studies

C ASE STUDIES PROVIDE INSIGHTS into how the real world works. They help formulate and test hypotheses, but cannot substitute for other forms of scientific research. In complicated areas of human endeavor, however, case studies can provide useful observations that go beyond the range of controlled experiments. They can indicate promising directions for further research and help demonstrate how many different factors—economics, biology, policy, and tradition—interact.

The committee commissioned these 11 case studies to expand the growing but still limited scientific literature on the range of alternative farming systems currently operating in the United States. U.S. agriculture is extremely diverse, and these case studies are only snapshots of certain agricultural sectors. These case studies were conducted during 1986. The committee is aware that a complete assessment of alternative farming in the United States would require a much larger number of case studies, with systematic data collection and analysis extending over several years. Moreover, the committee believes that a more comprehensive set of case studies should be developed and regularly updated as a way to track the evolution in the profitability of alternative agricultural systems. Nonetheless, the committee believes that these case studies provide a useful understanding of the range of successful alternative systems available to U.S. farmers.

Farmers and other innovators often develop, through their own creativity, new approaches to solving common farming problems. Examination of these approaches in case studies provides insights that may benefit others. Yet to draw valid inferences from a given farm, it is important to understand fully how a given production system works, what it accomplishes, and at what cost. Case studies per se should not be considered alone as ample evidence to judge the farming practices in question, nor should these farming practices necessarily be implemented on other farms. But case studies can broaden the perspective of profitable alternatives and help focus future research.

Working from its own experience, a survey of the literature, and discussions with alternative farming researchers, the committee compiled a prospective list of case study farms. Only the Thompson and Kutztown farms have been the subject of previous scientific research. The others have not been examined in detail prior to this project. Consequently, the case studies differ in level of scientific evidence, documentation, and analysis.

To complete a given case study, the committee relied on existing scientific literature and secondary data that added further insight to the direct observations made during on-farm visits. Such information was necessary in many cases to document or explain the feasibility of applying alternative farming practices or systems. In some instances, the biological and agronomic basis of special performance features on a case study farm was readily understood in light of current knowledge. Occasionally, the performance data reported by the case study farmer could not be fully explained or reconciled with current scientific knowledge or experience on similar farms in the same area. These cases are noted herein, and are often identified as areas for additional research.

A committee member or staff consultant visited each of the case study farms. Where possible, a local expert—a university researcher, cooperative extension specialist, or Soil Conservation Service professional—accompanied the visitor to the case study farm to provide verification and interpretation of procedures used, resources, performance, and other aspects of the special management features being examined on the farm. An outline was used to guide the interviewer, but the discussion always extended far beyond the questions anticipated.

Secondary data related to the climatic conditions, pest problems encountered in the locality of the case study farm, and procedures used by conventional farms were assembled from various sources. Published reports and verbal information were obtained from experts familiar with the location and type of farming in the area. As the case study draft was prepared, additional information was obtained by telephone and letter from various sources, including the farmer and local experts, to fill in details overlooked or not fully understood during the initial visit. Each farmer and local expert was asked to review at least one draft of the case study manuscript and to indicate any errors or significant omissions.

Committee members evaluated each case study according to several criteria. First, special features of the case study farm were reviewed to determine whether they were explicable with existing scientific knowledge and theory. If so, the scientific findings were documented and the committee assessed the applicability of these features to other farms. Where the available data permitted, the committee also examined the resource conservation, food safety, and environmental and financial impact of the special features.

OVERVIEW OF CASE STUDY FARMS

Crop and Livestock Farms

The Spray Brothers Farm near Mount Vernon, Ohio, encompasses 720 acres, including 400 acres of cropland. The farm enterprises in 1986 included 32 milk cows; 40 to 50 head of beef cattle; 88 acres of soybeans; 12 acres of adzuki beans; 100 acres of corn, of which 40 percent is sold off the farm and the remainder fed to livestock; and 100 acres of wheat and oats. The oats, soybeans, and adzuki beans are sold through specialty health food distributors, and the wheat is sold through normal marketing channels or as seed wheat. Some corn is sold as poultry feed at a premium price.

The BreDahl Farm near Fontenelle in southwestern Iowa is a relatively small farm of 160 acres producing 35 to 40 acres of corn, 35 to 40 acres of soybeans, 20 acres of alfalfa, and 20 to 30 acres of pedigreed oats for seed. It also produces the following animal products: lambs, wool, sheep breeding stock, cattle, and hogs. Special features include intensive, flexible management of on-farm resources—complex rotations, innovative ridge-tillage practices, and some spot spraying of herbicide when necessary. Strip cropping, terracing, and rotations have greatly reduced erosion on the farm. All fields are fenced so that livestock can be put on the fields after harvest to glean any crop residue. Turnips are double-cropped following oats and used as sheep forage. In addition, the turnips are grazed, a practice that provides excellent feed for the animals as well as improving soil tilth and fertility and reducing erosion.

The Sabot Hill Farm near Richmond, Virginia, is a diversified operation of 3,530 acres producing beef cattle, forage, and cash grain. Of special interest on this farm is the dramatic reduction in herbicide use as a result of harvesting weeds, notably Johnsongrass, as a crop. Erosion is substantially reduced through no-tillage planting, strip cropping, and pasturing practices. Improved pastures and rotational practices also allow animals to graze year-round with little supplemental feeding. These practices significantly reduce pest control, tillage, and feed costs.

The Kutztown Farm in eastern Pennsylvania is an "alternative" 305-acre mixed crop and livestock farm with field slopes up to 25 percent. Crop acreages are adjusted to meet the entire feed needs for finishing 250 to 290 head of beef cattle and 50 to 250 hogs per year. The farmer uses no commercial fertilizer except for a small quantity of liquid starter fertilizer. He uses a small amount of herbicide on corn and soybeans on about 45 percent of the acreage and crop rotation and cultivation control weeds on the remaining acreage. Some surplus hay is sold. The farm in past years has had a surplus of nitrogen. Estimated soil erosion is about 4.5 tons per acre per year, which is below county and state levels but above the rate at which soil is formed.

The Thompson Farm in central Iowa has 282 acres of tilled cropland with a 50-cow foundation beef herd and a farrow-to-finish hog operation of 90 sows. The farm's innovative ridge tillage system is effective in most years for controlling weeds and preventing soil erosion in its rotation of corn, soybeans, oats, and hay. The owners use synthetic chemical pesticides only in emergencies, and they apply raw manure and municipal sludge to maintain soil fertility. Livestock are raised with adequate sunlight and space to reduce stress and without growth hormones or subtherapeutic doses of antibiotics. The pigs are fed probiotics (beneficial bacteria) to prevent intestinal diseases.

Fruit and Vegetable Farms

The Ferrari Farm near Stockton, California, is composed of 223 acres and produces about 75 acres of various tree fruits, 126 acres of nuts, and 22 acres of fresh-market vegetables. Most of the farm is certified by the state as organic, and about two-thirds of the Ferrari crops are sold as organically produced, according to state law and local certification standards. The remainder of the crops are produced with an integrated pest management (IPM) program that includes the occasional use of pesticides. The Ferraris have been innovative in their pest control strategies, experimenting with such new biological controls as the codling moth granulosis virus to control codling moth in apples, pheromones to control the oriental fruit moth, and predaceous mites to control phytophagous mites. They apply compost for its nutritional value and in the belief that it helps control nematodes. They also eliminate disease-prone crops. The Ferraris sell to wholesalers, a few retailers, and directly to consumers at the San Francisco farmers' market. Indirect measures such as growth in acreage and capital stock and a low debt-to-asset ratio indicate that the farm is prosperous.

Four farms in south Florida producing fresh-market vegetables and using IPM were examined for this case study. John Hundley of Loxahatchee farms about 9,640 acres including 1,500 acres of sweet corn, 120 acres of cabbage, 3,000 acres of radishes, 1,600 acres of seed corn, and, 1,300 acres of field corn. The farm also includes 500 acres of pasture on which cattle are run, a 120-acre orange grove, and, 1,500 acres of sugarcane. Ted Winsburg of Palm Beach grows 350 acres of fresh-market peppers. John Garguillo of Naples raises 1,300 acres of tomatoes for the fresh market. Fred Barfield of Immokalee raises 1,000 acres of vegetables, primarily varieties of bell peppers, tomatoes, and cucumbers; he also has a 550-acre orange grove, a 1,000-head herd of purebred Beefmaster cattle, and a 1,200-cow commercial, mixed-breed herd. All four farms employ Glades Crop Care, Inc., to perform pest scouting as a means of reducing pesticide use.

Stephen Pavich & Sons is one of the nation's leading producers of fresh grapes. The Paviches produced about 1 percent of the nation's grapes in 1986 on about 1,125 acres of vineyard; in 1987 they purchased another 160 acres of grapes, a 14 percent increase in acreage. They use innovative vineyard management practices such as reliance on natural enemies to control mite and insect pests. The principal source of soil nutrients is the application of 2.5 to 3.0 tons of compost per acre. The farm maintains a permanent ground cover, which is periodically flail-chopped, to control weeds; weeds close to the vines are controlled by hand weeding and hoeing. Grape leafhopper is their primary insect pest. Most years the Paviches are able to produce all or nearly all of their grapes without using any chemical pesticides except sulfur, to prevent fungal diseases. They consider plant nutrition, especially avoiding an excessively high ratio of nitrogen to calcium in the plant tissue, to be an important element of pest control. The farm sells 97 percent of its grapes on the conventional market and the rest to specialty organic markets. A premium price of $1.00 to $2.00 per 23-pound box, a 12 to 25 percent premium, is charged for the organic grapes to cover the cost of certification and special handling and storage.

The Kitamura Farm near Sacramento, California, includes 305 acres, of which 160 acres are devoted to the production of processing tomatoes. The farm employs a modified version of an IPM program developed at the University of California that includes rotation of fields planted with tomatoes, scouting for insects, and precise irrigation management. By terminating irrigation 10 days earlier than the generally accepted practice, the Kitamuras incur a lower incidence of mold in their tomatoes. Through these techniques the farm has been able to virtually eliminate insecticide and fungicide use while maintaining high yields and fruit quality.

Other Farms

Coleman Natural Beef includes 2,500 cow-calf units on about 284,500 acres of mostly mountain rangeland near Saguache, Colorado, west of Denver. The Colemans have imported no livestock replacements for more than 25 years but instead select from their own herd, using artificial insemination of 300 cows to get replacement bulls. Pastureland acreage per cow is high compared to crop-livestock farms or beef feedlots, a practice that the Colemans say reduces the incidence of health problems in the cattle by avoiding crowding stress. Unlike other ranchers in the area, the Colemans apply no fertilizer or lime to their rangeland. Feeder cattle are fattened under contract in commercial feedlots using feeds tested for pesticide residues. The Colemans routinely administer inoculations but do not use growth hormones or subtherapeutic doses of antibiotics in their natural beef program. They receive a 25 percent premium price for 2,500 head of their own beef plus 12,500 head produced to their specifications by other ranches, all of which is nationally advertised and marketed as natural beef. The marketing enterprise is apparently much more profitable than the ranch.

The Lundberg Family Farms near Chico, California, is a 3,100-acre farm that includes 1,900 acres of rice produced with reduced rates of chemical pesticides and fertilizers and a 100-acre field experiment that produces rice with no chemical fertilizers or pesticides. The Lundbergs have been experimenting with the production of organic rice for 18 years. On all of their fields the Lundbergs substitute the decomposition and incorporation of their rice straw into the soil for the conventional practice of burning the straw, and they employ an early-season irrigation schedule to prevent crop damage by tadpole shrimp *(Triops longicaudatus)*. They have largely succeeded in the first objective of their experimental field—nonherbicidal weed control. Their remaining objectives include finding an acceptable way of providing nitrogen without using synthetic chemical fertilizers or increasingly scarce and expensive animal manures. In spite of a nearly 50 percent premium added to the price of their organic rice, the experimental rice acreage has not been profitable in most years. With what they have learned from the field experiment, however, the Lundbergs have been able to improve the financial performance of their other fields by significantly reducing herbicide use.

CASE STUDY

1

Crop and Livestock Farming in Ohio: The Spray Brothers

THE FARM OF THE SPRAY BROTHERS, Glen and Rex, is located in Morgan Township, Knox County, which is nearly in the geographic center of Ohio. The farm homes and some of the land are adjacent to State Highway 586 approximately 3 miles north of Martinsburg and 11 miles south of Mt. Vernon, the county seat of Knox County. The Sprays currently own 650 acres and cash-rent an additional 70 acres. They have farmed the rental acreage in the same manner as their own land for 15 years. They currently have 400 acres of cropland; the rest of the land is in permanent pasture (low depressional areas or steep sloped uplands) and woodland (7 to 10 acres). A 4-year rotation of 100 acres each of corn-soybeans-small grain-red clover hay is currently followed on the tillable cropland. The Spray brothers' farming operation is a full, equal partnership (Table 1).

Most other farms in the immediate vicinity are about 200 to 250 acres, although the average farm size in Knox County is 177 acres and one neighboring farm comprises 600 acres. The dominant farming system in the area is row-crop farming with continuous corn or a corn-soybeans rotation. Knox County has a high percentage of row crops planted with no-tillage equipment, a practice that markedly reduces the erosion potential for these soils. Dairy farmers in the area generally follow a corn-corn-soybeans-hay-hay-hay 6-year rotation. Beef farmers use a corn-soybean-hay-hay 4-year rotation.

The Sprays do not participate in any government programs except for the dairy diversion program.

GENERAL DATA

The Sprays operate a diversified animal and cash grain farm, as follows:

Milk cows*: The farm has 32 Holstein cows and 10 replacement heifers. Milk is sold to a local grade A market. Dairy cattle are bred using artificial insemination.

Beef cows: The Sprays have 40 to 50 Herefords; their calves are finished by feeding out in a 90-day finishing regime. Cattle are marketed at about 15 to 16 months of age. Some finished cattle are sold locally while the remainder are marketed through National Farmers Organization (NFO) markets as far away as Green Bay, Wisconsin.

Soybeans: Soybeans are a major cash crop, occupying 100 acres of cropland per year. The Sprays clean, bag, and market the entire crop of soybeans and sell it to organic tofu specialty markets at a premium price. In 1985 they marketed to tofu manufacturers in Cleveland and Worthington, Ohio, and in West Virginia at a price of $9.00 per bushel including transportation costs. The Sprays use a soybean cultivar with a white hilum that is desirable for tofu production. The soybean screenings, which contain cracked and broken beans and weed seeds, are fed to the dairy cattle and to beef cattle being finished for market.

Adzuki beans (Phaseolus angularis, wild): This crop was a new enterprise, occupying about 12 acres in 1985. These beans are sold to specialty health food companies for $42.00 per bushel and yield about 20 to 25 bushels per acre on this farm. The adzuki beans replaced 12 acres that would normally have been planted with soybeans.

Corn: A major cash crop as well as a cattle feed for on-farm consumption, corn acreage on the farm is 100 acres each year. Of this total, 40 percent is sold off the farm and 60 percent is fed to animals. The Sprays developed a specialty market for shelled corn as poultry feed for an Amish farm in Pennsylvania at a $0.50 per bushel premium price in 1985. The corn used for dairy and beef feeding on the farm is harvested with a picker because the cob is considered an important carbohydrate constituent for the cattle. The Sprays grind and mix the corn for animal rations directly on the farm. The seed corn used by the Sprays is a triple-cross hybrid instead of the more common single crosses. The advantages of the triple-cross hybrid are some prolificacy (multiears) and a savings of about 33 percent in the cost of seed.

Small grains: Wheat (50 to 70 acres) and oats (30 to 50 acres) occupy about 100 acres annually and serve as a nurse crop for the red clover used in the

---

*This enterprise was scheduled to be terminated by the U.S. Department of Agriculture Milk Production Termination Program in August 1987. This case study does not reflect that termination.

TABLE 1　Summary of Enterprise Data for the Spray Brothers Farm

Category	Description
Farm size	720 acres, 32 dairy cows, 40–50 beef cows
Labor and management practices	All enterprises are managed by the two brothers, Rex and Glen Spray. Glen Spray's son is a salaried employee. Student labor is hired during the growing season to help with haying and weed control. Hired labor costs are about $1,200/year for 300 man-hours.
Livestock management practices	Dairy cows are kept on pasture and fed roughages and supplements. Beef replacements are produced on the farm and marketed at 15–16 months of age.
Marketing strategies	Premium prices are received for corn, soybeans, wheat, oats, adzuki beans, and some of the beef because the farm is a certified organic farm. On-farm facilities are available for seed cleaning, bagging, and storage.
Weed control practices	No herbicide has been applied in 15 years. The farmers rotate corn, soybeans, small grain, and red clover. Two diskings in the early spring control weeds; late planting of corn and soybeans again uproots weeds; and corn and soybeans are rotary-hoed at emergence. Frequent cultivations (2–4 per season) also control weeds. Hand weeding of Jimson weed is performed in adzuki bean fields.
Insect and nematode control	No problems with insects or nematodes are apparent.
Disease control practices	Rotation and the use of disease-resistant varieties are cited as the reason for the absence of disease problems in farm crops. Soil microbial populations are also cited by the farmers as a disease-inhibiting factor.
Soil fertility management	The farmers use a corn-soybeans-small grain-red clover hay rotation. No lime or fertilizer has been purchased since 1971. Microbial fertilizer is applied once per 4-year rotation; the fertility benefits are not yet proven. Manure is applied to 100 acres/year.
Irrigation practices	None
Crop and livestock yields	Yields of corn exceed the county average by 32 percent, soybeans by 40 percent, wheat by 5 percent, and oats by 22 percent. (There is no county yield comparison for clover hay; the farm averages 6 tons/acre.)
Financial performance	The farm's overall economic viability is good. The value or sale of crops, livestock, and livestock products was $188,000 in 1985.

fourth year of the rotation. The wheat is all sold off-farm through normal marketing channels as grain or as seed wheat. (The availability of marketable seed wheat is another advantage of the farm's seed cleaning, bagging, and storing facility.) In 1985 some of the oats were sold for a premium price ($3.20 per bushel) to an organic market in Pennsylvania for processing into rolled oats.

Pasture: Almost 300 acres are unimproved permanent pastures made up of timothy, white clover, and blue grass.

TABLE 2   Normal Monthly Precipitation at Fredricktown Observation Station,
Knox County, Ohio

Month	Normal Precipitation (inches)
January	2.8
February	2.2
March	3.4
April	3.7
May	4.2
June	4.2
July	4.3
August	3.1
September	3.0
October	2.3
November	2.8
December	2.4
Average annual total	38.4

NOTE: The normal monthly precipitation is the average of the inches of precipitation for that month from 1941 to 1970.

SOURCE: National Oceanic and Atmospheric Administration. 1980. Climates of the States, 2d ed. Detroit: Gale Research Co., Book Tower.

Red clover: This crop occupies 100 acres of tillable land annually and is used as a hay crop as well as a green manure crop when incorporated in the fall. In some years the second crop is allowed to mature, and clover seed is harvested and sold. In 1986 the Sprays sold 100 pounds of clover seed at $0.65 per pound.

Climate

The climate in Knox County is typical of central Ohio, with warm and moderately humid days in summer and cold and cloudy winters. On the average there will be 15 days per year with temperatures above 90°F and 7 days per year with temperatures below 0°F. The growing season at the Fredricktown Observation Station (about 24 miles northwest of the Spray Brothers Farm) is 147 days. The growing season is longer than 170 days 10 percent of the time and shorter than 123 days 10 percent of the time. Showers and thunderstorms account for most of the precipitation during the growing season (Table 2). Snowfall averages 30 inches per year but varies greatly from year to year.

PHYSICAL AND CAPITAL RESOURCES

Soil

Knox County is on the outer edge of the glaciated region of Ohio. The last glaciation of the Wisconsin age completely covered western Knox

TABLE 3   Erosion Potential of Selected Soils Under Different Tillage
and Rotations

Soil	Annual Soil Loss (tons/acre)			
	Spray Brothers' Rotation (Reduced Tillage-Chisel Plowing)[a]	Continuous No Tillage[a]	Corn-Soybeans No Tillage	Corn-Soybeans Conventional Tillage[b]
Luray clay loam (0.2% slope)	0.62	0.14	0.36	1.7
Titusville silt loam (6.0% slope)	6.85	1.58	3.95	18.5

NOTE: All figures are based on an assumed 200-foot slope land. Figures are calculated using the Universal Soil Loss Equation.

[a]The rotation is corn-soybeans-small grain-red clover hay.

[b]Conventional tillage represents spring disking and normal field harrowing.

SOURCES: R. Adamski, communication, 1989. U.S. Department of Agriculture. 1984. Universal Soil Loss Equation Charts. Soil Conservation Service. Washington, D.C.

County. The soils map of the Sprays' farm is quite complex, with approximately 20 soil types or phases. The farm is founded on deep, well-drained to very poorly drained soils. These soils were formed in glacial outwash, alluvium and lacustrine deposits on terraces, floodplains, and glacial lakebeds. The farm land of the main farm (along Martinsburg Road, State Highway 586) fits this description closely.

The tilled land is in a flat basin, with rolling topography on the edges and rolling hills interspersed in the flat basin. All of the arable land on the farm has been tilled, and neither drainage nor wetness is a problem. The soil surface texture ranges from silt loam to silty clay loam. Some of the soils are underlain with sand and gravel and can be excessively well drained and slightly droughty. Although the soils are complex, four soil associations dominate the farm: (1) flat to nearly flat Fitchburg-Luray soils formed on lacustrine materials; (2) Chili-Crane-Homewood soils formed on glacial outwash; (3) Centerburg-Bennington soils formed on glacial till; and (4) Homewood-Loudonville-Titusville soils formed in loamy glacial till and residuum from sandstone.

The soils making up the tilled land are good soils for the region and, because of tile drainage, are reasonably easy to manage. The Homewood-Loudonville-Titusville soils on steep slopes could present a serious erosion problem if they are not properly managed. Problems could occur if the soils were tilled in an excessively wet state (Table 3).

Buildings and Facilities

The homes of both brothers are modern, substantial houses. Farm buildings are located at three locations and are modest but adequate. Farm building locations are neat and well managed. Primary and special facilities include a milking barn with a pipeline milking system and bulk tank, a new grain cleaning and storage facility for bagging and storing grains for off-farm sale (a $45,000 investment), storage bins for dried grain and shelled corn, an ear corn storage building, and a grain drying facility.

Machinery

The Spray brothers currently have nine tractors of varying ages and sizes. The largest is a 125-horsepower tractor mainly used for soil tillage with the farm's primary tillage tool, an offset disk. Other tractors are dedicated to specific tasks such as row-crop cultivation, manure loading, mowing and baling hay, and the like. None of the tractors was purchased new. All were purchased after being dealer demonstrators or otherwise used.

Other farm equipment includes a self-propelled combine with grain and corn headers, a four-row planter for corn and beans (36-inch row width), a grain drill, a two-row corn picker, two manure spreaders, a four-row field cultivator, two harrows, a rotary hoe, a sickle bar mower, a mower-conditioner for hay, a hay rake, and a square baler. Because grain handling is an important part of the farming operation, the Sprays own four gravity-feed grain wagons, two grain elevators (the auger type), and a portable hammer mill for grinding their cattle feed. Their estimate of total machinery inventory is $100,000.

MANAGEMENT FEATURES

Soil Fertility

The Spray brothers have not purchased any lime or chemical fertilizers since 1971. However, they do use microbial fertilizers that are applied one time in the rotation on those fields being planted with corn.

The Sprays use the manure that is available from their dairy and beef herds but do not consider this contribution of nutrients to be particularly important because fewer than 100 acres of their tillable land receive manure each year. They apply the manure to the clover fields in the fall and winter months because these are the fields to be planted with corn the next year. Thus, manure is applied only once in a 4-year rotation and then not to all the soils.

The committee engaged consultants to provide an estimate of nutrients supplied by the manure to the soil. The estimate assumed an application of 4 to 6 wet tons of manure per acre by a single pass. The average nutrient composition of beef manure is 18 pounds of nitrogen (N), 7 pounds of phosphorus (P_2O_5), and 9 pounds of potassium (K_2O) per ton; dairy manure is composed of 12 pounds N, 3 pounds P_2O_5, and 11 pounds K_2O per ton. Using an average of the two values and assuming 50 percent N mineralization from the manure in the first year, the 4-ton application rate would supply 30 pounds available N, 20 pounds P_2O_5, and 40 pounds K_2O in the first year after application and lesser amounts in subsequent years. These are not insignificant concentrations of nutrients.

The primary source of nitrogen for the rotation is the nitrogen provided by the incorporation and decomposition of the red clover green manure crop. This crop is generally harvested for hay in late spring or early summer. (A second hay crop may be harvested if needed or allowed to mature for clover seed.) The clover is then allowed to grow back and is partially incorporated by a late disking in October. The fall incorporation allows some decomposition and mineralization to occur before cold weather. In the spring the fields are disked twice more and harrowed before planting the corn in mid-May. The brothers estimate that incorporation of the red clover will provide about 125 pounds of nitrogen per acre with about 75 pounds or 60 percent available to the corn crop.

The Spray brothers do not believe in soil testing, and therefore only limited soil test data were available. The results indicated a soil pH of 6.2 to 7.2, 53 to 74 pounds per acre of available phosphorus, and 80 to 120 pounds per acre of available potassium. These data are insufficient to determine the pH, phosphorus, or potassium status of the entire farm; however, the yields obtained would seem to indicate that currently there is no soil fertility problem on the farm. A detailed and complete soil testing program over time would be required to determine if this fertility status can be maintained indefinitely. The Spray brothers attribute the favorable nutrient status to organic matter and microbial activity.

Tillage

The Spray brothers use chisel plowing as a form of reduced tillage. The primary tillage implement is the 12-foot offset disk. As indicated earlier the clover crop is disked once in October and then twice in the spring before planting corn. Likewise, the corn residues are disked once in October or November before frost and again once or twice in the spring before planting soybeans. The fields are harrowed before planting corn or soybeans. Soybean residues are also incorporated by disking, and wheat is planted immediately after tillage. The Spray brothers told the interviewer that they do not till deeper than 4 to 6 inches with the disk.

Weed Control

The Spray brothers have not used any herbicide for 15 years. They attribute the success of their weed control program to five factors: (1) the rotation they use, which includes a red clover hay crop; (2) the two diskings in early spring that kill off the sod as well as successive flushes of weed seeds; (3) a relatively late planting date for corn (near the middle of May) and for soybeans (in late May or early June); (4) the use of a rotary hoe at corn and soybean emergence; and (5) frequent cultivations during the growing season (two to four times as needed). In discussion, the brothers also noted that hired labor is used for cleaning up corn, soybean, and adzuki bean fields. Jimson weed is the major weed in the bean fields, and it is removed by hand. The soybean fields and corn fields were relatively free of weeds during the farm visit in September 1986. Weeds in the small grains on the Sprays' farm are not a problem because of the rotation.

Insect, Nematode, and Disease Control

Insecticides, nematocides, and fungicides have not been used on the Spray Brothers Farm in 15 years. The brothers stated emphatically that they did not have serious infestations of insects or diseases. The current and consistently high yields they obtain suggest that they are correct in their assumption. Field observation of mature or nearly mature soybeans and corn confirmed this point; there was no evidence of a disease or insect problem in September 1986.

The fields of small grain were not observed during the farm visit. The Sprays reported that they use an oat variety with a high test weight per bushel, an attractive feature in the marketplace. In 1986 this oat variety was not on the recommended list for Knox County, although it was on the recommended list in 1985. The loss of its earlier recommendation suggests that these oats may currently be susceptible to the fungus disease races prevalent in Ohio. The wheat variety used by the Sprays has good resistance to rusts, smuts, viruses, and powdery mildew and was on the recommended variety listing in 1986. In general, the Spray brothers did not feel that their small grain acreage was adversely affected by diseases or insects. They cited the benefits of rotation, an active soil microbial population, and good soil health as the reasons for reduced disease and insect occurrence, although there may be other factors that are important as well.

The Sprays were asked why alfalfa was not used in the rotation instead of red clover. Their answer is significant with respect to pesticide use; they do not use alfalfa because they believe that it would be necessary to use insecticides to control alfalfa weevil. Clover, on the other hand, does not have such an endemic insect problem and thus is a successful alternative legume whose use eliminates the need for insecticide. Also clover seed costs less than alfalfa seed and is a better leguminous crop for biomass production in the first year than the locally popular cultivars of alfalfa, a significant attribute for a legume that is used for only a single year in a rotation.

Labor

The farm's 720 acres and its multiple enterprises are operated by Glen and Rex Spray and Glen's son, who is paid a cash salary. During haying season and for weed control, the farm employs additional student labor. The Sprays hire about 300 man-hours of labor each year at a cost of about $1,200.

ANIMAL ENTERPRISES

Dairy Operations

The 32-cow dairy herd is kept on pasture near the milking barn and fed roughage (clover hay) in bunks near the barn. Ground concentrate is prepared on the farm, using farm-grown ear corn, soybeans, and the following supplements: vitamins, minerals, *Lactobacillus acidophilus* fermentation product, and corn germ meal (18 percent crude protein).

Replacement heifers are kept on pasture until they are ready to calve. Winter housing without stanchions is available.

Beef Operations

The Spray Brothers Farm currently has 40 to 50 beef cows, and the Sprays plan to add 25 more after the dairy herd is sold. They raise their own Hereford calves and grow them to market weight at about 15 to 16 months of age. The cows are pastured most of the year and are returned to the finishing area for concentrated feeding (mostly ground ear corn and a non-medicinal supplement) about 90 days before marketing. All cattle have free-choice minerals and salt. The beef animals being fattened are fed inside but are not confined.

PERFORMANCE INDICATORS

Crop Yields

The farm visit and additional discussions with the Knox County extension agent, Soil Conservation Service personnel, and some Ohio State University faculty members (from the departments of agronomy, agricultural engineering, and agricultural economics) confirm that the Spray brothers have a highly productive, well-managed farming operation. Their farm has a proven and accepted reputation in the area; and, although the outstanding yield data (Table 4) are not completely verified, the enterprise returns for 1985 (Table 5) are consistent with the Spray brothers' yield estimates.

Soil Fertility

It would seem that a crop rotation that includes clover and small grain for 2 out of 4 years would be a soil-conserving management system as compared with the production of only corn and soybeans. Yet, estimates made by the Soil Conservation Service computer program using soil loss equations and rainfall patterns in Knox County indicate that, although the Spray system is a better system than conventional (plowing) tillage, it is not as effective as a no-till system, which is used on 11.2 percent of the corn and 1.4 percent of the soybean acreage in Knox County. Erosion was not evident

TABLE 4 Yield Comparisons of Spray Brothers Farm and Knox County Averages, 1981–1985

Location or Soil Type	Corn (bushels/acre)	Soybeans (bushels/acre)	Wheat (bushels/acre)	Oats (bushels/acre)	Clover Hay (tons/acre)
Spray Brothers[a]	145–150	48.0	45.0	80.0	6.0
Knox County[b]	111.5	34.4	42.8	65.5	—
Soils inventory[c]					
Luray silty clay loam	125.0	40.0	45.0	80.0	—
Bennington silt loam	102.0	30.0	40.0	65.0	6.0
Homewood silt loam	110.0	30.0	45.0	90.0	6.0

[a]Yield estimates obtained from Glen and Rex Spray. Data based primarily on field harvest, with limited hand harvest data. The 1986 yields for corn, soybeans, wheat, and oats were 140, 48, 45, and 75 bushels per acre, respectively.

[b]Data obtained from Joseph Brown, Knox County Agricultural Extension Service, Mount Vernon, Ohio.

[c]Data from An Inventory of Ohio Soils, Knox County, Table 1, Report No. 70 (Ohio Department of Natural Resources, 1984). This is an estimate of yield by soil map unit when optimum level of management is imposed.

TABLE 5 Spray Brothers Partnership Income, 1985

Income Source	Amount (dollars)
Sale of animals and animal products	
Beef cattle	18,382.48
Dairy products	49,216.02
Government payment (milk diversion)	5,669.80
Cull dairy cows	5,243.45
Subtotal	78,511.75
Sale of agronomic crops	
Soybeans	56,726.34
Corn	14,791.59
Other grains (for example, wheat, oats)	23,084.78
Feed	14,725.83
Subtotal	109,328.54
Miscellaneous	
Custom machinery work	2,653.90
State gas tax refund	695.16
Sale of microbial fertilizer	15,461.27
Subtotal	18,810.33
Total	206,650.62

in the corn and soybean fields observed during the farm visit; fall disking and frequent cultivation of row crops on the sloping fields, however, could present an erosion hazard.

The Sprays' ability to supply all the nutrients needed for high yields of corn, wheat, soybeans, and oats using minimal off-farm inputs is a real accomplishment. The manure applications (once in the 4-year rotation) and the red clover green manure crop supply, at least for now, the nitrogen necessary to grow corn and small grains. How the Sprays manage to maintain an adequate pH balance and enough phosphorus and potassium in the soil—despite the fact that they have not added lime, phosphorus, or potassium in 15 years—cannot be readily explained. It may be due in part to previous additions of high concentrations of phosphorus and lime by the Sprays' father prior to 1972 and to the natural fertility of the soils. Detailed soil nutrient evaluations and further study of nutrient cycling on the farm would be useful and informative.

Both Glen and Rex Spray attribute much of their success in maintaining adequate soil fertility to the use of microbial fertilizers at least once in 4 years on all of their tillable acreage. On-farm research or comparisons have not been performed, however, and so no definite statement can be made about the reasons why the soil on the Spray Brothers Farm remains fertile.

Weed, Disease, and Insect Control

For the past 15 years the Spray brothers have been producing high yields of agronomic crops without the use of chemical pesticides, a considerable achievement that deserves further study. The clover in the rotation, later-than-normal planting dates for corn and soybeans, the tillage system, frequent cultivation, and some hand weeding are certainly factors in their success. Field observations in September 1986 confirmed that this alternative system is still working for them.

The Spray brothers did not stop using chemical herbicides because they were worried about the health and environmental risks associated with pesticide use. They stopped because herbicides were altering weed populations in such a way that weeds that had never been seen before were becoming problems. When this occurred, they stopped using herbicides and began to explore other weed control methods. The Sprays do not use chemical insecticides or fungicides because as a certified organic farm their products must be pesticide free to retain their organic designation.

Marketing Strategies

An important component of the financial performance of this operation is the farm's ability to market its corn, soybeans, wheat, oats, and adzuki beans, as well as some of its beef cattle, at higher-than-normal market prices. This kind of marketing is possible because the Spray Brothers Farm is an Ohio Ecological Food and Farm Association certified organic farm.

TABLE 6   Per Acre Costs of Field Operations Reported by Spray Brothers, 1982

Field Operations	Amount (dollars)	
	Corn After Sod	
	Unit Cost	Cost/Acre
Offset disk (3 times)	9.00	27.00
Field cultivate (3 times)	6.00	18.00
Spread fertilizer	1.75	1.75
Plant corn	7.50	7.50
Rotary hoe (1 time)	4.00	4.00
Cultivate (2 times)	5.00	10.00
Harvest	17.00	17.00
Total costs/acre		85.25
	Soybeans After Corn	
	Unit Cost	Cost/Acre
Chop stalks	6.50	6.50
Offset disk (2 times)	9.00	18.00
Field cultivate (3 times)	6.00	18.00
Plant beans	7.50	7.50
Rotary hoe (2 times)	4.00	8.00
Cultivate (2 times)	5.00	10.00
Harvest	18.00	18.00
Total costs/acre		86.00
	Wheat After Soybeans	
	Unit Cost	Cost/Acre
Spread fertilizer (1 time)	1.75	1.75
Disk (1 time)	6.00	6.00
Drill seed	5.50	5.50
Total costs/acre		13.25

TABLE 7   Reported per Acre and per Bushel Costs of Production for Corn and Soybeans in Ohio, 1982–1983[a]

Category	Corn (dollars)	Soybeans (dollars)
Variable costs		
Seed	15.00	7.00
Fertilizers	8.50	8.50
Other	85.25	86.00
Land charges	125.00	125.00
Total costs/acre[a]	233.75	226.50
Per bushel costs		
Low yield[b]	2.33	5.66
High yield[c]	1.94	4.53

[a]Costs reported by Spray brothers are not directly comparable to enterprise budgets from The Ohio State University Extension. The Sprays did not assess costs for labor. Procedures used for estimating equipment costs and land charges are not available.

[b]Forty bushels of soybeans and 100 bushels of corn.

[c]Fifty bushels of soybeans and 120 bushels of corn.

With this certification and judicious advertising in appropriate journals and meetings, the Sprays have achieved firm and regular specialty markets that add greatly to their profitability. It is clear, however, that not all midwestern farms could exploit such markets; if even a small percentage of farms shifted to organic methods, the market would become saturated and its premium prices would be greatly reduced.

The Spray brothers' recent investment in a seed cleaning, bagging, and storage facility complements this marketing strategy. The facility ensures a quality product and provides savings in cleaning and handling, some marketable wheat and clover seed sales, and a usable by-product (screenings for cattle feed). The facility also removes fines from shelled corn, wheat, oats, and beans, a capability that improves air flow and allows the Sprays to run an effective grain drying operation without supplementary heat.

Financial Performance

The overall economic viability of the Spray brothers' enterprise is strong at this time (see Table 5). Direct comparisons with the performance of similar farms using conventional methods would require whole-farm analysis of entire rotations, which is beyond the scope of the present study. However, per acre production costs for the Spray brothers are below the county averages (Tables 6 and 7). Yields are above the county averages (see Table 4).

The farm's effective management and the distribution of labor over the entire year is impressive. The brothers are obviously busy, yet they participate in community activities, host numerous visitors to their farm, give lectures to student groups at The Ohio State University, assist other farmers in establishing organic farms, and provide leadership for the Ohio Ecological Food and Farm Association. They also host a farm field day each year; on September 19, 1986, about 80 people attended this annual event. The Sprays are obviously proud of their operation and convinced that their system of farming is appropriate for them.

The Sprays and their methods are gaining some acceptance by certain faculty at The Ohio State University and by U.S. Department of Agriculture-Agricultural Research Service faculty at the nearby Coshocton station. The Sprays' immediate neighbors are not particularly supportive or enthusiastic about their operation, but the brothers are given credit for being good farmers by the Knox County extension agent. They are also willing to open their farm for further research and evaluation by The Ohio State University faculty, and there is currently some hope that this will occur.

Finally, the Spray brothers' response to inquiries about what, if anything, limits further acceptance of their farming system was enlightening. They felt strongly that, despite acknowledgment of their achievements, the major hurdle that their system faces is the reluctance of the state's Cooperative Extension Service to accept their methods and educate other farmers on crop rotations and alternatives to chemical fertilizers and pesticides.

2

A Mixed Crop and Livestock Farm
in Southwest Iowa:
The BreDahl Farm

T HE BREDAHL FARM IS LOCATED IN ADAIR COUNTY in southwestern Iowa, 4 miles south and 2 miles east of Fontanelle and about 60 miles southwest of Des Moines. The elevation is approximately 1,200 feet. The farm comprises 160 acres and has been in the BreDahl family since 1927 (Table 1). Clark BreDahl began operating the farm in 1974; he and his wife, Linda, currently cash-rent it from his mother. BreDahl has rented more land nearby, usually for hay and pasture, but until recently he felt that it would not be profitable to expand further, given the prevailing low commodity prices. In 1987, however, he planned to rent 160 acres of land on a crop-share basis for grain and forage production because he believed that it would add to his net income.

Like many farmers in Iowa, Clark BreDahl is an agricultural college graduate; Linda BreDahl teaches school full-time and, because of teaching schedules, is available for limited farming activities during most of the year and full-time farming in the summer.

GENERAL DATA

In some respects, Clark BreDahl farms as his father did, using crop rotations and strip cropping. He raises about 35 to 40 acres of corn, 35 to 40 acres of soybeans, 20 acres of alfalfa, and 20 to 30 acres of pedigreed oats for seed (the oats are sold as either registered or certified seed, depending on the opportunity).

The basic livestock enterprise is two flocks of sheep. One is a flock of 40 registered Rambouillet ewes, and the other is a flock of 130 commercial ewes of Rambouillet × Finn breeding (Finnish Landrace). These commercial ewes are bred to Suffolk rams to produce three-breed crosses for the slaugh-

TABLE 1 Summary of Enterprise Data for the BreDahl Farm

Category	Description
Farm size	160 acres, 300–500 ewes and lambs, 30–50 cattle, 10–20 sows
Labor and management practices	All management, which is intensive because of the highly diversified operation, is provided by the farmer. Farm labor is provided entirely by the family, except at haying, lambing, and shearing times, when additional laborers are hired. The farmer's wife works off-farm as a teacher.
Livestock management practices	Cattle and sheep glean fenced corn fields and feed on turnips in fall and winter.
Marketing strategies	Oats sold as seed bring double the feed price. Soybeans are sold on the regular market. Corn and hay are fed to livestock. Beef and sheep are sold on the regular market.
Weed control practices	The farmer uses ridge tillage, controlled burning, and rotary hoeing (both preemergence and weekly for 4 weeks). Disk cultivation is also used if weeds become a problem. Spot spraying with glyphosate or paraquat is used to control thistle or morning glory.
Insect and nematode control	The farmer reports no problems with insects or nematodes. Animals consume crop residues, and the fields are usually kept in a 5-year rotation.
Disease control practices	No disease problems were mentioned.
Soil fertility management	The farm's typical rotation is corn-soybeans-corn-oats-alfalfa (1 or 2 years). Turnips are sometimes planted with oats. A small amount of commercial fertilizer is applied to first-year corn after alfalfa (sometimes 20–40 pounds N, up to 30 pounds P_2O_5, and 30 pounds K_2O, closely following university recommendations). In addition, 80–120 pounds N are applied to corn after soybeans. Manure is applied prior to planting oats.
Irrigation practices	None
Crop and livestock yields	Corn yield averages are about 100–120 bushels/acre; soybeans yield 35–40 bushels/acre.
Financial performance	Costs are reduced by the use of on-farm resources (feeds, nitrogen fixation, operator labor) rather than relying on purchased inputs. The farm's cash flow obviates the need for borrowed capital. Net returns from the farm are adequate to support the family during most years.

ter lamb market. Some of the F × R crossbred female lambs are sold for breeding purposes, and usually 100 to 300 additional feeder lambs are purchased to finish for slaughter.

As opportunities arise, the BreDahls may purchase 30 to 50 feeder cattle annually. The swine enterprise on the farm consists of a herd of 10 to 20 white (usually Yorkshire) sows in a farrow-to-finish system. The BreDahls move in and out of swine farming as their economic and resource situations dictate. (They disposed of a cow herd recently because the farm was unable to sustain it.)

BreDahl is a careful livestock manager. He protects young sheep and swine in the livestock housing facilities and follows the best sanitation practices to prevent diseases. He also uses veterinarians frequently and will use any medication recommended for sick animals. However, he does not routinely use subtherapeutic levels of antibiotics in the animal feed. Protein supplements, prepared by the local cooperative, are fed along with home-grown grain. The various rations are determined according to extension service recommendations.

Climate

The climate of Adair County is typical of southwestern Iowa—subhumid and continental, with cold winters and hot, humid summers. The maximum mean daily temperature in July and August is 85°F. Average annual precipitation in nearby Greenfield (9 miles northeast) is 33 inches, mostly in the form of rain; 70 percent of the annual precipitation falls in the months of April through September (Table 2). The wettest months are between May and September, each with about 3 to 5 inches of rainfall. One year out of 5, the area will experience maximum daily temperatures of over 95°F in May and June, 99°F in July, and 97°F in August.

TABLE 2   Normal Monthly Temperatures and Precipitation at Greenfield and Corning, Iowa

Month	Normal Temperature (°F)			Normal Precipitation (inches)	
	Greenfield	Corning		Greenfield	Corning
January	19.4	19.3		0.89	0.90
February	25.6	25.2		1.21	0.92
March	35.7	35.1		2.33	2.14
April	50.7	50.0		3.34	3.15
May	61.9	60.9		4.11	4.06
June	70.9	69.8		4.72	4.55
July	75.6	74.8		3.69	4.09
August	73.3	72.6		3.96	4.90
September	65.0	63.9		3.87	4.28
October	54.3	53.1		2.32	2.33
November	38.7	38.3		1.47	1.58
December	26.3	26.2		0.87	0.89
Average annual	49.8	49.1	Average annual total	32.78	33.79

NOTE: The BreDahl Farm is 9 miles from Greenfield and 25 miles from Corning. The normal monthly temperature is the average of the normal daily maximum and minimum temperatures for that month from 1941 to 1970. The normal monthly precipitation is the average of the inches of precipitation for that month from 1941 to 1970.

SOURCE: National Oceanic and Atmospheric Administration. 1980. Climates of the States, 2d ed. Detroit: Gale Research Co., Book Tower.

PHYSICAL AND CAPITAL RESOURCES

Soil

Adair County soils were formed under the area's native vegetation, tall prairie grass. The terrain varies from nearly level upland ridges and bottom-land to moderately sloping uplands. The soils are moderately well-drained to poorly drained silty clay loams mostly formed on loess deposits. As water percolates down through the soil's loess mantle, it reaches a relatively impermeable paleosol developed in glacial till. Water tends to move along this less-permeable material and seeps out along the hillsides leading to areas that are usually too wet to till unless drainage tile is installed. The topography of the BreDahl Farm is moderately rolling.

The predominant soil type on the farm is Sharpsburg silty clay loam with less than 5 percent slope. Sizable acreages of relatively flat Macksburg and Winterset soil series are present along with Nira silty clay loam on the steeper slopes (5 to 9 percent). Average expected corn yields under excellent management on these soils range from 148 to 161 bushels per acre. The area's corn suitability rating (CSR), which is based on yield and intensity of production, ranges from 69 to 95—with 100 being the best in the state. Poorly drained upland Clearfield silty clay loam and Clarinda silty clay loam have CSR values of 50 and 30, respectively, and are not under row-crop cultivation. Likewise, the poorly drained alluvial Colo-Ely silty clay loam complex is not usually row-cropped.

The farm's soils are classified in land capability classes 1 through 8, depending on the degree of slope and other land-use capability factors. Fewer than 20 acres of the farm are in U.S. Department of Agriculture Soil Conservation Service (SCS) class 4, which has up to 13 percent slope. This land is used as permanent pasture by the BreDahls; under conventionally tilled row crops, this class of land would incur erosion probably exceeding 30 tons per acre per year. Even land with only 5 to 9 percent slopes would sustain excessive erosion if it were in continuous row-crop production.

BreDahl said that when his father bought the farm in 1927, it was run down and not very productive. This led him to use strip cropping and crop rotations. Today, much the same type of crop rotational system is being used. In the most recent soil survey of Adair County, the BreDahl Farm was found to be one of the few in the county on which the soils were not classified as moderately to severely eroded.

The farm is not irrigated, but there are four shallow wells for household and livestock use. A pond was constructed in 1974 with the help of the SCS and the local Agricultural Stabilization and Conservation Service (ASCS) office. ASCS shared the costs of building the 1-acre pond and buying the grass seed needed to establish a permanent vegetative cover around the pond.

Clark BreDahl approves of terracing in instances in which nothing else will work; but on most of his crop acreage, he feels that he can get the same soil conservation benefit at a lower cost by using no-tillage and ridge-tillage systems for row crops along with strip cropping. Another reason for his preference for these methods is that terracing interferes with some farming operations and is costly on these soils.

Buildings and Facilities

The BreDahls have a good set of outbuildings, barns, and feedlots. The barn, built in the 1950s, is 48-by-54 feet with a hayloft. Two open-front pole sheds (60-by-34 feet and 60-by-36 feet) complete the major general-purpose buildings. The farm is completely equipped to handle pig, sheep, and cattle raising. Corn is picked and stored as ear corn. Because the entire farm is fenced and cross-fenced, after the harvest, cattle and sheep can be turned into the fields to consume crop residues.

Machinery

BreDahl has been able to keep his equipment costs to a minimum. He estimates that the current market value of all his equipment would be approximately $25,000. He has two tractors (a 1967, 65-horsepower model and a 1966, 45-horsepower model), a 4-row no-tillage planter, a 4-row disk cultivator, a grain drill, a rotary hoe, a disk, and a plow. His newest piece of equipment is a 1980 square baler with hay mower, rake, and hay wagons. He uses a 2-row corn picker to harvest ear corn. His corn is stored in corn cribs because both harvesting and drying are less expensive and storage is simpler using this method. Much of the ear corn is ground on the farm and fed to livestock both by hand and in self-feeders.

MANAGEMENT FEATURES

Crop Rotation

The BreDahl Farm's crop rotation usually extends over 5 or 6 years, consisting of a corn-soybeans-corn-oats-alfalfa sequence. Sometimes alfalfa is kept for 2 years. Turnips are sometimes sown with the oats. By growing alfalfa for 1 or 2 years for hay, BreDahl said that most of the nitrogen (N) needed to raise corn (100 pounds N or more per acre) is supplied by the legumes in his rotation. He may add 20 to 40 pounds N with phosphorus and potassium fertilizer to sod ground going into corn, depending on the results of annual soil tests. Generally, 80 to 120 pounds N in the form of anhydrous ammonia is side-dressed on corn after soybeans.

Soil Fertility

Soil tests are generally in the high to very high range for potassium, and up to 30 pounds of potassium (K_2O) are applied per acre to maintain these levels. Phosphorus usually tests in the medium range, but is increasing. Up to 52 pounds of phosphorus (P_2O_5) per acre are applied from all sources (manure and fertilizer). Soil pH is maintained between 6.5 and 6.9. Soil fertility additions closely follow university recommendations.

Feedlot manure is spread on the soil following soybeans or corn in the rotation and before the field is planted to oats. In the BreDahls' operation, it is most convenient to spread the manure at this position in the rotation because it interferes less with row-cropping activities. Packed manure left on the soil surface interferes most with the use of a ridge-tillage planter and the cultivator. Most of the fields in the rotation receive manure once per rotation, that is, every 4 to 6 years. The manure from 300 to 500 lambs, 30 to 50 cattle, and up to 250 hogs is typically spread over an area of 20 to 30 acres each fall and spring. BreDahl tries to avoid lodging by the careful selection of oat cultivars following soybeans; he also plans to windrow the seed oats to alleviate this problem.

The farm's corn yield in 1985 was 140 bushels of corn per acre, the highest yield ever obtained. Average yields are generally in the 100- to 120-bushel range, except in 1977, when a drought reduced yields to less than 20 bushels per acre. BreDahl's ASCS corn base yield in 1986 was 107 bushels per acre, down 3 percent from the previous year. The county average corn base yield is 96 bushels per acre per year. All of the corn is fed to livestock; it is shelled for sheep and hogs but ground with the cob for feeder cattle.

The farm has had less success producing soybeans, which average 35 to 40 bushels per acre. In 1986, however, a farm record was set at 45 bushels per acre. The soybeans are sold on the open market, and the oats are sold for seed, which brings nearly double the market price. Soybeans and alfalfa are fertilized minimally. The nitrogen supplied by the soybeans and manure is adequate to make an excellent crop of oats. Yields over the last 10 years have ranged from 65 to 100 bushels per acre per year.

Turnips are often planted as a second crop on some of the oat acreage. After the oats and straw are harvested in July, BreDahl typically disks the ground, broadcasts 4 to 5 pounds of turnip seed along with 50 pounds dry N per acre, and then disks the ground again, lightly. (This practice is also used on government-diverted acres.) BreDahl prefers the standard garden variety of purple-top globe turnips, although he has tried other varieties. He has learned that both oats as a sole crop and turnips as a sole crop do better than if seeded together. Seeding the crops separately also makes the turnip pasture more timely: it comes to maximum production beginning in late September and sometimes lasts into the new year.

Government-diverted acres on the BreDahl Farm typically are on the ground previously planted in soybeans. Thus, instead of planting oats on all the soybean ground, some of the land is set aside to qualify for government support payments. Set-aside land is typically seeded with a cover crop of turnips prior to the end of June, and the oat acreage is planted in turnips as soon as possible after harvest in July.

TABLE 3   Adair County Estimated Soil Losses Under Various Tillage Methods, 1986

Tillage Method	Percentage of Cropland Prepared		Estimated Soil Loss (tons/acre/year)
	Corn	Soybeans	
Conventional (plowing)	41.6	23.0	15–20
Mulch tillage (chisel plowing)	45.7	75.4	12–15
Ridge tillage, strip-cropped	1.5	0.2	4–6
No tillage	11.2	1.4	4–6

SOURCE: R. BreDahl, Adair County extension director, communication, 1986.

Tillage and Planting Methods

BreDahl's row-crop planting methods alternate among no tillage, ridge tillage, and some moldboard plowing. Using his 4-row planter with 38-inch rows, Clark BreDahl plants 32 rows of corn or soybeans in eight passes. Then he plants the same-sized strip with alfalfa or oats. His fields are planted, therefore, in approximately 100-foot-wide strips. His disk cultivator matches his planter.

Depending on weather conditions, he may plant corn without tillage into an alfalfa meadow after killing the legumes with (2,4-Dichlorophenoxy) acetic acid (2,4-D) (applied at a rate of 1 to 2 quarts per acre) in the spring when the plants are 6 to 8 inches tall. A residual herbicide is usually applied at this time as well. If this system is used, BreDahl does not cultivate the ground and returns to the field only to harvest the crop. A side-dressing of fertilizer may be applied at planting. Alternatively, and preferentially, the alfalfa is plowed under in the spring, and weed control is by cultivation. After two cultivations, a ridge is formed that will be used to ridge-plant soybeans without tillage the following spring.

According to the Adair County extension director and the district soil conservationist, less than 13 percent of the corn and 2 percent of the soybeans in the county are grown using either no-tillage or ridge-tillage systems, even though these systems result in a significant reduction in soil loss (Table 3).

Clark BreDahl maintains organic matter in the soil through crop rotation, by incorporating the crop residue left by his livestock and by the addition of manure.

Weed Control

By adjusting his cultivator disks, BreDahl can create a ridge and cover weeds and crop residue. His ridge-tillage planter then levels the ridge and plants the seeds in the newly opened soil. In this fashion the corn gets a head start on the weeds. The rotary hoe is used 3 to 4 days after planting, before the crop has germinated and before weeds are visible. The same procedure is repeated once or twice, weather permitting. The rotary hoe does its best work before the weeds emerge, destroying them before they break the surface. If weeds become a problem, two pairs of disk hillers are set on the cultivator, one behind the other (one pair to throw the soil away from the plant and the other to move it back). The cultivator shields are always kept in place because even fairly large corn and soybean plants can be damaged or covered completely by the soil moved by the disk hillers.

BreDahl is not philosophically opposed to using herbicides, but he uses them sparingly, mainly on corn that has been no-tilled into alfalfa. His primary contention is that on an operation the size of his farm, mechanical weed control is just as effective and considerably cheaper than chemical weed control. The reduced risk of pollution is considered a side benefit.

PERFORMANCE INDICATORS

The BreDahl Farm is less than half the size of the average farm in Adair County (337 acres), but Clark BreDahl maintains that it is not the size of the operation that is important, but what is left after cash costs are subtracted. Careful cost containment is a characteristic of the BreDahl operation. A local farm management tax consultant, who handles a few hundred farms in the area, indicated that in one year recently the BreDahl Farm was in the bottom 25 percent of the area farms in terms of gross sales, but in the top 10 percent in terms of net income.

The successful operation of a farm like the BreDahls' requires a high degree of managerial ability, a quality the BreDahls seem to have. Their management strategy has been to minimize risks and to pay off all debts. They appear to have accomplished both goals. Recent sales of some of their beef and sheep herds, as well as feeder pigs, have eliminated their debt. Currently, they do not borrow for operational costs and do not owe money for equipment.

Data were not available for a detailed analysis of the farm's financial performance. The BreDahls report, however, that they have internalized most of their operational costs by minimizing purchased inputs and by reducing interest expenses for the farm to zero. They provide all of their own labor except at lambing and haying time; they hire custom shearing of the sheep—this item represents about two-thirds of their $1,600 annual labor bill. The variety of livestock raised on the farm provides market animals and income throughout the year. In addition, because the farm is cross-fenced, the BreDahls can make use of poor or damaged crops and crop residue as feed sources instead of letting the crops rot in the fields.

Clark BreDahl reports that in 1986 the average cash rental for corn land (row cropland in general) was $66.00 per acre in Adair County. The Bre-Dahls, on the other hand, pay $50.00 per acre (to Clark BreDahl's mother) for the whole farm they rent. Given that barely half the farm is planted with corn and soybeans, however, this appears to be a fair rental charge.

Although he has participated in the government's commodity support programs, BreDahl feels that the family farmer would be better off without them. He concludes that these programs have encouraged farmers to convert to a monocrop culture without livestock and that they have greatly reduced the number of farm families. He does not think that his own family's way of farming would be as feasible for a farm 10 times the size of their farm and spread out in various locations. He does feel, however, that by eliminating price supports, farmers with a good managerial strategy eventually could be better off.

The BreDahls' crop rotation system of 1 or 2 years in alfalfa, 2 years in corn, 1 year in soybeans, and 1 year in oats and possibly turnips has minimized purchased farm inputs of fertilizer and pesticides while maintaining high roughage production. Similarly, the alfalfa, whether plowed under or no-tilled, provides nearly all the nitrogen required by the corn crop. In addition, cattle and sheep glean the corn fields, thereby reducing the need for supplemental feeding and minimizing volunteer corn problems in the subsequent crop of soybeans. The farm's actual costs of production in 1986 (land and cash) were $1.52, $3.10, and $1.39 per bushel for corn, soybeans, and oats, respectively.

The double crop of turnips provides an excellent feed source for the BreDahls' sheep enterprise. Animals can begin grazing the turnips in 60 to 80 days, but they obtain optimum dry matter after 90 days. Both cattle and sheep have been grazed on the crop, but sheep tend to be more efficient, wasting less of the forage. The turnips provide a complete ration for sheep as well as 80 to 85 percent of their water consumption. The turnips are grazed from mid-September through early winter.

When the turnips reach full growth, three-quarters of the root will be aboveground. The sheep first eat this part of the root and the vegetative top and then graze the turnip down and cup out the tap root below the surface. The holes left in the soil after grazing fill with water, snow, and ice, which help to hold soil moisture and prevent runoff. The BreDahls have run up to 200 ewes for 10 days on an acre of turnips—that is, 2,000 ewe-days per acre. More commonly, they average 1,400 to 1,500 ewe-days per acre. Clark BreDahl estimates that this method provides maintenance at a current cost of less than $0.04 per day per ewe.

The BreDahls' system of farming could be used by others but only so long as such farms (1) diversified their operations enough to raise the livestock and to take advantage of all crop roughages and (2) maintained a small enough operation to be managed properly. The system works most effectively when a farm is in one contiguous unit and cross-fenced. Careful budgeting and management are also essential to success in operations such as this.

The family's financial goal, although not always achieved, has been to make a living from the farm and to save Linda BreDahl's teaching salary. This case study illustrates that, given the required management skills and a conservative investment strategy, a family can still make a living today on a 160-acre diversified farm in Iowa.

3

A Diversified Crop and Livestock Farm in Virginia: The Sabot Hill Farm

SABOT HILL FARM is located a few miles northwest of Richmond, Virginia, in Goochland County. Run by Sandy and Rossie Fisher, the farm is large in comparison with other farms in the eastern United States. The land is in one parcel of more than 3,000 acres, half of which is in forest. The Fishers purchased an adjoining farm of an additional 480 acres 4 years ago, making their total holdings 3,530 acres (Table 1). The average size of a farm in Goochland County is 200 acres.

GENERAL DATA

Sabot Hill is a diversified beef cattle, forage, and cash grain farm. The livestock enterprise consists of 500 head of cattle. Three hundred cows are bred to calve in the fall, and 200 are stocker cattle. The Fishers produce hay both as cattle feed and as a cash crop. They plant approximately 125 acres in an alfalfa-orchard grass mixture grown on Madison soils and normally harvest from 3.5 to 4 tons per acre. This hay is square-baled and sold primarily to neighboring horse owners. The Fishers grow orchard grass on another 65 acres, which yield 1.5 tons per acre. The rest of the hay acreage (300 acres) is a mixture of fescue, Johnsongrass, legumes, and millet, which is primarily used on the farm or sold as cattle roughage. Yields of 2 tons per acre are common (Table 2).

Corn and soybeans are major cash crops. The Fishers produce approximately 175 acres of corn each year, of which 75 are irrigated, and 150 acres of soybeans. They also participate in the Federal Feedgrains Commodity Program and raise the corn to maintain their corn base. The established yield at the farm is 78 bushels per acre. The Goochland County Agricultural Stabilization and Conservation Service (ASCS) corn base average, by com-

275

TABLE 1 Summary of Enterprise Data for the Sabot Hill Farm

Category	Description
Farm size	3,530 acres, 500 beef cattle
Labor and management practices	The family provides farm management and much of the labor, although there are several hired employees. Animal births are timed to coincide with the maximum amount of available labor.
Livestock management practices	The farmer raises 300 cow-calf units and 200 stocker cattle. A Hereford-Angus cross is predominant, with some other breeds raised on a trial basis (for example, Brahman). Feeder calves are grazed with no supplement used.
Marketing strategies	Cattle are sold in the Lancaster, Pennsylvania, market in 60-head lots.
Weed control practices	The farmer interseeds native Johnsongrass with legumes and harvests it as a crop until the grass is depleted. The farm uses a 2- or 3-year rotation of corn, hay, and soybeans. The estimated savings on chemicals is $20,000 per year.
Insect and nematode control	No insect or nematode problems are reported.
Disease control practices	Pinkeye (conjunctivitis) and lumpjaw (caused by eating foxtail) are the major livestock diseases. All cattle receive complete injections for bovine diseases. Calves are not fed antibiotics.
Soil fertility management	The farmer uses 2- or 3-year rotations with corn, hay, and soybeans sometimes double- or triple-cropped. Commercial fertilizer (20-20-20) is applied at corn planting; 30 pounds N is applied by side-dressing. No other crops receive commercial fertilizer.
Irrigation practices	The farmer irrigates 75–175 acres of corn with moving sprinkler guns (280-foot diameter coverage). Other crops and some corn are grown dryland.
Crop and livestock yields	Hay yields range from 1.5–4.0 tons/acre. Corn yields are 100 bushels/acre dryland and 165 bushels/acre irrigated.
Financial performance	The farmer reports annual savings of $20,000 on chemicals, largely because of the shift from chemical control of Johnsongrass to cultural practices (overseeding with competing legumes and harvesting as hay).

parison, is 61 bushels per acre. The Fishers' primary focus, however, is on maximizing forage production, the main feed input to their beef cattle.

PHYSICAL AND CAPITAL RESOURCES

Soils

The topography of Sabot Hill runs from flat cropland along the James River to rolling to steep hillsides, which are primarily in permanent pasture, hay, or forests. Most of the land is not well suited for row-crop production. The soils include Louisburg, Madison, and Pacelot varieties.

The Soil and Conservation Service (SCS) has placed Virginia's soils into four categories according to their productive potential for specific field and

TABLE 2   Crops and Yields on Sabot Hill Farm, 1986

Crop	Acreage	Yield/Acre
Cash grains		
Corn, dryland	100	100 bushels
Corn, irrigated	75	165 bushels
Soybeans	150	34 bushels
Forest	1,500	—
Hay		
Alfalfa-orchard grass	125	3.5–4.0 tons
Mixture (fescue, Johnsongrass,		
legumes, millets)	300	2 tons
Orchard grass	65	1.5 tons
Pasture	1,215	[a]
Total	3,530	—

[a]The carrying capacity is 0.5 animal units (cattle)/acre.

forage crops. The Louisburg and Pacelot soils found on Sabot Hill are in class 4, which is the least productive soils classification (their productive potential for corn is less than 90 bushels per acre). Madison soils are in class 3 for corn, having a productive potential of 90 to 110 bushels, but they are in class 2 for alfalfa hay production.

Approximately 400 acres of the farm lie along the floodplain of the James River. These soils are silt loam, mostly of the Monocan classification. Monocan soils are class 2 soils, having a productive potential for corn of 110 to 135 bushels per acre. The Fishers have invested nearly $600 per acre in a drainage tile system on 200 acres of river bottom ground now being used for crop production. The Monocan soils have less than a 2 percent slope.

The rest of the farm is considerably more hilly. Approximately half of this acreage is in woods that have slopes in the D and E category (12 to 25 percent slope) (Cook, 1962). This land is composed primarily of Louisburg soils. Much of the woodlands are fenced for cattle. The rest of the farm, which lies on the Piedmont Plateau, is composed of primarily Madison-type soils having B (2 to 6 percent) and C (6 to 12 percent) slopes.

Before the Fishers took over the farm, most of this land, over 800 acres, was conventionally farmed. Soil erosion on the more hilly ground was estimated to have exceeded 30 tons per acre; on some of the less sharply sloped ground, erosion was estimated at 15 to 20 tons per acre. Corn yields on these upland soils rarely averaged over 80 bushels per acre during good years, and the yield was half that during dry seasons.

Through strip cropping and converting much of the highly erodible land to permanent pasture and hay ground, and through the use of no-tillage planting equipment, the Fishers have been able to cut soil erosion to 4 tons per acre. Sandy Fisher has been a leader in the use of no-tillage equipment and conservation measures in the county and state. By shifting from conventional row-crop farming to a more sound, conservation-oriented operation with primary emphasis on forage and livestock production, the Fishers have significantly reduced the rate of soil erosion.

Indeed, the Fishers take a keen interest in their soils. They regularly have all of their fields soil-tested by the Brookside Farm Laboratory Association, which tests annually for nitrogen, sulfate, phosphate, calcium, magnesium, potassium, sodium, zinc, and boron.

The Sabot Hill Farm also includes a custom lime-spreading company, which spreads about 5,000 tons of lime per year on various other farms at $21.00 per ton. Because they own this company, the Fishers apply lime on their farm more frequently than most other farmers in the area. They also lime their pastures whenever necessary to maintain a soil pH of about 6.5.

Irrigation

The Fishers currently irrigate 75 to 175 acres of corn using a movable, self-coiling, single-head sprinkler gun. The moving gun sprinkler attached to the 6-inch retractable hose line can be extended 800 feet. The gun can irrigate a 280-foot-diameter field, pumping 600 gallons per minute. The pumping station is a movable 100-horsepower air-cooled engine with a 4 × 4 pump, burning from 2.5 to 3 gallons of diesel fuel per hour. The irrigated ground receives two applications of water at 2 inches each time.

Normal annual precipitation in the area is 42.6 inches of moisture, nearly all of which is in the form of rain (Table 3). The Fishers' irrigated corn crop on 75 acres yielded 165 bushels per acre during a severe summer drought in 1986; their nonirrigated bottomland corn crop yielded 100 bushels per acre that year.

Buildings and Facilities

The Sabot Hill Farm has an extensive set of buildings and facilities. There are over 20 houses on the farm, half of which house employees; the rest are rented out.

The inventory of equipment and machinery is extensive. The farm has a well-equipped modern shop in which much of the machinery is overhauled.

MANAGEMENT FEATURES

Crop Rotations

When the Fishers took over the management of this farm 9 years ago (they had spent more than 7 years previously farming and ranching in Colombia), the farm was planted with nearly 300 acres of corn and 200 acres of soybeans. Herbicide expenses for these crops were more than $26,000 annually, much of which was spent to control Johnsongrass in the cropland along the river. The farm was using EPTC plus R–25788 on 500 acres of cropland, applying 1 gallon of active ingredient per acre at $40.00 per

TABLE 3   Normal Daily Temperatures and Monthly Precipitation at Richmond, Virginia

Month	Normal Daily Temperature (°F)			Normal Degree Days		Normal Precipitation	
	Maximum	Minimum	Monthly Average	Heating	Cooling	Inches	Days With ≥0.01 Inches
January	47.4	27.6	37.5	853	0	2.86	10
February	49.9	28.8	39.4	717	0	3.01	9
March	58.2	35.5	46.9	569	8	3.38	11
April	70.3	45.2	57.8	226	10	2.77	9
May	78.4	54.5	66.5	64	111	3.42	11
June	85.4	62.9	74.2	0	276	3.52	10
July	88.2	67.5	77.9	0	400	5.63	11
August	86.6	65.9	76.3	0	350	5.06	10
September	80.9	59.0	70.0	21	171	3.58	8
October	71.2	47.4	59.3	203	27	2.94	7
November	60.6	37.3	49.0	480	0	3.20	8
December	49.1	28.8	39.0	806	0	3.22	9
Average annual	68.8	46.7	57.8				
Average annual total				3,939	1,353	42.59	113

NOTE: The normal daily maximum by month is the average of each day's (midnight to midnight) high temperature for every day in that month from 1941 to 1970. The normal daily minimum by month is the average of each day's (midnight to midnight) low temperature for every day in that month from 1941 to 1970. The normal monthly temperature is the average of the normal daily maximum and minimum temperatures for that month. The normal degree days heating are the sums of the negative departures of average daily temperatures from 65°F. The normal degree days cooling are the sums of the positive departures of average daily temperatures from 65°F. To calculate the normal degree days heating or cooling, multiply the difference between 65°F and the normal monthly temperature by the number of days in the month. The normal monthly precipitation is the average of the inches of precipitation for that month from 1941 to 1970.

SOURCE: National Oceanic and Atmospheric Administration. 1980. Climates of the States, 2d ed. Detroit: Gale Research Co., Book Tower.

gallon, for a total of $20,000; in addition, the farm spent approximately $5.00 per acre to prepare the land and apply the chemicals.

The Fishers said that when they first took over the farm, they tried many herbicides, especially EPTC plus R–25788, but the effectiveness of the herbicides appeared to be decreasing. The conclusion they reached was that either the Johnsongrass on the farm was building up a tolerance to the chemical or the organic matter in the soil was tying up the chemical.

Faced with this realization, the Fishers decided to change their primary goal from that of cash grain farming to forage production and to alter the farm's cropping system. Their experience in South America led them to believe that there might be some value in intensively haying Johnsongrass, especially if it could be interseeded with a legume. With this in mind, their focus shifted from trying to eliminate Johnsongrass by spraying to accepting it as a part of the forage mix. The main management strategy changed to one of maximizing the amount of food and feedstuff the farm could produce with a minimum of purchased inputs.

Today, the farm's acreage in corn and soybeans has been cut to 175 and 150 acres, respectively, and the chemical bill has been reduced to around $6,000 per year, a savings of $20,000. Most of the corn and soybeans are still sold on the cash grain market, but the focus of the Fishers' production and marketing strategy is now trained more intensively on getting as much feed value as possible from pasture for the cattle and selling roughage to the farm's neighbors. The feeding values of Johnsongrass and other species are given in Table 4.

The farm has a complete complement of haying equipment. But the key piece of equipment that assisted the switch in cropping systems was the purchase of a no-tillage grain drill. By using no-tillage planting, the Fishers avoid the high cost and severe erosion that would result from the tillage of pastureland prior to planting. No-tillage planting makes pasture renovation economically feasible. The farm currently owns two no-tillage drills. The drills, which cut easily through heavy corn stubble or heavy soil on pastureland, are heavy (5 tons when loaded) and are equipped with 14-inch hydraulically operated coulters.

Although the rotational system varies for different areas of the cropland, the Fishers use two main rotations on various fields: a 2-year rotation of corn, hay, and soybeans with three crops in 2 years; and a 3-year rotation of corn, hay, and hay-soybeans. In the 2-year rotation, the Fishers plant corn in early spring and harvest it in September or October. They drill a millet-pea-rye mixture into shredded corn stubble around October 15 and harvest two cuttings of hay the following spring before drilling soybeans in July. They plant corn in the spring (in conventional 38-inch row centers with a high plant population of 26,000 seeds per acre) on 175 acres of bottomland in fields from 20 to 40 acres in size that are plowed and disked before planting. The Fishers use a corn variety that requires 115 days from planting to maturity. This cultivar is resistant to viruses carried by the Johnsongrass but produces yields of from 10 to 15 percent less than other cultivars that do not have this resistance.

TABLE 4   Feeding Value for Ruminants of Selected Plant Species on Sabot Hill Farm (in percent)

Roughage	Total Dry Matter	Crude Protein	Total Digestible Nutrients	Calcium	Phosphorus
Dry					
Alfalfa and bromegrass hay	89.3	13.3	46.0	0.74	0.24
Alfalfa hay (all analyses)	90.5	18.4	54.0	1.33	0.24
Johnsongrass hay	89.0	8.5	48.0	0.87	0.26
Millet hay (foxtail varieties)	87.0	7.5	51.0	0.29	0.16
Mixed hay (good)—less than 30% legumes	88.0	8.4	49.8	0.61	0.18
Orchard grass hay (early cut)	89.0	13.4	58.0	0.35	0.32
Pasture grasses and clovers (mixed, from closely grazed fertile pastures; dried, from Northern states)	90.0	20.3	66.7	0.58	0.32
Pearl millet hay	87.2	7.0	49.8	—	—
Soybean hay (good, all analyses)	91.0	15.7	52.0	1.15	0.24
Sudan grass hay (all analyses)	91.0	7.3	51.0	0.50	0.28
Green					
Cabbage, entire	9.4	2.2	8.1	0.06	0.03
Clover/mixed grasses, hay stage	27.3	3.0	18.9	0.16	0.08
Corn fodder, dent (all analyses)	30.0	2.5	21.0	0.09	0.08
Johnsongrass pasture	25.0	5.2	15.6	0.22	0.07
Orchard grass pasture	23.0	4.3	17.0	0.14	0.12
Pearl millet pasture	21.0	1.8	13.0	—	—
Soybeans and pearl millet pasture	24.5	4.2	15.9	—	—
Turnips, roots	9.0	1.1	7.8	0.06	0.02
Turnips, tops	15.0	2.0	10.8	0.49	0.06

NOTE: A dash indicates that data are not available.

SOURCES: Morrison, F. B. 1949. Feeds and Feeding, 21st ed. Ithaca, N.Y.: Morrison Publishing Company. National Research Council. 1982. United States-Canadian Tables of Feed Composition, 3d. rev. Washington, D.C.: National Academy Press.

The Fishers also broadcast up to 500 pounds of nitrogen, phosphorus, and potassium (NPK) per acre in a 20–20–20 mixture and chisel the soil 11 to 12 inches deep before planting. After planting, they apply 2.5 pints of atrazine and 1 gallon of EPTC plus R–25788 with 60 pounds N over the top; they apply an additional 30 pounds N as a side-dressing at 50 days just after planting.

Hay follows corn in both the 2- and 3-year rotations. Various millets (German and pearl) are no-tilled into the ground along with black-eyed peas (or Austrian winter peas) and rye. These plantings are usually done after October 15, as soon as the corn is harvested. No fertilizer or herbicides are used. The planting rate is one bushel of rye, one-half bushel of winter peas, and one-half bushel of millet per acre. The total seed cost is less than $20.00 per acre. The Fishers claim that this mixture provides an excellent cattle feed. Peas provide 14 to 15 percent protein, and millet provides 5 to 6 percent. Using this mixture, about 147 tons of round bales were harvested in 1986 in two cuttings from 50 acres (or 2.9 tons per acre).

Depending on the weed populations, hay can be cut until fall, and another cutting can be harvested the following spring before no-tillage planting of soybeans. Alternatively, the hay can be cut only twice, and then no-till soybeans can be planted in July. If there are too many weeds, especially Johnsongrass, the Fishers will keep the field in hay for another year before planting soybeans. In this case, they will redrill the hay mixture in the fall. If Johnsongrass is relatively sparse, however, they will drill soybeans after the second hay cutting in May, no-tillage planting the soybeans in 7-inch rows. If weeds become a problem in the field, they may harvest the soybeans along with the weeds as a hay crop. The Fishers normally do not apply herbicides to soybeans.

In summary, the unique feature of the Fishers' cropping system is their view of Johnsongrass and other weeds: they no longer focus on trying to eliminate them but instead cultivate them as sources of feed for the livestock operation. In learning more about Johnsongrass—especially its feed value and how it grows—the Fishers have developed the strategy of interseeding millet and various legumes in infested fields and putting ground with limited potential for row crops back into productive use.

They chose millet because it competes well with Johnsongrass and has comparable feeding value (see Table 4). Even fields heavily infested with Johnsongrass can eventually be made to produce: the grass can be harvested as hay three or four times for a year or two to weaken its root system, which makes it easier for grains or soybeans to grow there in subsequent years without herbicides. By incorporating Johnsongrass into a hay crop, the Fishers derive economic benefit from land that would otherwise have little value. (Similar land infested with Johnsongrass rents for as little as $5.00 to $10.00 per acre.)

Last year the Sabot Hill Farm sold about 1,500 tons of regular bales of hay as well as about 600 tons of large round bales. Forage yields per acre vary considerably depending on whether the hay ground is harvested only twice and then planted with soybeans or whether it is hayed all season long.

Pasture Renovation and Interseeding

By using a no-tillage planter the Fishers estimate that they have more than doubled the productive capacity of many of their pastures. For seed costs of under $20.00 per acre (plus a cash operating cost for planting of less than $5.00 per acre), they have been able to field-graze their cattle until February, compared with grazing only until September or October for un-improved pasture. This extension permits them to sell most of their round bale hay crop to neighbors short on feed, thus increasing the farm's profits.

The following outlines a typical approach to pasture renovation at Sabot Hill. After the pastures have been grazed down around September 1, the Fishers drill in 1 bushel to the acre of wheat, barley, or rye, mixed with 0.5 bushel of Austrian winter peas. They also recommend including 10 to 20 pounds of orchard grass or some endophyte-free fescue per acre. According to Sandy Fisher, fall is the best time to lime and fertilize, if necessary, because the slots created by the no-tillage drill allow the lime to penetrate the soil more easily. After seeding, animals are kept off the pasture until late fall and then are allowed to graze it down two or three times during the winter and early spring.

The Fishers have also tried other methods such as no-tillage planting of bluestem grass (which is native to Kansas) into fescue pasture, allowing it to grow throughout the summer and fall, and then letting the cattle graze it in March. By using a combination of pasture renovation methods, the Fishers find it possible to graze their cattle throughout the year, requiring little supplemental feeding most years.

The Fishers maintain that by using this system approximately every 2 to 3 years, they enhance the fertility of the pasture for subsequent years. They observe that the vigorous growth of the small grain root systems actually serves much like a shallow subsoiling, loosening up the sod to allow im-proved water penetration. They believe that the animals also get improved nutrition from the seeded crop (compared with unimproved pasture).

In other areas of the farm a similar method is followed using oats and turnips. The Fishers drill oats at a rate of 0.5 bushel per acre and turnips at a rate of 1 pound per acre. Turnips provide valuable feed and improve the condition of the soil (see the BreDahl case study). The key to success with this practice is to fertilize the crop heavily, promoting vigorous growth (10 to 14 inches high); graze it quickly with a large number of cattle; and then rotate the stock off so that regrowth can take place. The Fishers are also considering seeding broccoli in place of turnips on the pastures to take advantage of broccoli's fast vegetative growth and its capacity to withstand light frosts.

An added benefit of the Fishers' system is that because the beef cattle are on pasture constantly, the pastureland requires no manure handling except dragging (with a section of chain-link fence) every few months. The Fishers clip some of the fields for weeds, but they look on such practices as an indication of a failure in pasture management.

Cattle

The farm is currently running 500 head of cattle. Of that number, 300 are cows bred to calve from September through November 15. The Fishers breed for calving in the fall to avoid labor shortages that might be incurred in spring calving. It is also believed that cooler fall temperatures minimize pest problems for the newly born animals. The Fishers keep the calves until the following fall and then sell them as stockers weighing 750 to 800 pounds.

The farm's predominant breeds are Black Angus and Hereford. The Fishers prefer the F_1 generation cross between the Hereford and the Angus, which produces Black Baldies, and plan to keep heifers from this cross as replacement cows. They are also considering what breed of bulls would be best to cross back to these cows; at present, they are trying out two Brahman or Zebu bulls.

Feeder calves do not get a grain supplement but are allowed to graze on the farm's improved pastures, which have been interseeded with various legumes and grains. The calves' primary health problems have been pinkeye and lumpjaw. They receive a complete set of shots for various bovine diseases.

When ready for sale, the cattle are sorted into 60-head trailer loads and sold in Lancaster, Pennsylvania, which is more competitive than the local market, justifying the additional cost of shipping the animals. As of July 1986 the cattle sold by the Fishers brought an average of $0.58 per pound liveweight for heifers and steers combined, or about $450.00 per head.

PERFORMANCE INDICATORS

Environmental Impact

The Fishers are deeply committed to soil conservation practices and are willing to experiment with new methods. They try to use as few purchased inputs as possible on their farm to maximize the land's potential without harming the soil or polluting the environment. During the interview for this study, Sandy Fisher, who is president of the Virginia Soybean Producers Association and a member of the local soil and water conservation district board, observed that in Goochland County, few farmers are willing to cut back on corn because they are reluctant to lose their corn base, on which corn price support payments are computed.

What the Fishers have accomplished on their farm is similar to the goals established by the Conservation Reserve Program in the 1985 Farm Bill, which is designed to take highly erodible land out of row-crop production. Ironically, early innovators such as the Fishers, who have already achieved this goal, are not eligible for government payments on the acreage that they have voluntarily removed from the production of price-supported crops.

Financial Performance

Cost and returns data for the Sabot Hill Farm are not available for presentation in this case study. Yet some generalizations can be made. Cash operating costs are low in comparison with those of farms dependent on purchased feed, and the Fishers have reduced their herbicide costs by about $20,000 through cultural practices. The farm's acreage is large in relation to that of other farms in the county. The relatively large size of the farm may provide the Fishers with enough economic stability to experiment with alternative production systems. Finally, the farm is a diversified operation, with its primary sources of revenue being the sale of stocker cattle and hay, with some income from the sale of corn and soybeans.

REFERENCE

Cook, R. L. 1962. Soil Management for Conservation and Production. New York: Wiley.

4

A Mixed Crop and Livestock Farm in Pennsylvania: The Kutztown Farm

THE KUTZTOWN FARM is a 305-acre mixed crop and livestock farm located near Kutztown in east-central Pennsylvania. The farm is located in Berks County, which has some large cash grain farms and many family-operated crop and livestock farms and is among the top five counties in the state for crop production and agricultural cash receipts. The principal source of income on the case farm is a beef-feeding operation; most of the crops grown on the farm are used to support this enterprise.

The farmland consists of rolling hills with some bottomland broken down into 98 fields averaging about 3.4 acres (Figure 1). Most of these fields are laid out on the contour, commonly in strips 100 to 200 feet wide. Soil pH, nutrient levels, and physical conditions are measured prior to selecting the crops each year. The 98 individual fields give the farmer great flexibility in fitting the crops to the conditions of each field (Culik et al., 1983).

The family owns 72 acres and has rented about another 173 acres from the Rodale Research Center since 1973; the family also rents 60 acres from neighbors. In comparison, the average cropland harvested per farm in Berks County during 1982 was 105 acres.

GENERAL DATA

Rodale Research Center scientists (Culik et al., 1983) studied the Kutztown Farm over a 5-year period, presenting their work in such a way as to protect the privacy of the Mennonite family that operates it. This farm is probably the most thoroughly studied alternative farming operation in the country. Reports on it have appeared in numerous publications, and it has been the subject of extensive comparisons with state and county average production. One Ph.D. dissertation in agronomy (Wegrzyn, 1984) and two

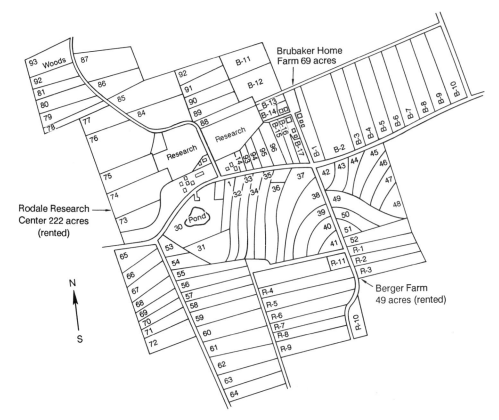

FIGURE 1 Farming operation including home farm and rented land near Kutztown, Pennsylvania, in 1978, 1979, and 1980. SOURCE: Wegrzyn, V. A. 1984. Nitrogen Fertility Management in Corn—A Case Study on a Mixed Crop-Livestock Farm in Pennsylvania. Ph.D. dissertation, The Pennsylvania State University, University Park.

M.S. theses in agricultural economics (Dabbert, 1986; Domanico, 1985) have been completed at The Pennsylvania State University using data obtained from this farm. This case study draws heavily on past research and recent interviews with the director of the Rodale Research Center at the time.

The family raises all the grain, hay, and silage used on the farm. Most of the farm's crops are used for feed and bedding, although the farmer sells some alfalfa and red clover hay. In addition to providing income, these crops are grown to balance rotations and enhance soil fertility. In recent years the family has increased the acreage of grains and reduced hay acreage. The livestock operation mainly involves finishing purchased beef cattle, but the family also raises hogs and laying hens. All livestock and other products of the farm are sold through conventional market channels (Table 1).

Acreages were distributed as shown in Table 2 from 1978 to 1982 and in

TABLE 1 Summary of Enterprise Data for the Kutztown Farm

Category	Description
Farm size	305 acres, 250–290 beef cattle
Labor and management practices	This is a family-operated farm with one full-time hired man and occasional help from relatives. There are complex management duties because of the diversity of the enterprise. The father manages the beef enterprise; his son manages crop production and machinery maintenance. The variety of crops grown on the farm results in a relatively even distribution of labor needs throughout the year.
Livestock management practices	Feeder cattle are purchased from Virginia. They are fed corn, silage, hay, roasted soybeans, and small grain supplements. The size of the hog herd varies from 50 to 250, depending on the prices of feeder and finished hogs.
Marketing strategies	Beef and hay are sold through conventional markets; no premium prices are obtained for alternative farming methods.
Weed control practices	Crop rotations and multiple cultivations of row crops are used. Imported chicken manure (not composted) is a suspected source of weed seeds. Rain at cultivation time often results in poor weed control. Herbicides are applied to approximately 45 percent of the land.
Insect and nematode control	Pest build-up is avoided in field crops by rotation. There are no reported insect problems in animal operation.
Disease control practices	The farmer uses prophylactic application of sulfa-type drugs to purchased beef feeders while in quarantine immediately after purchase.
Soil fertility management	A variable rotation with corn, soybeans, small grains, and hay is used. Manure (10 tons/acre) is applied twice in a 5-year rotation; on-farm beef manure and imported chicken manure are used. Starter fertilizer is applied to corn in proportions of 3.6 pounds N, 7.2 pounds P, and 3.6 pounds K per acre once in a 5-year rotation.
Irrigation practices	None
Crop and livestock yields	Crop yields exceed county averages for soybeans, hay, wheat, and corn grain; yields are lower for corn silage and rye.
Financial performance	Expenditures for fertilizers and agricultural chemicals per acre are substantially below county averages. Investment in machinery is very low because of the age of the equipment; repair costs are high (mostly for parts). Economic analysis indicates the Kutztown Farm is somewhat less profitable than a comparable conventional farm.

1986. Corn silage is the most prevalent crop, currently occupying 29.5 percent of the land; another 26.2 percent is used for hay production.

Climate

The county has a fairly moderate, humid continental climate. Average annual precipitation is 42.5 inches (Table 3); it is normally well distributed throughout the year (2.8 to 4.4 inches per month) with the most monthly

TABLE 2 Kutztown Farm Crop Acreages, 1978–1982 and 1986

Crop	Acreage, 1978–1982		Acreage, 1986	Percentage of Total	
	Range	Mean		1978–1982	1986
Hay					
Alfalfa	30–63	51.7	40	17.6	13.1
Red clover[a]	32–70	45.9	40	15.6	13.1
Subtotal		97.6	80	33.3	26.2
Small grains					
Barley	9–36	18.1	20	6.2	6.6
Oats[b]	13–29	20.9	20	7.1	6.6
Rye	17–26	23.4	30	8.0	9.8
Wheat	8–34	23.2	20	7.9	6.6
Subtotal		85.6	90	29.2	29.5
Row crops					
Corn, grain	14–30	25.0	20	8.5	6.6
Corn, silage	52–78	63.9	90[c]	21.7	29.5
Soybeans	14–42	21.8	25	7.4	8.2
Subtotal		110.7	135	37.5	44.3
Total		293.9	305	100.0	100.0

[a]Red clover hay includes other hay.
[b]Oats includes spring barley and oat mix.
[c]60 acres high-moisture ear corn plus 30 acres regular silage.

SOURCE: Culik, M. N., J. C. McAllister, M. C. Palada, and S. L. Rieger. 1983. The Kutztown Farm Report: A Study of a Low-Input Crop/Livestock Farm. Regenerative Agriculture Library Technical Bulletin. Kutztown, Pa.: Rodale Research Center.

precipitation occurring in July and August. During the summer, precipitation of 0.1 inch or more occurs on an average of 10 days per month.

Maximum daily temperatures in nearby Allentown range from 35.7°F in January to about 85.4°F in July. Minimum mean daily temperatures range from 19.8°F in January to 62.7°F in July. Culik et al. (1983) report that the average growing season is 194 frost-free days.

PHYSICAL AND CAPITAL RESOURCES

Soil

Most of the farm's cropland is on steep, shaley hills, with slopes of up to 25 percent that are somewhat eroded from past cropping. The surface of the soil is covered with flat, shaley pebbles. The soils are of shale, silt loam, sandstone, gneiss, and limestone origins (Table 4). The Berks and Weikert soil series are inceptisols, and the Fogelsville and Ryder soils are alfisols. Their productive capacity depends on their age and weathering status.

The predominant soil type (on nearly two-thirds of the farm) is Berks shaley silt loam soil described as moderately deep, well-drained, medium-textured, shaley soils that have formed in material weathered from gray

TABLE 3   Normal Daily Temperatures and Monthly Precipitation at Allentown, Pennsylvania

Month	Normal Daily Temperature (°F)			Normal Degree Days		Normal Precipitation	
	Maximum	Minimum	Monthly Average	Heating	Cooling	Inches	Days With ≥0.01 Inches
January	35.7	19.8	27.8	1,153	0	3.02	11
February	37.9	20.9	29.4	997	0	2.78	10
March	47.7	28.5	38.1	834	0	3.61	11
April	61.3	38.5	49.9	453	0	3.79	11
May	71.7	48.4	60.1	190	38	3.78	12
June	81.0	57.9	69.5	21	156	3.47	11
July	85.4	62.7	74.1	0	282	4.36	10
August	82.8	60.6	71.7	6	214	4.18	10
September	75.9	53.4	64.7	85	76	3.59	9
October	65.6	42.5	54.1	344	6	2.73	8
November	51.7	32.8	42.3	681	0	3.59	10
December	38.7	22.6	30.7	1,063	0	3.59	11
Average annual	61.3	40.7	51.0	5,827	772	42.49	124
			Average annual total				

NOTE: The normal daily maximum by month is the average of each day's (midnight to midnight) high temperature for every day in that month from 1941 to 1970. The normal daily minimum by month is the average of each day's (midnight to midnight) low temperature for every day in that month from 1941 to 1970. The normal monthly temperature is the average of the normal daily maximum and minimum temperatures for that month. The normal degree days heating are the sums of the negative departures of average daily temperatures from 65°F. The normal degree days cooling are the sums of the positive departures of average daily temperatures from 65°F. To calculate the normal degree days heating or cooling, multiply the difference between 65°F and the normal monthly temperature by the number of days in the month. The normal monthly precipitation is the average of the inches of precipitation for that month from 1941 to 1970.

SOURCE: National Oceanic and Atmospheric Administration. 1980. Climates of the States, 2d ed. Detroit: Gale Research Co., Book Tower.

TABLE 4 Soil Types on Kutztown Farm[a]

Soil Type	Number of Fields	Acres	Percentage of Total
Berks shaley silt loam	61	193.4	63.4
Fogelsville silt loam	25	76.0	24.9
Ryder silt loam	5	17.1	5.6
Weikert-Berks shaley silt loam	4	9.5	3.1
Other[b]	3	9.2	3.0
Total	98	305.2	100.0

[a]Number of fields and total acreage vary slightly from year to year.
[b]Other includes 5.5 acres of Litz shaley silt loam and 4.2 acres of Melvin silt loam.

SOURCE: Culik, M. N., J. C. McAllister, M. C. Palada, and S. L. Rieger. 1983. The Kutztown Farm Report: A Study of a Low-Input Crop/Livestock Farm. Regenerative Agriculture Library Technical Bulletin. Kutztown, Pa.: Rodale Research Center.

shale and siltstone. These soils are gently sloping to very hilly (Culik et al., 1983). The soils are easily tillable, with a topsoil horizon usually about 9 inches thick and a subsoil extending to about 24 inches. They have a low available moisture capacity and are very prone to drought. Soil erosion is a moderate hazard, and crop rotations that feature frequent row crops (corn or soybeans) are not recommended.

The Fogelsville soil series (on about 25 percent of the farm) is described as "deep, well-drained, nearly level to sloping silty soils that have formed in material weathered from shaley limestone or cement rock" (Culik et al., 1983). These soils are easily tilled and also easily eroded, with topsoil about 8 inches thick and a substratum extending to about 38 inches.

The farm contains 17.6 acres of highly erodible land with over 11 tons of potential soil erosion per acre. At the other extreme, 31.7 acres have an erosion potential of less than 3 tons of soil erosion per year (Table 5 and Figure 2). The average soil erosion for the entire farm is estimated as 4.5 tons per acre per year (Culik et al., 1983).

Buildings and Facilities

One large barn houses the farm's cattle and hogs. The beef barn includes a quarantine area in which purchased feeder stock are kept for 3 to 4 weeks before they are housed with the other beef animals. Chickens are kept in a small chicken house. There is also a machine shed and a shop where the farmer repairs the machinery.

Machinery

The farm uses conventional tillage and frequent cultivation, averaging eight machinery operations per field per year. Because the farmer is so expert mechanically, he is able to use older equipment and keep it repaired.

TABLE 5   Kutztown Farm Estimated Soil Erosion, Based on Segments of the Farm, 1978–1982

Segments	Number of Acres		Estimated Soil Loss (tons/acre/year)
1	8.8		11.27
2	18.0		4.00
3	8.8		13.84
4	8.2		6.61
5	44.3		3.64
6	11.2		0.77
7	12.6		6.79
8	46.3		4.00
9	32.0		3.32
10	12.2		3.19
11	28.9		4.94
12	44.1		4.77
13	20.5		2.67
Total	295.9	Farm average	4.53

SOURCE: Culik, M. N., J. C. McAllister, M. C. Palada, and S. L. Rieger. 1983. The Kutztown Farm Report: A Study of a Low-Input Crop/Livestock Farm. Regenerative Agriculture Library Technical Bulletin. Kutztown, Pa.: Rodale Research Center.

He is also able to adapt or fabricate parts of equipment, a talent shared by many other successful alternative farmers (see the Ferrari and Coleman case studies).

The farm's 1982 machinery inventory included six tractors and a combine, plus equipment for planting, cultivating, haying, making silage, and spreading manure, with a total market value of about $67,000. (Purchased new, an inventory like this—which would not ordinarily be found on a farm of this size—would have cost more than $286,000 in 1982.) Because of the age of the equipment, depreciation and other costs of ownership are lower. However, this cost saving is somewhat offset by repair costs (replacement parts, engines, and so forth) and losses that result from a lack of timeliness of operations when the aging machinery breaks down. Although available data do not permit direct comparisons of the machinery inventory on the Kutztown Farm with that of a comparable farm using conventional practices, it is clear that this farmer is substituting his mechanical craftsmanship (in repairing old machinery) for the capital that would be needed to purchase newer equipment.

The timeliness of field operations is critical on farms. Equipment age and other factors influence the timely completion of operations. Machinery breakdowns occur on the Kutztown Farm. Usually, however, the farmer can quickly fix the problem. Back-up tractors are available when needed. Equipment and labor needs are divided among eight crops produced during a 9-month period (March through November); consequently, the demand on equipment is spread out. This means that, when a certain piece of equip-

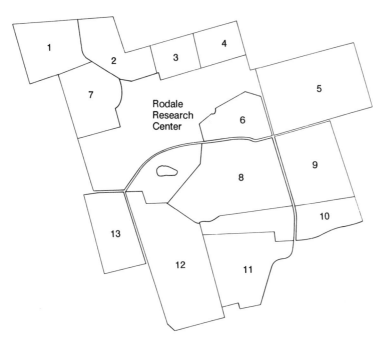

FIGURE 2 Kutztown Farm segments for erosion calculations. Segmented according to relatively uniform sod type and slopes. SOURCE: Culik, M. N., J. C. McAllister, M. C. Palada, and S. L. Rieger. 1983. P. 30 in The Kutztown Farm Report: A Study of a Low-Input Crop/ Livestock Farm. Regenerative Agriculture Library Technical Bulletin. Kutztown, Pa.: Rodale Research Center.

ment breaks down, it can be repaired while minimizing production losses (Culik et al., 1983). The machinery inventory is much the same in 1986 as it was in 1982.

MANAGEMENT FEATURES

Labor

The farm is operated primarily by the farmer with one full-time hired man and occasional help from his wife and other family members or other relatives as necessary. Since 1978, and particularly in recent years as the father's health has declined, the son in the family has taken on the primary role of managing and producing the crops and repairing and operating the farm equipment and machinery. For the purposes of this report, the son is considered the farmer; his father now focuses primarily on managing the beef feedlot. Culik et al. (1983) measured crop labor input during the 1982 season (Table 6). These data may or may not represent average labor input in other years. Nevertheless, labor is distributed much more evenly throughout the year, largely because of the variety of crops grown.

TABLE 6   Kutztown Farm Labor Requirements, 1982

| | Hours/Acre/Season | | | | |
Crop	March–May	June–August	September–November	December–February	Total
Alfalfa hay	0.0	8.0	3.0	0.0	11.0
Barley	0.0	7.2	2.8	0.0	10.0
Corn, grain	1.6	1.1	3.0	0.0	5.7
Corn, silage	1.6	1.1	3.0	0.0	5.7
Red clover hay	0.0	8.0	0.0	0.0	8.0
Rye[a]	0.0	7.2	2.8	0.0	10.0
Soybeans	1.6	1.1	1.5	0.0	4.2
Spring barley/oats	1.8	7.2	1.0	0.0	10.0
Wheat	0.0	7.2	2.8	0.0	10.0

[a]No budget for rye is given (Dum et al., 1977). Labor requirements for rye are assumed to be the same as those for barley.

SOURCES: Culik, M. N., J. C. McAllister, M. C. Palada, and S. L. Rieger. 1983. The Kutztown Farm Report: A Study of a Low-Input Crop/Livestock Farm. Regenerative Agriculture Library Technical Bulletin. Kutztown, Pa.: Rodale Research Center; Dum, S. A., F. A. Hughes, J. G. Cooper, B. W. Kelly, and V. E. Crowley. 1977. The Penn State Farm Management Handbook. University Park, Pa.: College of Agriculture, The Pennsylvania State University.

Culik et al. (1983) estimated that the Kutztown Farm's labor requirements exceeded those of conventional farms by 10 to 30 percent. Comparable data for conventional farms are not available, however. The labor requirements in Dum et al. (1977), which have been used for such estimates, are known to be obsolete and inaccurate (V. Crowley, Penn State University Farm Management Extension Director, interview, 1987).

Tillage and Crop Rotations

When the family first began farming the Rodale land in 1973 (on a special lease requiring that agricultural chemicals not be used and other management provisions), the yields were described by the farmer's father as disastrous for several years (interview, 1982). No crop yield data are available from that phase of the operation. Beginning in 1978, however, detailed yield information was collected by the staff of the Rodale Research Center for 5 years. The farmer recalls that crop yields were "very low until after the first plow-down of a legume," when yields increased substantially. He observed another increase in yields following the second plow-down of a legume (the second rotation), but since then yields have not increased with subsequent plow-downs.

Crop production on the Kutztown Farm includes alfalfa and red clover hays, barley, oats, rye and wheat, corn for grain and silage, and soybeans. The acreages planted in each of these crops varied during the study period, but the cropland usually was apportioned into about one-third hay, one-third small grains, and one-third row crops (corn and soybeans) (see Table

2). Currently, the farmer has increased the farm's corn acreage to about 36 percent and decreased its hay acreage in part because he became aware that nitrogen availability was more than adequate (Wegrzyn, 1984) and in part because of the declining price of hay. (Hay prices were reported to be $70.00 per ton in 1985, compared with $110.00 per ton during the period from 1978 to 1982 covered by the Culik study.) With the exception of the hay crops, all of the crops are used for the livestock raised on the farm. From 1978 to 1982, approximately two-thirds of the hay was sold off the farm; today, that proportion has dropped to one-half. The farmer uses certified seed for most crops, although occasionally he uses some home-grown red clover, timothy, small grains, or soybean seeds.

Culik et al. (1983) report that during their study a complex crop rotation was used throughout the farm that involved the consideration of many factors before a crop was selected for an individual field. The standard cropping sequence included small grains used for establishing leguminous hay crops, followed by corn, soybeans, or more corn, and, again, small grains. The hay crops included alfalfa, alfalfa-timothy, red clover, or mixed species.

The farmer generally keeps hillsides in alfalfa hay in longer rotations, with shorter rotations used on the less sloping fields. During the Culik study, the species used for hay included pure alfalfa, pure clover, and alfalfa or clover seeded either with timothy or bromegrass. Currently, only alfalfa or a mix of red clover and timothy is used.

The farm's use of different hay crops has been a deliberate management strategy to spread out the harvest dates, thus avoiding the need to hire additional labor or purchase additional machinery. Each year the alfalfa is cut first, followed by the clover and the clover mixes, thus spreading the haying time—and its accompanying labor and machinery requirements— over about a month in the spring. Spreading the timing of the hay harvest also reduces the risk of rain damage, although every spring some hay is lessened in quality by rain.

In the establishment year of the hay rotation the crops are seeded with a small grain. At the first hay harvest the residual straw from the small grain is mowed and baled with the hay; a second cutting is normally obtained late in the summer. After the establishment year the alfalfa or alfalfa-timothy hay is usually cut three times; red clover and timothy are only cut twice. The clover hay stand is sometimes plowed down after just 1 full year of production.

Currently, the farmer reports that he is still using a rather flexible approach to his rotation, depending on weather and other conditions. He says his typical rotation is as follows. After plowing down a legume, he plants corn for 1 or 2 years (occasionally 3), followed by 3 years of small grain (usually rye, then barley, then wheat). He may underseed the wheat with a legume mixture (timothy and clover) if conditions permit or wait until the next spring to plant oats and alfalfa together. Timothy is always combined with the red clover.

The alfalfa is grown for about 3 years. The farmer says the alfalfa crop in the mix is nearly depleted in 3 years and must be plowed down. A former director of the Rodale Research Center observed that the stand is typically 50 to 75 percent alfalfa when it is plowed down (correspondence, 1986). The farmer observed that if he were to use chemicals it would be possible to slow the stand depletion caused by diseases and insects. The conventional practice in the area is to apply carbofuran in the spring to control alfalfa weevils and dimethoate later in the season to control potato leafhopper, which tends to do severe damage. In this way an alfalfa stand will produce heavier yields and can be maintained for as long as 6 years. Conventional farmers often still plow down their alfalfa after 3 years, however, because of their rotation or other farm management considerations.

During the study period monitored by Culik et al. (1983), another third of the farm was in row crops (corn or soybeans). Currently, corn is grown on about 44 percent of the land. Corn is always grown on plowed-down hay fields, usually for 2 and occasionally for up to 3 years. Animal manures (from poultry or cattle) are usually applied to the second- or third-year corn. The farmer considers weather, weed, insect, and nutrient factors when deciding the number of years of corn production. Normally, corn is grown for 1 or 2 years, making use of the residual nitrogen from the leguminous hay plow-down and the animal manure. On the farm's more fertile soils, however, corn is occasionally grown for 3 years.

The production practices used for both the corn grain and the silage fields are similar. The corn is grown in 38-inch-wide rows with populations averaging about 17,000 plants per acre; typical corn plant populations in the area are 18,000 to 20,000 plants per acre for corn grain and 20,000 to 24,000 plants per acre for silage. Wegrzyn (1984) attributes the relatively low plant population on the farm to the fact that an old, well-worn, 4-row planter was used. Although higher plant populations are considered a standard agronomic practice for weed control (because they provide a heavy canopy early in the season, shading the emerging weeds), the Kutztown Farm achieves above-average yields with below-average plant populations.

The farmer plants hybrid corn seed, usually of several varieties and sometimes mixed in the same field. The corn is rotary hoed at least once to control the early weeds and then cultivated two or three times. Because rainfall of 0.01 inch or more occurs on the average every third day (see Table 3), however, sometimes the farmer is unable to cultivate at the optimum time for weed control.

In 1986, the corn was harvested by chopping two rows for silage in the normal manner and then picking eight rows. The picked ears were ground in the field with a picker-grinder before being blown into the wagon with the silage. The benefit of this method is that, by alternating silage chopping with the ear picking, the feed value of the silage is increased, and about 80 percent of the corn residue is left in the field, providing abundant organic matter and preventing soil erosion. Compared with the erosion that occurs when all corn stalks are cut for silage, a method that leaves virtually no ground cover, the farmer's practice of alternating two rows of silage and eight rows of grain is estimated to reduce soil erosion by 36 percent on the more erosive category of soils and 30 percent on the less erosive soils. (These estimates are based on data in Domanico, [1985].)

Adding the extra ear corn at silage harvest time also reduces expenses later when grain corn would normally be added to the feed ration. The addition of wet ear corn also seems to help with silage packing and subsequent preservation. Currently, only about 20 of the farm's 110 acres of corn are harvested (in the fall) as grain and stored for feeding. Of the remaining 90 acres, about 22.5 acres are harvested as regular silage and about 67.5 acres are harvested as high-moisture ear corn and combined with the silage to increase its feeding value.

Soybeans, which are grown on about 20 acres, are roasted (by custom hire) and used as feed for the cattle, hogs, and chickens. Prior to 1985, soybeans planted on the non-Rodale land were drilled on a 7-inch row spacing, and herbicides were used for weed control. On the Rodale land, soybeans were planted in 38-inch rows and cultivated 2 or 3 times for weed control. Since 1985, all soybeans are planted in 30-inch rows, and the farmer reports that yields are similar on both lands. Scientific comparisons are impossible, however, because of varying cropping histories and soil types. Soybeans are never grown in any field 2 years in succession because of the farmer's concern about the risk of disease or excessive erosion.

Roughly 29 percent of the land is typically planted in small grains (wheat, rye, barley, and oats). The farmer grows barley and rye as much for the straw (for livestock bedding) as for grain. Yields from these grains are generally lower than yields from wheat, but production of straw is greater. The farm uses oats, the first crop planted in the spring, as a backup crop when untimely rains prevent a fall planting of winter wheat. About 90 percent of the oats are underseeded with alfalfa or clover (correspondence with the farmer, 1986).

The farmer reports that small grains are grown for 1 to 3 years before returning to a leguminous hay. The economic implications of including this combination of four small grains in the rotations are discussed later in this case study.

Soil Fertility

The farmer views the management of soil nutrients over the whole farm as particularly important. During the period of the Culik study (1978–1982), the staff of the Rodale Research Center frequently performed soil tests and plant tissue tests for each of the farm's 98 fields. The farmer had access to the test results (Table 7). During the transition from chemical to nonchemical farming methods on the Rodale-rented land beginning in 1973, the farmer grew a higher-than-normal percentage of alfalfa or clover hay. By 1978 some fields had higher-than-optimal levels of nitrogen, according to nitrogen response trials conducted by Wegrzyn (1984). When the farmer

TABLE 7   Trends in Soil Sample Test Results, 1978–1982[a]

Item	1977	1978	1980	1981	1982
Soil pH	6.8	6.7	6.9	6.6	6.7
Phosphorus (pounds/ acre)[b]	181	276	191	213	230
Potassium (pounds/acre)	188	211	227	313	274
Magnesium (pounds/acre)	296	321	364	406	496
Calcium (pounds/acre)	2,085	2,967	2,927	3,048	3,048
Cation exchange capacity	8.0	10.6	9.4	11.8	11.2
Organic matter (percent)	—	—	2.2	2.3	—

NOTE: A dash indicates that data were not reported.

[a]Data are averages for selected fields.
[b]Levels over 101 pounds of phosphorus per acre are considered high in this area.

SOURCE: Culik, M. N., J. C. McAllister, M. C. Palada, and S. L. Rieger. 1983. The Kutztown Farm Report: A Study of a Low-Input Crop/Livestock Farm. Regenerative Agriculture Library Technical Bulletin. Kutztown, Pa.: Rodale Research Center.

saw the results of those tests (Table 8), he increased his grain plantings and reduced the legumes in the rotations.

A more serious challenge than the level of nitrogen is the regulation of all nutrients. The major vehicle for such regulation is the application of manure. The farmer applies about 10 tons of manure (including bedding material) per acre twice during the 5-year rotation, causing a bimodal fluctuation of potassium and nitrogen in the fields over the 5 years. A former director of the Rodale Research Center observes that after a manure appli-

TABLE 8   Generalized Nitrogen (N) Budget for Corn on Kutztown Farm

Major N Supplies (available N)	N/pounds/acre/year	Percentage of Total N
Forage legume residue	2,800	0.36
Soil N pool	2,207	0.28
Steer manure	1,526	0.19
Poultry manure	1,457	0.17
Total available N supplies[a]	7,635	1.00
Crop requirements[b]	6,449	
Measured crop uptake[c]	6,530	
N balance[d]	+1,104	

[a]Does not include contributions from soybean residue, precipitation, autotrophic N fixation, crop residues older than 1 year, or manure residue older than 2 years.
[b]Based on 1978 Pennsylvania State University Soil Testing Service calculations for 40 acres of corn on 28 separate fields.
[c]Based on 1978 Kjeldahl analyses of whole plant samples from 28 separate corn fields.
[d]Total N supplies minus measured crop uptake equals the N balance.

SOURCE: Wegrzyn, V. A. 1984. Nitrogen fertility management in corn—A case study on a mixed crop-livestock farm in Pennsylvania. Ph.D. dissertation, The Pennsylvania State University.

TABLE 9   Comparison of Fertilizer and Other Agricultural Chemical
Expenditures, Kutztown Farm Versus Berks County Estimated Average

| | Expenditures per Acre (dollars) | |
Item	Kutztown Farm, 1978–1982	Berks County, 1982
Fertilizers	13.85[a]	47.17[b]
Other agricultural chemicals	4.28[c]	17.49[b]

[a]From U.S. Department of Commerce. 1982. 1982 Census of Agriculture, Vol. 1. Geographic Area Series, Pt. 38, Pennsylvania State and County Data, Table 6. Washington, D.C. A mean of 599 gallons of starter fertilizer at $3.15/gallon plus 181 tons of chicken manure at $12.00/ton, divided by 293 acres.

[b]Calculated from Culik, M. N., J. C. McAllister, M. C. Palada, and S. L. Rieger. 1983. The Kutztown Farm Report: A Study of a Low-Input Crop/Livestock Farm, Tables 20 and 23. Regenerative Agriculture Library Technical Bulletin. Kutztown, Pa.: Rodale Research Center. Mean expenditures per farm were divided by mean acreage of cropland harvested per farm. This procedure ignores fertilizers added to land not harvested and may overstate mean expenditures for the county.

[c]Estimated by dividing total expenditures for chemicals on this farm by the number of non-Rodale acres. This procedure slightly overstates the cost per acre to which chemicals were applied.

cation there is a gradual drawdown of available potassium, especially during the alfalfa portions of the hay rotation, and then a jump in potassium as manure is applied to the hay. This jump is followed by another gradual reduction until the small grain is planted, with another jump as manure is applied again. A somewhat similar pattern occurs with nitrogen.

The farmer supplements the nutrients provided by legume rotations and beef manure produced on the farm with imported nutrients: chicken manure purchased under a contract with a local egg producer and a small quantity (4 gallons per acre) of liquid starter fertilizer (9-18-9) for use on the corn. These materials provide a total of 3.6 pounds of nitrogen (N), 7.2 pounds of phosphorus (P), and 3.6 pounds of potassium (K) per acre.

Culik et al. (1983) reported that from 1978 to 1982, the mean quantities of manure and fertilizer purchased were as follows: chicken manure with wood shavings, 181 tons at $12.00 per ton ($2,172), and liquid starter, 599 gallons at $3.15 per gallon ($1,887)—a total of $4,059 per year or $13.85 per acre compared with a county average of $47.17 per acre (Table 9). Currently, 40 tons of chicken manure are delivered to the farm every 6 weeks (320 tons per year). Most of this manure is stockpiled (uncovered) until spring, when it is applied to certain fields.

Culik and his coworkers reported that the chicken manure supplied 30 pounds N, 14 pounds P, and 7 pounds K per ton of fresh manure (that is, an analysis of 1.5-0.7-0.35). However, from 30 to 90 percent of the nitrogen in manure can be lost through ammonia volatilization when the manure is left exposed (Vanderholm, 1975). Nearly 50 percent may be lost in the first 24 hours (D. Pimentel, correspondence, 1987).

To maintain a soil pH of 6.5 to 7.0, lime was applied to many fields in 1978; the use of legumes in the rotation requires that soil pH be maintained at or near neutral.

During the Culik study, soil tests and plant tissue tests were performed for each of the 98 fields of the farm. Soil magnesium, calcium, and cation exchange capacity remained fairly constant during the study. Soil organic matter, measured in 1980 and 1981, was about 2.2 to 2.3 percent, a level similar to that in other fields in the area (Culik et al., 1983). Available phosphorus, calcium, and potassium in the top 6 inches of the soil profile remained high enough so as not to limit crop production. In fact, the levels of these nutrients tended to increase from year to year, an increase that cannot be explained by the amounts of these nutrients applied in the manure. Although this phenomenon is not well understood, the Rodale Research Center scientists speculate that deep-rooted sod crops in the rotation may be drawing nutrients upward from deep in the soil profile.

Wegrzyn (1984) estimated the nitrogen budget for corn in a typical year on the Kutztown Farm. The largest source of nitrogen was found to be forage legume residuals (see Table 8).

Weed and Insect Control

The Kutztown Farm largely avoids weed and insect problems by using intensive, yet flexible, crop rotations. Corn, in particular, is rarely grown in a given field for more than one or two seasons in succession as a means of breaking the reproductive cycle of corn root worm.

Weed control on the Rodale land is accomplished primarily through cultivation and rotations. The corn, for example, is rotary hoed at least once for early weeds and then cultivated two or three times. From 1978 to 1982, the farmer used herbicides (atrazine, alachlor, butylate, and linuron) on corn and soybeans on the non-Rodale land. Currently, the farmer applies a mixture of atrazine and metolachlor to control yellow nutsedge on the non-Rodale land.

The farmer has reported an increasing problem with control of velvetleaf in fields in which herbicides are used: "Velvetleaf weeds don't seem to be a problem in organic fields, but we do have weed problems that change from year to year. If we have wet weather during critical cultivating time, weeds can take over" (correspondence, 1986).

The total cost of chemicals during the period studied by Culik and colleagues ranged from $354.00 to $1,029.00 (the mean was $565.00); these costs were primarily for herbicides applied to the non-Rodale land (132 acres) and work out to an average of $4.28 per acre of non-Rodale land. A small and unknown fraction of the $565.00 average chemical cost was used to purchase sprays for barn insects. The comparable expenditure for other farms in Berks County in 1982 was $17.49 per acre (see Table 9).

No weed control of any kind—neither cultivation nor herbicides—is used on the small grains, and very few weeds are observed in these fields. Culik

TABLE 10   Livestock Sales, 1978–1982[a]

Commodity	1978	1979	1980	1981	1982
Beef cattle (head)[b]	164	173	204	248	276
Eggs (dozen)[c]	3,100	3,634	4,531	4,967	5,235
Hogs (head)[d]	168	150	139	39	55

[a]Data do not include two to four cattle, several hogs, and eggs consumed annually on the farm.

[b]Feeder cattle (of many breeds) are purchased at 650–700 pounds. Finished weight is 1,100–1,150 pounds. In 1985, about 290 head were sold.

[c]Number of laying hens in 1978 was about 200, increasing to almost 300 in 1982. Hens are kept for an egg-laying period of 14 months, after which they are butchered for consumption by the families of the three men operating the farm (the farmer, his father, and the hired men). The flock size is currently 20 hens.

[d]Hogs are purchased at 45–50 pounds and sold when finished, usually after 90 days. The hog herd is now 50 head.

SOURCE: Culik, M. N., J. C. McAllister, M. C. Palada, and S. L. Rieger. 1983. The Kutztown Farm Report: A Study of a Low-Input Crop/Livestock Farm. Regenerative Agriculture Library Technical Bulletin. Kutztown, Pa.: Rodale Research Center.

et al. (1983) reported that during the 5 years of the study crop rotations controlled weeds, insects, and diseases.

In 1986 weed control in the corn fields was excellent. However, because weather conditions interfered with the timing of cultivations, the soybean fields had severe weed problems in 1986 despite rotary hoeing and cultivating. Weeds are sometimes a serious problem when untimely rain prevents cultivation, while in other years cultivation controls weeds better than herbicides do. The farmer suspects that the chicken manure he uses contained weed seeds (telephone interview, 1986).

Animal Enterprises

The farm gave increasing emphasis to its beef cattle finishing operation from 1978 to 1982. Cattle sales increased 68 percent (Table 10) at the same time hog production declined by 67 percent. Egg production also increased 69 percent. Since 1982, the chicken and hog enterprises on the farm have been reduced. Currently, the farm has 20 laying hens (for family use) and 50 hogs. The farmer reported that he increases the number of hogs in production when the price of feeder hogs declines. The number of cattle sold increased slightly to about 290 in 1985.

Animals are confined but have small exercise yards; they are occasionally grazed in the fall on one field that is fenced. Antibiotics are used only to treat acute disease problems as they arise. Newly purchased feeder cattle are isolated until they have stabilized and are fed antibiotics (sulfa and chlortetracycline) for the first 3 to 4 weeks after shipping. Otherwise, drugs are not used prophylactically or as subtherapeutic growth promoters. The farmer reports that feeder cattle purchased in Virginia seem to have fewer disease problems than locally purchased animals. The decline in the size of the hog herd noted above was partly due to disease problems in animals purchased at a local livestock auction.

The cattle are confined in the barn, and urine and droppings are caught in the bedding, which helps to keep the barn somewhat dry underfoot. The farmer mentioned that poor ventilation in the barn in which the cattle are fed sometimes causes health problems (telephone interview, 1986). There is essentially no runoff or effluent from the barn, and except for ammonia volatilization, virtually all the nutrients excreted by the animals are caught in the bedding. The straw bedding is a high-carbon material, and it is reasonable to assume that losses of nitrogen are reduced. There is little smell of ammonia, even when the manure is dug out, but some nitrogen losses are inevitable. With a high carbon-nitrogen ratio in the bedding and manure, it is reasonable to expect that when they are applied to the field, some soil nitrogen is temporarily immobilized by soil bacteria while they are breaking down the cellulose in the straw.

Cattle are fed approximately the following amounts per head per day, for 200 to 240 days: corn silage, 15 pounds; a barley, oats, wheat, and rye mix (processed in a roller mill), 5 pounds; roasted soybeans, 1 pound; and ground, high-moisture ear corn, 7 pounds (wet basis). In addition, the cattle are fed leguminous hay, vitamins, and minerals. Younger stock receive more hay than do animals near finishing. The average weight gain is about 2.3 pounds per head per day, with some animals gaining up to 2.5 pounds. When they reach 1,100 to 1,150 pounds the cattle (often 2 to 4 head per week) are sold to local butchers or meat packers.

Hogs and chickens are fed the same feed ration: corn, oats, barley, wheat, and rye (in proportions of about 75 percent corn to 25 percent small grains); vitamins and minerals; and roasted soybeans mixed with the grains (in a ratio of 1:5). The hogs are allowed in the barn with the cattle but are fed (ad libitum) separately from the beef cattle. When finished, the hogs are sold to local butchers; local markets buy the eggs, except for the meat and eggs used by the two families on the farm and the hired man, who receives room and board in addition to a wage.

PERFORMANCE INDICATORS

Soil Conservation

Culik et al. (1983) estimated that the soil erosion on the Kutztown Farm (based on the Universal Soil Loss Equation) ranged from a low of 0.8 tons per acre per year in one 11-acre area to 13.8 tons per acre per year on the most erodible 8.8-acre area. The Soil Conservation Service, which has estimated that soil erosion on some farms in Berks County is as high as 18 to 40 tons per acre per year, put the tolerable soil loss levels on the Kutztown Farm between 3 and 5 tons per acre per year. Pimentel et al. (1987) estimated that this tolerable level exceeds the rate of soil formation by a factor of 10 times. The average soil erosion on the Kutztown Farm was estimated at 4.5 tons per acre per year when moldboard plowing in combination with contour and strip cropping was used. (If moldboard plowing was used without contour and strip cropping, however, it was estimated that the average soil erosion on the Kutztown Farm would more than triple to 14.7 tons per acre per year.) As discussed earlier, the levels of various soil nutrients on the Kutztown Farm increased from 1977 to 1982.

Yield Performance

Crop yields on the Kutztown Farm are generally equal to or slightly higher than state or county averages (Table 11). The notable exceptions are barley and rye yields. These grains have been substantially below average most years because the farm uses long-stemmed varieties to provide ample straw for bedding, not dwarf varieties, which are typically grown for higher yields.

In addition to the lower yield effect of selecting long-stemmed cultivars, the farm sometimes has a peculiar problem with small grains, especially rye: excess nitrogen in the soil can cause lower grain yields but even higher yields of straw. For example, 1981 was a year with normal rainfall following a very dry year; alternative systems are very responsive to moisture, and in a dry year the nitrogen in the soil is not completely used but instead accumulates in a mineralized form. In 1981 the farmer reported that he applied the usual manure before rye, not realizing that there had been considerable mineralization of the nitrogen released from the organic matter that had been applied during the previous dry summer. In 1981 excess nitrogen was released not only from the manure applied that year but also from the mineralized nitrogen left from the year before; the rye grew 6 to 8 feet tall with such heavy stems that they lodged (bent to the ground); and very little of the grain was recovered at harvest.

In 1981 the farm averaged less than half of the average state and county yields of barley. The farmer observed, "I think the poor barley yield was mostly due to winter kill. This problem seems to be worse with the early fall seeding and weather patterns such as heavy freezing with bare ground" (correspondence, 1986). William Liebhardt (correspondence, 1986) has suggested that the early fall-seeded barley may also have had disease problems.

Corn yields on the Kutztown Farm averaged 28 percent higher than the county average and 17 percent higher than the state average from 1978 through 1982. In 1980, a very dry year, the farm's corn yield was 47 percent higher than the county average. This result is consistent with the findings of several studies (see, for example, Lockeretz et al. [1984]) indicating that, under dry weather conditions, farming systems based on crop rotations have relatively higher yields than conventional farms. The likely cause for the better dry weather performance is better soil tilth and moisture-holding capacity. For other crops, yields for Kutztown Farm corn silage equaled

TABLE 11   Kutztown Farm Crop Yields per Acre Compared With County and State Averages

Crop	1978	1979	1980	1981	1982	Mean
Alfalfa hay (tons/acre)						
County	3.2	3.1	2.4	3.3	3.2	3.0
Kutztown	—	—	2.4	3.8	3.8	3.3
Corn grain (bushels/acre)						
County	95.7	95.0	52.0	92.0	92.0	85.3
Kutztown	121.3	124.4	76.6	121.3	96.6	108.0
State	95.0	95.0	75.0	96.0	97.0	91.6
Corn silage (tons/acre)						
County	17.0	13.8	10.6	15.4	14.9	14.3
Kutztown	17.6	—	9.3	15.3	15.0	14.3
State	15.5	15.0	12.6	16.2	15.2	14.9
Other hay (tons/acre)						
County[a]	2.0	1.6	1.9	1.8	2.2	1.9
Kutztown[b]	—	—	1.3	3.6	3.1	2.7
State[a]	1.8	1.8	1.8	1.9	2.0	1.9
Rye (bushels/acre)[c]						
Kutztown	23.6	24.0	39.6	30.3	27.3	29.0
State	32.0	27.0	31.0	33.0	34.0	31.4
Soybean (bushels/acre)[c]						
Kutztown	38.8	36.8	27.9	44.0	42.8	38.1
State	31.5	32.0	24.5	31.0	32.0	30.2
Wheat (bushels/acre)						
County	35.0	33.0	41.0	41.0	39.0	37.8
Kutztown	—	33.1	37.0	41.8	34.6	36.6
State	33.0	31.0	37.0	36.0	36.0	34.6

NOTE: Oats could not be compared directly because the Kutztown Farm grew a spring barley and oats mix. A dash indicates that data are not available.

[a]Includes red clover and mixed hays.
[b]Includes red clover and timothy hay.
[c]County average data are not available.

SOURCE: Culik, M. N., J. C. McAllister, M. C. Palada, and S. L. Rieger. 1983. The Kutztown Farm Report: A Study of a Low-Input Crop/Livestock Farm. Regenerative Agriculture Library Technical Bulletin. Kutztown, Pa.: Rodale Research Center.

county and state averages, and soybean yields averaged 26 percent above the state mean.

Financial Performance

Any assessment of the financial performance of the Kutztown Farm is complicated by a lack of comparable data for conventional farms. Culik et al. (1983) used a number of simplifying procedures to facilitate an economic comparison of the Kutztown Farm with a conventional comparison farm. One of their key procedures was substituting certain Kutztown Farm data for the comparable items in the *Penn State Farm Management Handbook* (Dum et al., 1977) in calculating the costs for the comparison farm—in particular, variable machine costs, which ignores labor cost, depreciation, and other overhead and fixed costs. Culik et al. (1983) estimated that the Kutztown Farm incurred a somewhat lower cost for producing various crops—for example, 1 percent lower costs per acre for producing corn and 20 percent lower costs for alfalfa. As a result, they estimated that the Kutztown Farm earned a 5 percent higher net cash income than a comparison farm ($69,430 versus $65,987).

When the Culik team's assumption of equal variable machine costs is relaxed, however, and the *Penn State Farm Management Handbook* costs are used without that adjustment, and when differences in yields are taken into account, the cost comparisons are drastically different. The cash operating cost per bushel (or the variable cost) of producing corn grain was found to be 6 percent higher, and alfalfa costs 45 percent higher, on the Kutztown Farm than on the comparison farm; the costs of producing all other crops were also significantly higher on the Kutztown Farm; some were more than double the comparison values. The farm's variable costs per bushel of small grains were found to be particularly high relative to those of the conventional comparison farm (because its grain yields are quite low, for reasons explained earlier).

There are some problems with the assumptions in this comparison, however. A conventional farm probably would not produce the same combination of crops as the Kutztown Farm. For example, farmers might choose the more profitable option of purchasing straw rather than committing such a high proportion of their land to the production of small grains, especially rye and barley, that typically produce low grain yields.

To provide a more direct comparison, researchers at The Pennsylvania State University (Dabbert, 1986; Dabbert and Madden, 1986; Domanico, 1985; Domanico et al., 1986) used economic simulation in conjunction with linear programming, relying on the Culik team's descriptions of the physical characteristics of the Kutztown Farm and its resource requirements and yields, together with comparable data from the *Penn State Farm Management Handbook* and elsewhere. Studies that compare actual operating farms using alternative methods of production with other standards (such as county averages) or matched nearby farms have been criticized for their lack of statistical controls and for uncontrollable differences among ostensibly comparable farms (Lockeretz et al., 1984).

The economic simulation approach also has inherent limitations, including the risk that the mathematical combination of management practices may appear to be reasonable but in reality may be unworkable. (In addition, in this particular case the analysis conducted by the Penn State researchers assumed that the farm could be operated with about the same complement of equipment under either conventional management or the mixed conventional-alternative procedures employed on the Kutztown Farm, a questionable assumption.) Consequently, the findings of this type of analysis must always be interpreted cautiously. A strength of this approach, however, is that it has the advantage of holding constant the resource base and certain other factors that would otherwise tend to confound the comparisons.

The Penn State analysis calculated income in terms of net return over cash operating (variable) costs, ignoring energy utilization and most of the externalities (except soil erosion). The conventional comparison farm was not assumed to produce the same combination of crops as the Kutztown Farm. Instead, it was assumed that both the Kutztown Farm and the conventional comparison farm would be optimally organized; that is, they would produce the most profitable combination of enterprises, subject to the limitations of the resources available and the technologies used. Specifically, the analysis was designed to provide directly comparable results from the Kutztown Farm versus alternative scenarios featuring the use of other technologies (including conventional practices, overseeding, no-till, and other options) in the context of specific assumptions regarding the level of soil erosion permitted, the rotations appropriate for alternative farming systems, and the use or nonuse of chemical pesticides and fertilizers.

The economic analysis postulated both a single-year planning horizon (see Domanico et al., 1986) and a multiple-year transition from conventional to organic farming, defined as a farming system compatible with the U.S. Department of Agriculture definition of organic farming (see Dabbert, 1986). Only Domanico et al.'s (1986) findings are discussed here.

Domanico et al. (1986) found that when soil erosion is not limited, the profit-maximizing conventional farm plan is 3 percent more profitable (in terms of net return over cash operating costs) as compared with an optimally organized alternative farm plan with the same resources. Soil erosion was estimated to be 9.7 tons per acre per year for the optimally organized conventional comparison farm compared with 5 tons per acre per year for the Kutztown Farm (Domanico et al., 1986). But when soil erosion is limited to a 5-ton-per-acre average across the farm, the conventional option is 1 percent less profitable than the alternative option. When soil erosion is limited to 3 tons per acre, the alternative option is estimated to yield a $3,200 (10.8 percent) higher profit than a conventionally operated farm. (Of course, the comparative financial performance of the Kutztown Farm under conventional and alternative management would also vary with different prices of farm commodities and inputs.)

The management and labor requirements of the Kutztown Farm would be likely to exceed those of a conventional alternative because of the farm's reliance on cultivation for weed control (on the Rodale land) as well as the complexity of the crop rotations and the large number (98) of small fields necessitated by the contour strip-cropping system. The magnitude of difference in management and labor requirements cannot be determined at present, however, because of data limitations.

REFERENCES

Culik, M. N., J. C. McAllister, M. C. Palada, and S. L. Rieger. 1983. The Kutztown Farm Report: A Study of a Low-input Crop/Livestock Farm. Regenerative Agriculture Library Technical Bulletin. Kutztown, Pa.: Rodale Research Center.

Dabbert, S. 1986. A Dynamic Simulation Model of the Transition from Conventional to Organic Farming. M.S. thesis, The Pennsylvania State University.

Dabbert, S., and P. Madden. 1986. The transition to organic agriculture: A multi-year model of a Pennsylvania farm. American Journal of Alternative Agriculture 1(3):99–107.

Domanico, J. L. 1985. Income Effects of Limiting Soil Erosion Under Alternative Farm Management Systems: A Simulation and Optimization Analysis of a Pennsylvania Crop and Livestock Farm. M.S. thesis, The Pennsylvania State University.

Domanico, J. L., P. Madden, and E. J. Partenheimer. 1986. Income effects of limiting soil erosion under organic, conventional, and no-till systems in eastern Pennsylvania. American Journal of Alternative Agriculture 1(2):75–82.

Dum, S. A., F. A. Hughes, J. G. Cooper, B. W. Kelly, and V. E. Crowley. 1977. The Penn State Farm Management Handbook. University Park, Pa.: College of Agriculture, The Pennsylvania State University.

Lockeretz, W., G. Shearer, D. H. Kohl, and R. W. Klepper. 1984. Comparison of organic and conventional farming in the corn belt. In Organic Farming: Current Technology and Its Role in a Sustainable Agriculture, D. F. Bezdicek, and J. F. Power, eds. Madison, Wis.: American Society of Agronomy, Crop Science Society of America, and Soil Science Society of America.

Pimentel, D., J. Allen, A. Beers, L. Guinand, R. Linder, P. McLaughlin, B. Meer, D. Musonda, D. Perdue, S. Poisson, S. Siebert, K. Stoner, R. Salazar, and A. Hawkins. 1987. World agriculture and soil erosion. Bioscience 37(4):277–283.

Vanderholm, D. H. 1975. Nutrient losses from livestock waste during storage, treatment, and handling. Pp. 282–285 in Managing Livestock Waste. Proceedings of the International Symposium on Livestock Wastes. St. Joseph, Mich.: American Society of Agricultural Engineers.

Wegrzyn, V. A. 1984. Nitrogen Fertility Management in Corn—A Case Study on a Mixed Crop-Livestock Farm in Pennsylvania. Ph.D. dissertation, The Pennsylvania State University.

CASE STUDY

5

Crop-Livestock Farming in Iowa:
The Thompson Farm

RICHARD AND SHARON THOMPSON'S FARM is in eastern Boone County, Iowa, at an elevation of about 1,000 feet. It is located in Jackson Township on sections 16, 17, and 21, which is about 8 miles west of Ames and 4 miles due north of the Iowa State University Agronomy and Agricultural Engineering Research Center. The farm has a total of 300 acres, all of it owned by the Thompsons and all of it tillable. Of this total, 282 acres are tilled, which is about the average farm size for the state of Iowa as a whole and for Boone County.

GENERAL DATA

The Thompsons have a diversified farming operation, which is no longer the norm in their area (Table 1). Statewide, about 55 percent of farmland is used to grow corn each year, and roughly one-third of the land area is devoted to soybeans. For Boone County, the corresponding figures are 40 percent (127,000 acres) and 35 percent (117,000 acres), respectively, proportions that are typical of north-central and central Iowa. There are only 11,000 acres of oats and 9,000 acres of hay in the county. The most common crop sequence in the vicinity is corn-soybeans-corn-soybeans. The Thompsons grow corn on 33 percent of their land and soybeans on 15 percent, on average. Specifically, the farm grows approximately 100 acres of hybrid field corn, 50 acres of soybeans, 50 acres of oats, and 50 acres of hay; another 32 acres are in pasture. The Thompsons keep 50 cows in the foundation herd and raise the calves through finishing; the farrow-to-finish hog operation has 90 sows.

Dick Thompson received an M.S. degree in animal production from Iowa State in the 1950s and started farming conventionally. For a 16-year period

TABLE 1 Summary of Enterprise Data for the Thompson Farm

Category	Description
Farm size	282 acres of tilled cropland, 50 cows, 90 sows
Labor and management practices	The farm's diversified operation spreads labor requirements throughout the year. It is managed and operated by Dick and Sharon Thompson and their son, with one full-time hired man who takes care of the swine. Dick Thompson spends considerable time doing on-farm research.
Livestock management practices	The farm has a 50-cow foundation herd of cattle (Angus-Hereford cows and exotic breeds of bulls); it also has a 90-sow, farrow-to-finish hog operation (1,300–1,400 pigs finished/year).
Marketing strategies	Most crops and livestock are sold through ordinary commercial markets with no price differential for methods of production. The exception is about 15 percent of beef animals, which are sold directly to individuals at a $0.10/pound premium, less the transportation cost to the locker/slaughter plant.
Weed control practices	Ridge tillage and high plant populations, in conjunction with crop rotation and cultivation with rotary hoe, disk hillers, and sweeps, are used. Small grains in the rotation disrupt weed reproductive cycles. If rain delays cultivation and weeds threaten crops, postemergence herbicides will be used.
Insect and nematode control practices	No particular pest problems were reported. Crop rotations and ridge tillage, plus a diversity of plant species, are credited for this situation.
Disease control practices	No antibiotics are used except to treat illness. Various measures are used to build resistance in the hog herd (for example, probiotics, transfer of manure from farrowing units to gestation pen). Cattle are not vaccinated. The farmer limes the pens to keep the pH unfavorable to pathogens, and uses isolation, sunlight, and special feed rations (for example, steamed rolled oats) to prevent scours and other diseases.
Soil fertility management	Municipal sludge and manure (18 tons/acre) are applied to corn and soybeans. Urea (30 pounds N) is applied to corn and oats at planting; 30 pounds K_2O is applied to corn and soybeans at planting. The farmer uses 5- to 6-year rotations with corn, soybeans, oats, meadow, and green manure in various combinations.
Irrigation practices	None
Crop and livestock yields	Corn yields are 130–150 bushels/acre versus the county average of 124; soybeans, 45–55 bushels/acre versus 40; oats, 80–100 bushels/acre versus 67; hay, 4–5 tons/acre versus 3.4. Pigs are finished and sold. The average number of pigs sold per sow is 14.4–15.6 versus an average of 14.8 for a group of 270 Iowa Swine Enterprise Record members.
Financial performance	Municipal sludge is provided free of charge; only a limited number of farms can receive this free resource. Costs are kept low by the use of on-farm resources, such as N_2 fixation and labor. Corn and soybean production costs are lower than for conventional farms. Farm cash flow is adequate to meet operating costs without borrowing, to maintain and enhance the capital stock of machinery and facilities, and to support the farm family.

(1967–1983), however, the farm was organic in the sense of using no purchased fertilizer or herbicides. The only off-farm nutrient input to the system during this time was through feed purchased for the livestock. More recently, sewage sludge from nearby Boone (population 13,000) has been used.

Thompson has always been an innovative farmer, and his operations change—to a greater or lesser extent—from year to year. Because he is still not satisfied with all the procedures used on the farm, he conducts on-farm research and demonstration trials involving tillage, weed control, fertility, rotations and cropping systems, cover crops and interseeding, hybrid and variety comparisons, and livestock management.

Commodities Produced, Used, and Marketed

Soybean yields on the Thompson Farm are 45 to 55 bushels per acre; 7-year county and state averages are 40 and 37 bushels per acre, respectively. (The state and county data are from "Iowa Agricultural Statistics," compiled and issued by the Iowa Crop and Livestock Reporting Service of the Iowa Department of Agriculture. The published attainable mean yield for the farm's best soils is 50 bushels per acre and 37 for the poorest.) All of the soybeans are sold, thus far only through standard market channels.

Corn yields on the Thompson Farm are now in the range of 130 to 150 bushels per acre. County and state averages (1979–1985) are 124 and 115 bushels per acre, respectively, and the highest attainable yield figures are 150 bushels per acre on the best soil and 115 on the poorest. The grain is fed to the livestock, supplying their needs for about 6 months out of the year; the stalks are used for bedding the animals.

Oats generally yield 80 to 100 bushels per acre for the Thompsons (although the yield was 127 bushels in 1985). The 7-year county and state averages are 67 and 62 bushels per acre, respectively. The oats are fed to hogs, calves, and yearling feeder cattle.

Hay harvest (three cuttings) on the farm yields 4 to 5 tons per acre per year. County and state averages from 1981 to 1985 were 3.4 and 3.3 tons per acre, respectively. The hay is all fed.

Statewide, 1983 and 1984 were drought years in which yields of row crops were severely depressed in some regions but not in central Iowa. The yields that are given for the Thompson Farm apply to more nearly normal cropping years. Also, the farming methods used by the Thompsons have been evolving over time; these numbers represent what can be produced now and not what the yields on the farm have been since 1979. Still, production on the Thompson Farm compares well with that of the best conventional farmers and the published attainable productivity of its soils.

The hogs produced on the farm are crossbreeds. Annually, the Thompsons finish 1,300 to 1,400 pigs that are then sold through normal market channels. Boars are purchased to match the sow herd, which is replenished by keeping replacement gilts produced on the farm.

The cows are Black Baldies, an Angus-Hereford cross; the two bulls are of mixed exotic European breeds. Replacement heifers and bulls are purchased from herds known to be healthy. About four dozen cattle are finished each year. Of these, six to eight head are typically sold to individuals on a carcass basis for a premium of $0.10 per pound. Transportation to the locker comes out of this premium.

PHYSICAL AND CAPITAL RESOURCES

Soils

The landscape of the area is gently rolling, with slopes of up to 10 percent. The soils are young mollisols derived directly or indirectly from glacial till of the Wisconsin glaciation, from which the ice receded only about 12,000 years ago. The farm is in the Clarion-Nicollet-Webster soil association area and is typical of the swell-swale topography that constitutes over one-fifth of the state. These soil types predominate and are very good to excellent for crop production in Iowa. Subsoils are calcareous, and in poorly drained parts of the farm the topsoils are calcareous also as a result of secondary calcium carbonate deposition. The Canisteo, Okoboji, and Harps soil series occur in such potentially wet places on the property. The soils are all loams and clay loams and have relatively deep A horizons of 1 to 2 feet, with organic matter contents now characteristically 2 to 6 percent. Clay accumulations in the B horizons of the subsoil restrict internal drainage. Prior to the installation of drainage tile, and in some places canals, this part of the state was marshland.

Buildings and Facilities

Buildings and facilities on the farm include a building that contains a machine shop. The ability to repair, modify, and construct equipment at home is an important survival skill for any farm. In the case of the Thompson Farm, such a capability is absolutely essential because of the amount and kind of equipment needed for farming operations and the modifications created by the Thompsons.

A manure bunker (48-by-176-by-12 feet) is currently being constructed and will be an important element in the improvement of manure handling on the farm. The bunker will also receive municipal sludge from the city of Boone; the farm used 270 dry tons in 1986 at 80 percent moisture. The sludge contains 2.5 percent nitrogen and 1 percent phosphorus on a dry basis and is monitored for chromium because of a tannery that contributes to the Boone waste flow.

The Thompsons currently dry their corn on the ear. They plan to build another crib, narrow and oriented to catch the winter winds, that they hope will allow them to harvest ear corn at 25 percent moisture if necessary. In addition, two metal grain bins (each with a 7,000-bushel capacity) have recently been added, which will expand the farm's options for marketing soybeans.

The hog operation uses a system of do-it-yourself insulated, prefabricated units with open fronts that offers sunshine, fresh air, and isolation and costs less than a confinement unit of comparable size. The farrowing isolits cost $937.00 per unit for 30 units in 1979; a confinement building at the time would have cost around $2,000 per unit. The nursery units cost $34.00 per head; an enclosed building would have required an investment of $80.00 per head. The Thompsons' finishing facility cost them $37.00 per head out-of-pocket, compared with the cost of a building for the purpose at about $145.00 per head.

Open-front housing requires a good windbreak to keep snowdrifts out of the area. The farrowing isolits are equipped with both liquid propane infra-red heaters and electric heat pumps. Two truck mud flaps are hung over the lower part of the nursery doors in winter, stopping the wind but allow-ing enough air circulation to prevent humidity build-up in the units. Nipple waterers provide clean water on demand throughout the year.

Machinery

The Thompsons own three tractors. A 3-year-old tractor with 120 horse-power and front-wheel assist is used with the manure spreader, the baler, and the feed grinder and to pull the disk and cultivator. A 55-horsepower tractor is also 3 years old; it has a front-end loader and is used to handle manure, to mow and condition hay, and to cultivate. A 60-horsepower tractor is at least 20 years old and is used for planting and cultivation. At planting time, all three tractors are often in operation at once.

Other farm equipment includes a 14-ton-capacity manure spreader; a 30-foot flexible rotary hoe; a heavy-duty offset disk; a stacker-baler, now used only for corn stalks; a baler that makes large, round bales; a 12-foot wind-rower-conditioner; a hay turner; an oat windrower; a grinder-mixer; a grain drill; a 4-row ear corn picker; a 4-row combine for soybeans and oats; and a water wagon for manure tea.

Climate

This grain-producing region is characterized by a continental climate with cold, dry winters (December, January, and February) and a warm-to-hot, humid growing season. Except for late July and early August the area's average precipitation equals or exceeds the evapotranspiration of the pre-dominant crops. The root zone of soils holds about 10 inches of plant-available moisture. Even so, moisture stress is not uncommon, and major drought years occur every 18 to 20 years.

The average length of the growing season is 189 days from April 14 to October 20. The soil is usually warm enough to plant corn by the first week of May, with soybeans generally planted around the middle to the end of

TABLE 2   Normal Monthly Temperature and Precipitation at Boone, Iowa

Month	Normal Temperature (°F)		Normal Precipitation (inches)
January	17.1		0.74
February	23.4		0.95
March	33.0		2.07
April	49.5		3.40
May	61.1		4.37
June	70.1		5.11
July	74.0		3.45
August	71.1		3.89
September	63.5		3.21
October	52.8		2.31
November	37.0		1.33
December	24.3		1.28
Average annual	48.2	Average annual total	32.11

NOTE: The normal monthly temperature is the average of the normal daily maximum and minimum temperatures for that month from 1941 to 1970. The normal monthly precipitation is the average of the inches of precipitation for that month from 1941 to 1970.

SOURCE: National Oceanic and Atmospheric Administration. 1985. Climatological Data Annual Summary—Iowa, 1985, Vol. 96, No. 13.

May. The average first frost (32°F) occurs around the first week of October. The mean growing degree-day accumulation for the cropping season through September is about 3,100 (50 to 86°F). Mean temperature and precipitation for the region over the course of the year are shown in Table 2.

MANAGEMENT FEATURES

The current working philosophy of the Thompson Farm is to limit or find substitutes for off-farm inputs wherever possible to reduce costs and promote the health of livestock and people. For example, herbicides and antibiotics are not routinely employed, although these inputs are used when a crisis occurs, as in the case of treating a sick animal. Or, if there are thistles in a pasture, the individual plants will be sprayed with herbicide. For the most part, balance and diversity give the Thompsons' operation a certain resilience, qualities that are manifest in, for example, the mix of species in the pastures and in the gut of the livestock and in the early-season weeds in the row crops.

Rotations

There are five outlying fields on the farm that use a 5-year rotation of corn-soybeans-corn-oats-meadow. Manure and sludge are spread on these fields just before they are planted with corn and soybeans. Four smaller fields near the homestead that alternate as pastures use a 6-year rotation of corn-soybeans-oats-meadow-meadow-meadow. Manure and sludge are applied prior to planting corn and soybeans on these four fields as well.

Clearly, raising animals is not an option for every farmer. Recognizing this, and wishing to demonstrate other farming options, the Thompsons have allocated land for two rotations that are relevant to the typical cash grain operation. One is the corn-soybean rotation that is so common in the Midwest; the other is a 3-year oats with green manure cover crop-corn-soybean rotation. The Thompsons' purpose in these rotations is to investigate alternative methods of production and show ways in which growers with no manure or sludge can still limit production inputs purchased off the farm.

Tillage and Planting Methods

Ridge tillage is a form of reduced or conservation tillage used in the Midwest that has gained some popularity, particularly in areas with heavy soils that warm slowly in spring. In this practice the new crop is planted directly into the ridge remaining where the previous crop grew; no prior working of the soil is needed to prepare a seedbed. Most ridge-tillage farmers plant on the top of the ridge and usually apply herbicide in a band over the row. Since 1980, however, the Thompsons have planted on ridges without using any herbicides at all.

In the Thompson modification of the standard ridge-tillage program, the planter shaves off the top two inches of the ridge, throwing soil, weeds and weed seeds, and cover crops into the middle or interrow zone. This method accomplishes two things: it helps to incorporate the manure, which has been applied just ahead of the planter, and it provides a planting strip in soil that is unoxygenated and fairly free of weed seed and that has not been exposed to sunlight. Weed seeds from the previous year may have fallen onto the surface of the ridge, but these are thrown into the interrow zone. In addition, the strip prepared for planting is in soil lacking the environmental cues—oxygen, light, and warmth—that signal dormant weed seeds to germinate.

Soybeans are planted at a rate of 12 seeds per foot of row instead of the 9 or 10 customarily recommended. The Thompsons use a tall, fast-emerging variety, one of the benefits of which is that it quickly establishes a small canopy over the row itself for within-row weed control. Weeds between the rows are easily cultivated. The Thompsons also plant a tall corn hybrid at a relatively high rate (24,000 to 26,000 plants per acre), again, to assist in weed control.

The basic planting unit used by the Thompsons is a 4-row ridge-tillage planter, which is set up for 36-inch rows (planter costs are about $2,000 per row). A number of alterations have been made, however, to adapt the planter to this planting method. The two drive wheels were rotated to the back of the planter, where they are less likely to pick up manure, and a soil scraper was added to each. The sweep that cuts off the top of the ridge was extended by 4 inches on each side to throw more soil into the middle. A metal plate was added on the rear of the sweeps to extend them back to the trash rods, and these rods are also covered with plates to keep soil and weed seeds from falling back into the row. With these changes the planter can be set deep enough to ensure a clean strip on the ridge for planting. Flexible plastic hoses on the planter are mounted to deliver fertilizer from tanks on the tractor. Disk openers enable this material to be placed 2 inches below and 2 inches to the side of the seed; in addition, the planter shoe can dispense starter fertilizer with the seed itself.

Every effort is made to rotary hoe all row crops at least twice, which is another key element of the Thompsons' weed management strategy. The depth of rotary hoeing can be controlled both by the three-point hitch and by gauge wheels on the implement. The field is first hoed 3 or 4 days after planting, before the crop emerges. The purpose of this hoeing is to turn up tiny, germinating weeds while they are still in the vulnerable white root stage. The second pass with the rotary hoe occurs about 7 days later, after the crop has emerged. (The soybeans should be showing their first true leaves.)

Crop losses as a result of hoeing are quite small; in most cases, only very shallow penetration of the soil is required. The operator can drive through the field at a brisk 10 to 12 miles per hour. The benefit to the crop in terms of weed control is great. Thompson maintains that when he is able to rotary hoe twice, his weed problems are well under control. With the 30-foot rotary hoe, Thompson can cover 150 acres in a single day, minimizing the system's vulnerability to changing weather conditions.

The 4-row cultivator costs about $1,200 per row. A mirror mounted low and forward on the body of the tractor allows the driver to position the cultivator precisely while still facing forward. Deep, adjustable shields ride over the crop at any desired height, protecting young plants from clods thrown by the cultivator and keeping weed seeds out of the row. The cultivator is also equipped to deliver a side-dressing of fertilizer nitrogen, although the Thompsons customarily apply fertilizer earlier in the season. Herbicide boxes with rubber flights are used to dispense cover crop seed, through more plastic tubing, to the row just ahead of the disk hillers at the last cultivation.

A pair of disk hillers and a sweep are used in each interrow zone for the first and second cultivation of soybeans. At the third and last cultivation, only a sweep is used with a ridging V behind it to create a firm ridge for the next year's crop. On the first two cultivations of corn the cultivator is mounted with a set of disk hillers and a sweep. On the last cultivation two pairs of hillers are used; they are turned to throw soil into the row, thus rebuilding the ridges. When ridges are constructed at the last cultivation, in late June or July, weed seeds have a chance to germinate and are then choked out by the growing crop and limited moisture. Experience has shown that in years when ridges cannot be made until late fall or the next spring, the following row crop has more weeds.

Weed Control

The main element of weed control on the Thompson Farm is the use of the modified ridge-tillage planting and cultivation system just described. In addition, the Thompsons choose varieties of soybeans and corn that are tall and that do well at higher population levels. If, using the ridge system, the weeds should exceed the economic threshold, Thompson would suggest banding postemergence herbicide over the soybeans or corn. Thompson believes that herbicide use helps select the particular weed species that proliferate on a farm. Years ago, on an atrazine program with continuous corn planting, his big problems were milkweed and ground cherry. He maintains that the velvetleaf (buttonweed) now so widespread in Iowa is a product of the grass herbicides used in recent years.

The inclusion of small grains and hay in the rotation helps to disrupt the weed cycle. The Thompsons value cover crops, such as rye, for their ability to inhibit weeds directly. Whether this inhibition occurs through allelopathy, direct competition, or a combination of the two, weed populations appear to be low or very low in fields in which rye is growing or has recently been grown.

Some suggest that an herbicide-free weed control program can only succeed in operations in which there are cover crops and small grains in the rotation to prevent the build-up of weed populations, particularly those of perennials. An extended field study on the farm has evaluated weed levels in the bean years of a corn-soybean rotation under three weed control systems: (1) ridge tillage without herbicides; (2) ridge tillage with the grass herbicide metolachlor broadcast before planting soybeans and corn; and (3) conventional tillage without herbicides.

The last of these treatments employs the method of weed control used before the development of herbicides: several diskings performed at intervals before planting to allow weed emergence followed by weed destruction. The results indicate why herbicides are now considered indispensable by many farmers. When it rains a lot, weed problems can be severe using this system.

Weed infestation, over time, became worse in the conventional tillage treatment. Broadleaf weeds also increased in the ridge-tillage-plus-metolachlor treatment. (There was no statistical difference between the soybean yields obtained in the two ridge-tillage treatments, although yields in the nonherbicide treatment tended to be a bushel or two higher.) Most importantly from the standpoint of the cash grain farmer, there was no increase in weeds in cases in which this form of herbicide-free ridge tillage was applied to a corn-soybean rotation.

Pest Control

There have been no particular pest problems on the farm. In springs in which cutworms are numerous, the Thompsons' fields do not seem to suffer more or less damage than their neighbors' land. This observation is perhaps surprising given the weedy appearance of the fields in spring; cutworm moths often seek out such weeds. In terms of leafhoppers, which can plague alfalfa, 1986 was a bad year. But there was no major damage from the insects—in particular, no yellowing of alfalfa—on the farm. Possibly, the diversity of plant species in the pastures and hay can be credited for this effect: the seeding mixture contains alfalfa, red and alsike clovers, timothy, and orchard grass. There is soybean cyst nematode in the county, although it has not appeared on the farm; hairy vetch is said to be among the many alternate hosts for this pest.

Labor and Costs

The Thompsons' diversified operation tends to spread the demands for labor—four full-time people—over the whole year. The farming is done mainly by Dick Thompson, his youngest son, Rex, and a hired man, employed full-time; occasionally, Sharon Thompson will also help, although her primary responsibilities are as secretary, recordkeeper, receptionist, accountant, and gardener. Rex Thompson is responsible for all machinery, feed grinding and preparation, and the field operations; the livestock, and especially the swine, are the responsibility of the other employee.

In 1984, when the demands of the farm's more than 200 field research plots and the many speaking requests for Dick Thompson became too numerous, the Regenerative Agriculture Association made it possible for the Thompsons to hire a farmhand. The association also pays for soil and leaf tissue testing; the farming operation itself, however, receives no outside financial support. Dick Thompson estimates that 2.5 full-time persons are employed in farming, and 1.5 persons do the research and demonstrations.

In terms of trips across the field, a conventional farmer might perform the following operations to grow a crop of soybeans: one pass with a combination chisel plow-disk in the fall prior to the cropping year; a pass in the spring to disk again and apply herbicide; two trips with a field cultivator to incorporate the herbicide; a planting trip; one rotary hoeing after emergence; two cultivations; and a final trip with an herbicide wick or spray nozzle to get the remaining broadleaved weeds. The operations require nine transits of the field. The Thompsons, on the other hand, pull the manure spreader over the field, plant, rotary hoe twice, cultivate three times, and occasionally weed their soybeans with hand hoes, for a total of seven or eight trips (Table 3). These weed control practices substitute labor for capital and represent money kept within the operation as opposed to the purchase of inputs. The expenses associated with the practices, such as the cost of diesel fuel used in cultivation, are out-of-pocket costs. Rather than taking

TABLE 3   Thompsons' Time per Task Labor Costs for Corn and Soybeans, 1986

	Minutes per Acre	
Field Operations	Corn	Soybeans
Spread manure and sludge	30	30
Plant	15	15
Rotary hoe (2 times)	15	15
Cultivate (3 times)	45	45
Harvest	30	30
Hoe weeds	—	15
Shred stalks	15	—
Total time per acre	2:30	2:30
Cost per acre ($6.00/hour)	$15.00	$15.00

out a loan in the spring to get the crops planted, the Thompsons are able to distribute their expenses over the growing season and operate on cash flow completely. Tables 4 and 5 compare the costs of production for the Thompson operation with a conventional, cash grain, corn-soybean operation.

Soil Fertility

During the period when no fertilizers were purchased, soybeans were nevertheless being sold off the farm; the only nutrient inputs were through

TABLE 4   Cost Comparisons for Corn and Soybean Production Using Thompson Methods and Conventional Methods

	Corn (dollars)		Soybeans (dollars)	
Category	Thompson	Conventional	Thompson	Conventional
Cost per acre				
Machinery	73.30	83.20	61.15	47.05
Seed	20.00	22.10	15.00	11.00
Chemicals	11.40	91.40	4.20	64.95
Labor ($6.00/hour)	15.00	19.20	15.00	16.80
Land	100.00	100.00	100.00	100.00
Total cost/acre	219.70	315.90	195.35	239.80
Cost per bushel[a]				
150 bushels of corn	1.46	2.11		
50 bushels of soybeans			3.91	4.80

NOTE: A rigorous comparison of the Thompson Farm with Iowa State University (ISU) estimates would require an economic analysis of entire rotations. Such analysis is beyond the scope of this present study. The cost of vetch seed for the cover crop in corn and soybeans is omitted from the Thompson data.

[a]These estimates reflect average yields and expenses. ISU's estimated costs of production for these crops are somewhat higher, indicating that the Thompsons' operation is profitable. Details of the Thompsons' labor expenditures and production costs are presented in Tables 3 and 5.

SOURCE: Iowa State University. 1986. Estimated Costs of Crop Production in Iowa—1986. FM-1712.

TABLE 5   Variable Production Costs for Corn and Soybeans on Thompson Farm, 1986 (in dollars)

Operations and Materials	Corn (150 bushels/acre)	Soybeans (50 bushels/acre)
Ridge-tillage planting	9.70	9.70
Seed	20.00	15.00
Spread manure ($1.00/ton)	18.00	18.00
Purchased fertilizer		
30 pounds N at $0.24/pound	7.20	
30 pounds K_2 at $0.14/pound	4.20	4.20
Herbicides	0	0
Rotary hoe (2 times at $1.75)	3.50	3.50
Cultivate (3 times at $2.95)	8.85	8.85
Corn picker	23.20	—
Combine	—	17.75
Transport grain	10.05	3.35
Dry grain	0	0
Labor	15.00	15.00
Land charge[a]	100.00	100.00
Total cost per acre	219.70	195.35

[a]This cost is for comparative purposes. The Thompsons actually own the land.

purchased livestock feed. Dick Thompson has calculated that the theoretical net gain-loss to the system per acre per 5-year rotation was: nitrogen, +101 pounds; phosphate, +112 pounds; and potash, −336 pounds. In reality, additional leaks in the system also occurred. Soil tests for phosphorus showed a steady increase over this period, whereas potassium remained in the medium range.

Leaf tissue analysis, although an additional cost, is a good way to determine the nutrient status of a farm's crop. Soil tests, on the other hand, indicate only the probability of response to additional fertilizer. In humid parts of the Midwest, testing for soil nitrogen is still controversial because all of the variables involved are not yet fully understood. Tissue tests are thus useful in taking some of the guesswork out of a fertility program.

When the Thompsons began tissue testing, they found that both nitrogen and potassium were below adequate levels. Their short-term solution was to purchase moderate amounts of both, in the form of chemical fertilizers, and to monitor nutrient levels through soil and leaf tissue tests. The tests had shown that by June, the soil had as much as several hundred pounds of nitrogen; at corn planting, however, around the first week of May, there was very little available nitrogen. The Thompsons now apply 18 tons of mixed sludge and manure per acre at planting of both corn and soybeans, but the substantial amendments of green manure and livestock manure do not begin to benefit the crop until the soil warms, allowing the microbial breakdown of the added substrates. Currently, 30 pounds of nitrogen (N) per acre, in the form of 28 percent N urea solution, is applied at the planting of both oats and corn. In 1986, 30 pounds of potash per acre were also applied to corn and soybeans at planting.

The farm's long-term strategy is to plug the nutrient leaks in the operation so that purchased inputs can be reduced or eliminated. For a number of years the mixed livestock manure and bedding were composted before they were applied to the land. Dick Thompson reported that this practice has been discontinued for a number of reasons:

- The success of the weed control program has made it less critical that weed seeds be destroyed by composting;
- Nitrogen in the compost was apparently stabilized to such an extent that it could not be mineralized fast enough to supply the corn crop;
- In the composting process itself, nitrogen was volatilized and potassium was lost in the liquid expressed from the compost windrow; and
- Finally, the process required a year's delay between the collection of the raw material and the application of compost to the field.

Dick Thompson also sees some evidence that compost is a less attractive substrate for soil fauna and flora than the mix of bedding and raw manure: the estimated earthworm population was significantly lower after the application of compost than after raw manure or sludge.

The current plan for manure is to haul material from the livestock pens directly to the new manure bunker where it is added to the sewage sludge. In the bunker, the manure and municipal sludge will be kept cool and anaerobic. Any liquid that collects in the bunker will be pumped off and used as a starter fertilizer in the spring.

The municipal sludge is delivered to the bunker at no cost to the Thompsons. Sludge deliveries began in 1984 and now amount to about 1,200 wet tons per year. Sludge is typically low in potassium, but the use of cement kiln dust as a precipitating agent may increase the potassium content of the material in the future. A tannery near Boone is responsible for the chromium found in the sludge. Chromium levels are being monitored by the city, however, and although there are no precise guidelines, it appears that the metals now found in the sludge can be applied to the farm's land for many years without causing problems.

The Use of Cover Crops

Cover crops are grown wherever possible on the Thompson Farm, both for soil conservation and for soil improvement. The district conservationist for the Soil Conservation Service has calculated annual erosion rates for the farm's soils, estimating them to average 4 tons per acre per year for the corn-soybeans-corn-oats-hay rotation and 1 ton per acre for the fields that are in a corn-soybeans-oats-meadow-meadow-meadow rotation. (These estimates do not factor in the additional erosion control effect of the interseeded and overseeded cover crops, which can be sizable when the cover crops provide a high degree of ground cover.) The maximum tolerable annual level of erosion in this area of Iowa is considered to be 5 tons per acre. In Boone County, land in a corn-soybean rotation loses an average of 8 to 10 tons of soil per acre per year. All of the Thompsons' fields except the cash grain experiments are in hay or small grains 2 years out of 5 or 3 years out of 6.

Winter ground cover is established after soybeans by aerially applying the seed as the beans approach senescence. As the soybean plant leaves fall to the ground, they cover the seeds, forming a good environment for germination. The cost of the service in this area is about $4.50 per acre, but it costs considerably less in other parts of the country. Dick Thompson uses the following application rates, which are fairly typical: 1 bushel of oats per acre (currently priced at $2.50 per bushel for cleaned seed) with 20 pounds of either hairy vetch (prices vary widely by locality, $0.38 to $0.70 per pound) or rye (about $3.50 per bushel). Oats die in the winter in central Iowa, an advantageous characteristic in that they provide ground cover without interfering with the following year's crop. Rye is exceptionally hardy and will grow vigorously the following spring. At seeding rates of greater than 20 pounds per acre, however, rye can dry out the soil through increased transpiration, hinder planting of the succeeding crop, and immobilize soil nitrogen. The soybean harvest is too late in the year to allow more than a few inches growth of a cover crop, but even such a late seeding helps to hold the snow on the ground.

Cover crops are also seeded into corn at the time of its last cultivation. Hairy vetch has been the best performer in this capacity, although its winter hardiness is unpredictable. One stand of vetch on the Thompson Farm in the spring of 1986 contained 40 pounds of nitrogen per acre at the end of April and 75 pounds per acre by the end of May. These rates were determined by excavating and analyzing the vetch in square-yard quadrants every 20 rows along a transect across the field; plant samples were then subjected to Kjeldahl analysis for nitrogen. Although the vetch roots were profusely nodulated, there is no way of knowing how much of the nitrogen was fixed from N_2 and how much was simply accumulated from the soil. The carbon:nitrogen ratio of this plant material was 10:1 or 12:1, so there should have been no immobilization of additional soil nitrogen as the green manure decomposed.

The cash grain rotations on the Thompson Farm do not receive applications of manure or sludge. The green manure crop in the 3-year rotation is grown only for nitrogen fixation and nutrient accumulation. In both this and the corn-soybean rotation, seeds of rye, oats, or hairy vetch are aerially seeded just before leaf fall of the soybeans. At the last cultivation, hairy vetch is seeded into corn in both rotations for green manure the following spring.

LIVESTOCK SYSTEMS

Feeds

A complete description of the feed rations used on the farm is available from the Thompsons. The foundation beef cow herd ration is hay, oats, and ground ear corn. Fattening cattle receive a protein supplement but no growth hormones. There are five mixes for the hogs: one for gestation, one for lactation, a pig starter, a nursery grower, and a finishing ration. The sows get ground ear corn, oats, and purchased lysine and minerals. The other pigs are given ground, shelled corn (some of which comes from on the farm, some from outside), ground oats, and mineral and protein supplements.

The cattle previously raised by the Thompsons were a large, exotic type that in winter required a great deal of corn just to maintain weight. The smaller, hardier Black Baldies that the Thompsons currently raise do well on mostly hay. (The Thompsons have changed from spring to fall calving, which has solved the problem of calving in the yard before cows get out to pasture in the spring and so helped to eliminate scouring in the calves.) Similarly, the hogs used to be of a tall, narrow body type, but eventually the Thompsons concluded that a medium-framed animal with more lung capacity was better suited to the outdoor environment that the farm maintains.

Disease Control

The Thompsons do not use antibiotics routinely in their livestock operations, and the cattle receive no vaccinations. Diatomaceous earth is added to the feed and is dusted on the cows once or twice a year for external and intestinal parasites.

The isolation, sunlight, and generous amounts of bedding the Thompsons use in the hog operation help to lessen the pressure from disease organisms. Agricultural calcium carbonate (fine barn lime) is spread on the floors to keep the pH above the range favored by potential pathogens. The pens are cleaned every 2 weeks with the front-end loader tractor, but the facilities have never been sterilized. Manure is moved weekly from the farrowing units to the gestation pen so the sows are exposed to the same microflora found in farrowing stalls. In this way, pregnant sows build immunity to any new microbe in the environment, and the piglets begin life with the corresponding passive immunities.

Antibiotics create a "vacuum" in the gut of an animal, a vacuum into which resistant pathogens may move with relatively few constraints. No amount of sterilization can keep a sow's microflora away from her piglets. Rather than add antibiotics to the hog rations, the Thompsons add one or more of a group of products referred to as probiotics. These additives contain live cultures of bacterial strains, prominently *Lactobacillus* species but also strains of *Streptococcus, Bacillus,* and probably other genera.

The goal of their use is to create a favorable and stable balance in the hog's gut through this selective diversity. Some of these probiotics seem to be effective in preventing scours, which is also avoided through the inclusion of oats and ground ear corn for bulk in the sows' gestation ration. As a further preventive measure, piglets get steamed rolled oats in a ration that contains no added sugars and only 16.5 percent crude protein.

All of the hogs are chased out of their hutches early every morning to discourage them from dunging inside. If the number of pigs in a unit is sufficiently high, they tend to defecate outside rather than foul the building in which they sleep. As the pigs grow, walls are removed to increase the size of their sleeping space or the number of pigs per unit is reduced or both. Iron shots are the only injection the Thompsons customarily administer in the hog operation.

TRANSFER OF TECHNOLOGY

Communication between the academic community and limited-input farmers like the Thompsons has improved over the years. There are probably at least several reasons for this improvement:

- Limited input farmers rely on sound management and agronomic principles rather than adhering to specific ideologies;
- The practices used on these farms are supported by empirical data, and incorporate many proven agricultural methods such as crop rotations;
- Individual scientists and farmers have worked to develop the dialogue; and
- Circumstances in the farm economy and the environment have led farmers to consider the philosophy of optimization rather than maximization in ways that might not have been foreseen a decade ago.

Every summer, several hundred people are drawn to the Thompson Farm field days. A number of researchers, teachers, extension agents, and administrators have visited the farm at one time or another and been struck with its accomplishments and successes. So far, most state universities have not moved to develop and promote input-efficient farming per se, as has been done, for example, with no-tillage cropping. For such research to occur on a more systematic basis, the Thompsons believe that funding must be available.

During discussions at the Thompson Farm field days, questions were raised about three major problems confronting agriculture: (1) the farm credit crisis, (2) the oversupply of grain, and (3) the environmental effects of production. In addressing these problems, the Thompsons stress the links among farm management decisions (such as what to plant or which tillage systems or disease control practices to use), economic performance, and reduced environmental degradation. The Thompson Farm represents one possible integrated solution to all three problems.

Experience from around the country has shown that for the methods used in the Thompson operation to be more widely adopted, farmers must first see them working in their own neighborhoods. They tend to view advice from leaders in government, universities, and the private sector more skeptically. As a result, demonstration farms, such as the Thompson Farm, play an important role in technology transfer. The Thompsons believe that the growing crowds at their field days and the desire of research scientists to conduct more in-depth studies of the family's farming system are positive signs.

Tree Fruits, Walnuts, and Vegetables in California: The Ferrari Farm

THE FERRARI FARM consists of 223 acres and is located near the town of Linden, California, east of Stockton. It is located in an alluvial plain near the confluence of the Sacramento and San Joaquin rivers, at an elevation of just a few feet above sea level.

GENERAL DATA

The Ferraris grow 22 acres of vegetables, including, currently, onions, broccoli, sweet corn, cabbage, and squash; 126 acres of nuts, including 111 acres of walnuts (of which 41 acres are produced organically) and 15 acres of almonds; and 75 acres of various tree fruits, including 12 acres of apples, 10 acres of plums, 7 acres of apricots, 1 acre of Asian pears, and about 42 acres of peaches and nectarines combined. The operators (George Ferrari and his son, Wayne) attempt to use organic methods on all of their crops, both as a matter of personal preference and out of concern for the health of consumers and those working in and around the orchards.

Most of the farm is certified as organic by the California Certified Organic Growers. Currently, about two-thirds of the total value of crop sales are sold as organic ($300,000 of the total $450,000); the remainder is produced using an integrated pest management (IPM) program that includes some use of pesticides. Produce from this remaining acreage is sold in conventional markets.

This case study points out several areas in which the Ferraris have taken innovative approaches: (1) the use of nonpesticide insect control in fruit production on a commercial scale and, specifically, the use of experimental biological controls, a pheromone and codling moth granulosis virus; (2) the successful use of an IPM scouting and advisory service; (3) a successful fertility program using no chemical fertilizers in most of their orchards; and (4) diversification of the species grown on the farm and the marketing strategies necessary to sell them (Table 1).

Climate

The average precipitation in the Stockton area is about 14 inches (Table 2). Almost no rainfall occurs from June through September; more than 2 inches per month normally fall in December through February. Temperatures in the area are hot in the summer (with monthly normal maximum temperatures exceeding 80°F in May through September); the winters are mild (with monthly normal minimum temperatures above 35°F). This climate is excellent for growing tree fruits and nuts, as long as irrigation can be provided during the growing season. According to Wayne Ferrari, however, spring frost, which sometimes occurs as late as April 30, has damaged crops on occasion.

PHYSICAL AND CAPITAL RESOURCES

Soil

Soils in the area are highly productive (class 1) Wyman clay loam soils. The topography is flat, facilitating gravity irrigation.

Buildings and Facilities

The Ferrari Farm has an extensive set of buildings, including the following:

- a 50-by 75-foot repair shop containing a full set of metal-working equipment and machine maintenance and repair facilities;
- a cooling plant with two cooling rooms (32-by-32-by-12 feet and 32-by-20-by-12 feet) for immediately cooling the fruit at harvest and storing the fruit after packing and prior to shipping; these rooms can hold 200 and 125 storage bins, respectively (each bin is 4-by-4-by-2 feet);
- a packing house (with a 40-by 100-foot new addition plus an older facility) containing various kinds of sorting and packaging machines and processing facilities for shelling nuts and drying fruits;
- a 40-by 60-foot roof shed;
- a 30-by 60-foot walnut cracking shed;
- a 36-by 80-foot storage building;
- a lean-to building for tractors and other equipment; and
- an office in Wayne Ferrari's home.

Machinery

The Ferrari Farm has an extensive inventory of machinery and equipment, including the following major items: a crawler tractor; three utility tractors

TABLE 1   Summary of Enterprise Data for the Ferrari Farm

Category	Description
Farm size	223 acres
Labor and management practices	The farm is operated by the Ferrari family (4 adults working full time and 2 teenagers working part time) plus 12 regular year-round hired workers and 8 regular seasonal hired workers. Additional seasonal workers are hired as needed. The farm raises a diverse combination of fruits, nuts, and vegetables, and requires intensive management. Wayne Ferrari manages the orchards; his wife does the bookkeeping; and his father manages the vegetable production. Both men and Wayne Ferrari's mother share in the management of packing and marketing the produce.
Marketing strategies	About 10 to 15 percent of the farm's crops have been sold at the San Francisco farmers' market for the past 30 years; there are many repeat customers. Premium fancy-grade produce is sold to wholesalers, mostly organic specialty markets; a 5 percent premium price is charged on organic walnuts. This produce is sold at about a constant price throughout the season in an effort to introduce some stability into the market.
Weed control practices	Strip-spraying with herbicides is used in some orchards, including glyphosate for Johnsongrass. In organic orchards, weeds are flail-chopped, disked, or hand-hoed.
Insect and nematode control practices	Codling moth granulosis virus (CMGV) is used successfully on 5 acres of apples with 2 to 3 percent worm damage on Red Delicious, 1 percent on Granny Smith apples. It has not yet been effective against codling moths in walnuts. Other organically approved substances have not been as effective as CMGV or chemical pesticides used under the advice of an IPM pest control adviser. The application of compost is credited for controlling nematodes. Pheromone materials used on an experimental permit have been found to be effective against oriental fruit moth. CMGV has been found to require very thorough coverage of trees and more frequent applications than chemical pesticides. The application of predacious mites to fruit trees has been reported to be effective against phytophagous mites.
Disease control practices	Bordeaux solution is applied weekly during high humidity to control blight. Disease-prone crops are withdrawn from production and replaced with disease-resistant species.
Soil fertility management	The Ferraris apply 275 pounds N/acre to conventional walnuts; gypsum is added when soil tests indicate a need for calcium. Vetch green manure is used on certified organic acres, as is 2.7 tons/acre composted steer manure. Supplemental foliar spray with a kelp fertilizer is used when crops are stressed by pests.
Irrigation practices	Flood irrigation of orchards and vegetable fields is used. Water comes from six wells, pumped from a 130- to 150-foot depth.
Crop and livestock yields	Yields vary by crop. Detailed yield data for six kinds of tree fruits (several varieties of each) and walnuts, almonds, and fresh vegetables were not available.
Financial performance	Cost and return data on individual enterprises and on the farm's overall operation were not available. Unobtrusive measures indicate the farm is prospering: approximate doubling of packing shed facilities in the past 5 years financed internally (there is no debt on machinery or buildings); the acreage owned has increased 9 percent since 1982; and the farm is supporting two families.

TABLE 2   Normal Daily Temperatures and Monthly Precipitation at Stockton, California

Month	Normal Daily Temperature (°F)			Normal Degree Days		Normal Precipitation	
	Maximum	Minimum	Monthly Average	Heating	Cooling	Inches	Days With ≥0.01 Inches
January	52.8	36.3	44.6	632	0	2.91	9
February	59.0	39.2	49.1	445	0	2.11	8
March	64.8	40.6	52.7	381	0	1.96	8
April	72.4	44.8	58.6	214	22	1.37	6
May	80.3	50.0	65.2	67	73	0.42	3
June	88.1	55.4	71.8	15	219	0.07	1
July	94.7	58.7	76.7	0	363	0.01	[a]
August	92.8	57.8	75.3	0	323	0.03	[a]
September	88.8	55.3	72.1	0	217	0.17	1
October	78.1	48.9	63.5	88	42	0.72	3
November	64.2	41.5	52.9	363	0	1.72	7
December	53.3	37.9	45.6	601	0	2.68	6
Average annual	74.1	47.2	60.7				
Average annual total				2,806	1,259	14.17	52

NOTE: The normal daily maximum by month is the average of each day's (midnight to midnight) high temperature for every day in that month from 1941 to 1970. The normal daily minimum by month is the average of each day's (midnight to midnight) low temperature for every day in that month from 1941 to 1970. The normal monthly temperature is the average of the normal daily maximum and minimum temperatures for that month. The normal degree days heating are the sums of the negative departures of average daily temperatures from 65°F. The normal degree days cooling are the sums of the positive departures of average daily temperatures from 65°F. To calculate the normal degree days heating or cooling, multiply the difference between 65°F and the normal monthly temperature by the number of days in the month. The normal monthly precipitation is the average of the inches of precipitation for that month from 1941 to 1970.

[a]Less than one-half.

SOURCE: National Oceanic and Atmospheric Administration. 1980. Climates of the States, 2d ed. Detroit: Gale Research Co., Book Tower.

(60 to 70 horsepower); three row-crop tractors, ranging from very small to 80 horsepower; five forklifts; three pickups; a 1-ton flatbed truck; a 16-foot van; a refrigerated, 28-foot semi-truck and trailer; a backhoe; a sweeper (for nut harvesting); a nut harvester (tree-shaker); a bulk trailer with a nut drier; a compost spreader truck; a pull-type compost spreader; a row-crop sprayer; a speed sprayer; a nut huller; various packing house equipment; a biomass burner that generates up to 500,000 BTUs of heat for the packing shed using walnut shells as fuel following a 90-second warmup with propane; and miscellaneous other machinery and equipment.

MANAGEMENT FEATURES

Soil Fertility

The Ferraris use different fertility programs on the conventional and certified organic portions of their farm. For example, 275 pounds of nitrogen are applied each year (125 pounds in the spring and 150 pounds in the fall) to the conventionally grown walnuts. In a wet spring, calcium nitrate is broadcast in the orchards; in a dry spring, ammonium sulfate is used. Urea is used in the fall. If calcium is found to be deficient based on the results of soil tests, gypsum may be applied; in 1985, for example, the Ferraris used 2 tons of gypsum per acre.

A much more complex fertility program is used on the certified organic acres. About 2.7 tons of composted steer manure are applied at a cost of $93.00 per acre compared with about $70.00 per acre for the conventional chemical fertilizer ammonium sulfate, which has less nutrient value than compost. Spreading compost (with a spreader truck) requires about 16 minutes of labor per acre, compared with about 4.8 minutes per acre for labor to apply chemical fertilizer, according to Wayne Ferrari.

The compost used on the farm is purchased from various local firms specializing in its production. Ferrari also reported that the analysis of compost provided by one of these firms is as follows: 1.7 percent nitrogen, 1.6 percent phosphorus (P_2O_5), 2.5 percent potassium (K_2O), 2.4 percent calcium, and 1.3 percent magnesium. Purple vetch (*Vicia benghalensis*) is used as a green manure crop in areas that are not overlain with permanent sod.

A material made from kelp is applied during times of plant stress when pest populations are expanding rapidly. Farmers who use this method contend that the foliar spray does not reduce pest populations but instead stimulates plant growth so that the damage done by the pests invokes less stress on the crop. The relationship between foliar feeding and pest damage has not been experimentally established, however (A. Berlowitz, correspondence, 1986).

Fertility management practices differ for the various crops, depending on soil conditions, the health of the trees, and other factors. Because of the large number of different species being produced and the resulting complexity of the fertility program, such activities require a high level of management.

Tillage, Crop Rotations, and Irrigation

No special features were noted in the methods used by the Ferraris for planting the orchards or for tilling cropland except for the rather high level of crop diversification noted earlier. The diversified cropping pattern is primarily a risk management strategy. By diversifying the Ferraris reduce the risk of crop loss from various pests, adverse growing conditions, and changes in the market. The family is constantly identifying areas of its orchards that appear to be unprofitable and then replanting new varieties. When trees are removed from an orchard, vegetables are usually grown on the bare ground for a period of 2 to 3 years prior to replanting. Moreover, vegetable production is often continued in the young orchard before the trees reach maturity. In this way, some income is earned from that land before the trees begin bearing.

The Ferraris produce more than one-third of their walnuts without the use of pesticides; these nuts are sold in the organic markets. All of the farm's other crops are produced with as few pesticide applications as possible. Whenever it becomes necessary to apply a synthetically formulated chemical pesticide to prevent the loss of a crop, this acreage is removed from the organic market for 2 years in compliance with state law. The state does allow certain other chemicals, such as Bordeaux solution (lime, water, and copper sulfate), to be used in organic orchards. Currently, about two-thirds of the Ferrari tree-fruit orchards and vegetable crops are produced without synthetically formulated chemical pesticides. The exact acreage varies from year to year.

The Ferraris irrigate their alternative orchards either with sprinklers or through gravity irrigation; they use sprinklers on all of their conventional orchards. Wayne Ferrari says that he applies less water than suggested by his hired pest control adviser because he prefers to save money on the cost of pumping the water. (Irrigation water for the farm is pumped by electric power from a depth of 130 to 250 feet from six irrigation wells ranging in depth from 350 to 500 feet.) He also reports that the water table is subsiding because of the intensity of irrigation in the Stockton area and observes that "every few years we have to add another 10 or 20 feet of column to the pumps." The sustainability of this practice is a concern.

Weed Control

In their conventional orchards, the Ferraris spray herbicides (primarily glyphosate) to control weeds, most notably Johnsongrass. In their alternative orchards, weed control methods include flail chopping and disking between the rows of trees and hand hoeing of weeds growing close to the trees.

Insect and Nematode Control

The Ferraris control insects and nematodes on their conventionally grown and alternative acreages with a variety of methods. A pest control adviser is hired to scout the entire farm and provide advice on the timing and necessity of spraying as pest populations approach or exceed economic threshold levels. Wayne Ferrari decides which material is to be applied. On the conventional acreage, phosalone is applied to control codling moths and aphids, especially on apples that have been grown on certain susceptible varieties of root stock; methidathion is used to control scale; and propargite is used to control mites.

Nematodes are not considered a problem in walnut production. The Ferraris use compost in the belief that it may help control nematodes on their other acreage, although the efficacy of compost application for nematode control in orchards has not been established.

On their alternative acreage the Ferraris use a number of biological controls and organically acceptable pesticides, as well as other methods of insect and mite control.* Occasionally, beneficial predators are released on the advice of the pest control adviser. The Ferraris also apply various materials approved by the state law governing organic farming, including dormant oils, pheromones, and various biological control materials. For example, the Ferraris are using a pheromone, available on a U.S. Environmental Protection Agency (EPA) experimental use permit, to control oriental fruit moth. This material is distributed through small wirelike devices, four of which are attached to each tree.

The pheromone emitted by the wires saturates the chemical receptors on the antennae of the male oriental fruit moth, making it difficult for him to find the female and breed (Weakley et al., 1988). The Ferraris have found this material to be very effective in controlling the oriental fruit moth. The labor cost for attaching the wires to the trees is approximately $0.25 per tree; the cost of the material under full-scale commercial production is not yet known.

Another biological control measure used by the Ferraris is the codling moth granulosis virus (CMGV), which is also used under an EPA experimental use permit. Scientists at the University of California are performing the safety tests necessary for EPA registration of CMGV.

Indications to date are that the virus is highly specific and innocuous to anything but the codling moth and some closely related insect species (Kurstak, 1982). Several problems that must be overcome, however, include the development of a sunscreen material to prevent CMGV from degrading

---

*Many materials permitted under the California organic farming legislation are, in fact, pesticides—for example, sulfur, Bordeaux solution, *Bacillus thuringiensis*, ryania, and so on. However, these materials are distinguished from synthetically formulated chemical pesticides because they are derived from naturally occurring substances.

in sunlight and practical ways to reduce the number of applications required. (These are common formulation challenges faced by developers and manufacturers of agrichemicals.)

When it has been necessary to mix CMGV with chemical insecticides, tests have shown that CMGV is compatible and does not deteriorate when mixed with most of the chemicals that are commonly used. As a result, it can be applied along with other materials, thereby reducing the number of separate sprays that would be required if CMGV had to be applied by itself (A. Berlowitz, correspondence, 1986).

The Ferraris report that they applied CMGV to 5 acres of Payne English walnuts (on black walnut root stock) and 6 acres of apples (2 acres of Red Delicious and 4 acres of Granny Smith). They observed 2 to 3 percent worm damage in the Red Delicious and 1 percent damage in the Granny Smiths. The material was not effective in controlling codling moth damage in the walnuts, but they attribute this failure to poor methods of application. (They say that they were not as careful as they should have been in obtaining complete coverage in the walnuts.)

In places where CMGV could not penetrate (because it is not a chemical pesticide with fuming and contact action), such as between tightly clustered apples, worm damage was higher. There were three flights of the codling moth; CMGV was applied three times per flight. The applications were timed to occur 7 to 10 days prior to the peak, at the peak, and just beyond the peak population of the moths. Applied at the proper times using a method that achieves total coverage of the foliage, CMGV has been found to be highly effective in controlling codling moths on experimental blocks of operating farms (Falcon et al., 1985).

Falcon et al. (1985) compared the effectiveness of CMGV with a widely used alternative method (oil) and a conventional method (the application of chemical pesticides) in an orchard near the Ferrari Farm. The results of this field experiment suggest that CMGV provides approximately the same protection against codling moth at about the same cost for materials, but that it also requires more frequent applications.

Another element of the Ferrari pest control program on the farm's alternative acreage is the periodic release of predacious mites. These mites prey on phytophagous mites that feed on the leaves of fruit trees, sometimes defoliating the trees. The beneficial predator mites are introduced to the orchard in a novel way: the Ferraris purchase bean plants (each about 12 inches tall) infested with predaceous mites (about 20 to 30 per plant) from an insectary. Wayne Ferrari and his workers then place one of these bean plants on a branch at the northeast side of each tree (about 100 trees per acre). The bean plants are placed at about chest height.

The predaceous mites are bred to be resistant to sulfur and miticides. Wayne Ferrari reports that the efficacy of this procedure is "fantastic" and that the predators effectively control the phytophagous mites. The cost of this material ($25.00 per acre) is about the same as the cost of chemical miticides such as propargite; however, Ferrari estimates that the labor cost of applying the infested bean plants would be somewhat higher (data are not available) than spraying miticide on the orchard.

In addition to applying CMGV material to an experimental plot of apples, the Ferraris also used this material on 3 acres of their conventional apples. Before applying CMGV, they had found the population of codling moth to be increasing very rapidly in one of the orchards just a few days prior to harvest. They were faced with a choice of either picking the crop early, thereby sacrificing the optimum sugar content of the apples, or spraying with the chemical pesticide phosalone, which would have required a 2-week delay before the toxicity of the pesticide had subsided sufficiently to legally permit workers to enter the orchards for harvest. This delay would have meant postponing harvest beyond the optimum stage of maturity.

Instead of choosing either of these options the Ferraris applied CMGV to this orchard. Because there is no reentry delay time with CMGV, it was possible to send harvest crews into the orchard at the optimum harvest time. Ferrari has observed that a major advantage of CMGV is that, as in the above instance, it can be used close to harvest; it is one of the few pesticide materials currently available for such use. He recognizes, however, that conventional producers are not likely to favor CMGV for general use during the growing season because of the greater number of applications required as compared with chemical insecticides.

Disease Control

Blight is one of the major diseases of walnuts. Late-flowering varieties (such as Hartley) tend to escape blight infection because humidity is typically low later in the season. Early varieties, however, are treated with a 1 percent Bordeaux solution every 7 days as long as the humidity is high and the walnuts are small. When the humidity is low or the nuts are large, blight is no longer a problem.

Labor

The labor force on the Ferrari Farm consists of four adult family members (Wayne; his wife, Irene, who takes care of the bookkeeping; his mother, Italia, who works in the packing shed and takes sale orders from buyers; and his father, George, who oversees the packing shed and vegetable production); two teenage children who work in the packing shed during the summer months and on weekends; 20 regular hired workers (reduced to 12 in the winter months); and miscellaneous seasonal workers as needed.

Wayne Ferrari provides the bulk of the management of the orchards, including irrigation scheduling, pest control, fertility management, and planting and harvesting scheduling. Wayne and George Ferrari jointly select the cultivars of crops to be planted, deal with buyers, and make other day-to-day decisions. Both of these men work virtually year-round on the farm, with their most intense schedule occurring during harvest.

PERFORMANCE INDICATORS

Yield Data

Because of the large number of crops grown and the diversity of cultivars of each crop, it is not practical for the Ferraris to maintain accurate production and yield data on the various orchards and segments within orchards or in the vegetable plots. Consequently, no specific yield data are available.

Further research is needed before the yield impact profitability of CMGV will be verified for various growing conditions and locations. It is well established, however, that CMGV can be effective and profitable—but only when applied with precise timing in a thorough coverage of trees, as a part of a comprehensive IPM program that ensures viability of various natural predators, and in combination with other aspects of good management.

Financial Performance

Although cost and return data are not available for the Ferrari Farm as a whole, several specific items of information were gathered and generalizations can be made on the basis of interviews conducted with the family in 1982 and 1986. In general, costs and returns vary depending on the crop grown and on whether conventional or alternative methods are used. For example, the Ferraris pay dues of 0.5 percent of the gross value of sales from all acreage certified as organic (currently about $1,500) to the certifying organization, California Certified Organic Farmers. Approximately one-third of the walnut acreage of a total 111 acres are certified as organically grown; the Ferraris receive a price premium of $0.04 per pound on the organic walnuts sold in the shell. Shelled walnuts bring a $0.10 per pound (5 percent) premium.

The Ferraris use three market outlets for their fruit: (1) wholesale outlet firms, which handle the premium quality produce; (2) the San Francisco farmers' market and a few other direct marketing outlets, at which they sell produce that does not meet the premium grading standards; and (3) the dried fruit processing facility, where cull fruit (particularly insect-damaged fruit) is dried for marketing. The Ferraris sell their premium produce to several different wholesalers, particularly in the Los Angeles, San Francisco, and Oregon markets. Most of these wholesalers specialize in organic produce; the Ferraris sell the fruit from their conventionally produced acreage to conventional wholesalers.

Approximately 10 to 15 percent of the Ferraris' farm products are sold through the San Francisco farmers' market and six small stores; currently, there are no restaurants included in their direct marketing network. Each Friday the Ferraris load their refrigerated van with various fruits and vegetables and, early on Saturday morning, they drive to the San Francisco farmers' market to sell this produce. They also make deliveries to a few stores that have telephoned orders to them in advance. The Ferraris report that they have been marketing their produce in the San Francisco area for approximately 30 years, with a large number of repeat customers. Because the produce that they are selling is slightly below the grading standards for premium products, they are able to give these customers a good quality product at a bargain price.

Over the years, Wayne Ferrari reports that he has made decisions to plant additional varieties of fruits and vegetables in response to questions and requests from his customers in the San Francisco market; he has also based such decisions on evidence in the wholesale markets suggesting expectations of profitable enterprises in years to come. His management strategy is extremely diversified in every aspect: the number of crops he produces, the farm's marketing outlets, sources of compost, fertility management methods, and pest management strategies.

One of the underlying goals guiding the Ferrari marketing strategy is a desire to promote stability in the market. They prefer to see the prices of produce remain relatively constant throughout the season rather than exhibiting wide fluctuations from month to month. They also prefer to avoid the inevitable haggling required in dealing with wholesalers and other buyers when prices fluctuate widely. Consequently, the Ferrari pricing strategy is to set a price for each of their various products when harvesting begins and to try to maintain that price throughout the harvest season and for as long as the product is available in storage. They realize that at times in the season, when prices are abnormally high, the wholesalers make significant profits because of this pricing practice. This fact does not appear to bother the Ferraris, however; they are willing to allow such profits in the interest of encouraging some degree of stability in the market and avoiding the haggling over price.

The premium prices received for certified organic produce do vary throughout the season. For example, at some points the Ferraris may receive a premium of $2.00 to $3.00 per box above the conventional price for Granny Smith apples; at other times the price may be $1.00 to $2.00 per box below the conventional price.

Despite the lack of detailed accounting data, there are some indirect indications of the Ferrari Farm's financial performance. First, the farm is expanding modestly based on earnings and savings, without incurring debt. Since 1982 the acreage of the farm has expanded by 18 acres through the purchase of an additional field. Second, the capacity of the packing plant has approximately doubled, both in floor space and in the number and sophistication of the machines which it contains. Each new item of machinery has been purchased with cash rather than credit. Expansion financed by earnings constitutes real growth, which is one of the most reliable indicators of good financial performance.

REFERENCES

Falcon, L. A., A. Berlowitz, and J. Bradley. 1985. Progress report on pilot studies designed to demonstrate how pear zone growers may improve pest and disease control and frost protection through better timing and management. Department of Entomological Sciences, University of California, Berkeley, December 28.

Kurstak, E. 1982. Microbial and Viral Pesticides. New York: Marcel Dekker.

Weakley, C. V., P. Kirsch, and R. E. Rice. 1988. Control of oriental fruit moth by mating disruption in California peach and nectarine orchards. Pp. 541–549 in Global Perspectives on Agroecology and Sustainable Agricultural Systems: Proceedings of the Sixth International Scientific Conference of the International Federation of Organic Agriculture Movements, Vol. 2, P. Allen and D. Van Dusen, eds. Santa Cruz, Calif.: Agroecology Program. University of California.

7

Florida Fresh-Market
Vegetable Production:
Integrated Pest Management

F OUR FARMS IN SOUTH FLORIDA that produce fresh-market vegetables are the subject of this case study. The common element linking these farms is that they are all served by the same integrated pest management (IPM) pest scouting service, Glades Crop Care, Inc. (Table 1).

GENERAL DATA

Hundley Farms of Loxahatchee: The farmer, John Hundley, grows 1,500 acres of sweet corn, 120 acres of cabbage, 3,000 acres of radishes, 1,600 acres of seed corn, and 1,300 acres of leafy vegetables. He also has a 120-acre orange grove and 1,500 acres of sugarcane; he runs cattle on 500 acres of pasture.

Ted Winsberg of Palm Beach: This farm consists of 350 acres of irrigated sandy soil. Winsberg has raised fresh-market peppers on all of this land continuously for over 10 years.

John Garguillo of Naples: The Garguillo Farm is located south of Ft. Myers, on the west side of the state. John Garguillo raises 1,300 acres of staked tomatoes for the fresh market.

Fred Barfield of Immokalee: Fred Barfield raises 1,000 acres of vegetables, primarily bell peppers (green, red, purple, and yellow), tomatoes, and cucumbers. He has also grown eggplant and yellow squash. The farm includes a 550-acre orange grove, a 1,000-cow purebred Beefmaster herd, and a 1,200-cow mixed-breed, commercial herd.

Glades Crop Care, Inc.: All four of these farms employ Glades Crop Care, Inc. (GCC), the largest IPM farm pest scouting service in south Florida. The GCC staff consists of about 20 field scouts as well as a backup staff. The scouts have at least a B.S. degree in an agricultural discipline and are supported by a technical staff (with M.S. or Ph.D. degrees) under the

TABLE 1 Summary of Enterprise Data for Four Farms in Florida

Category	Description
Farm sizes	350–9,640 acres
Labor and management practices	All four farms hire the services of an IPM scouting firm during all phases of crop growth. The firm provides frequent and extensive scouting. One grower (Winsberg) retains his labor force even if crop prices decline to the labor cost of harvesting.
Marketing strategies	Fresh produce is marketed through a regular packing plant owned by the farm firm (Garguillo), a cooperative (Hundley), or a vegetable exchange (Winsberg).
Weed control practices	Plastic mulch over seed beds smothers and shades the weeds, preventing emergence. Herbicides are used where plastic mulch is inappropriate.
Insect and nematode control practices	An IPM scouting service is used by all four farms, which greatly reduces pesticide usage. The long, hot growing season, however, necessitates chemical control: endosulfan and fenvalerate in peppers; and methomyl, fenvalerate, and endosulfan in tomatoes. Methyl bromide is used as a fumigant for nematodes. Pesticide usage has been substantially cut in all cases.
Disease control practices	The farmers use soil fumigation and rely on several applications of fungicides and bactericides to control plant diseases.
Soil fertility management	Commercial fertilizers are used to supply N, P, K, Ca, and trace elements.
Irrigation practices	Fields are subirrigated with seepage from parallel ditches 80 feet apart.
Crop and livestock yields	No yield impacts were reported.
Financial performance	All four farms appear to be financially sound. The farmers report per-acre cost savings of as much as $400 from the use of IPM pest scouting and ensuing reductions in the frequency of pesticide applications.

direction of H. Charles Mellinger. The committee's interviewer was accompanied on the farm visits by Madeline Biemueller Mellinger, president of GCC.

The fundamental concept of IPM is that only when a pest reaches an economic damage level—that is, when the expected decline in the value of revenue from sale of the crop exceeds the cost of spraying—will treatment (usually a pesticide) be employed. For an IPM program to be effective, the pest scout must be completely familiar with the cultural practices being used on the farm: field preparation, bed fumigation and formation, fertilizer application, transplanting or seeding, and irrigation.

Scouting begins at the transplant greenhouse for some crops to ensure that diseases and insect problems are not spread to the fields. The IPM monitoring program continues in the production fields through the harvest; during this stage, scouts monitor pest populations and evaluate any diseases that are present and their severity. GCC has developed extensive field manuals that assist their scouts with pest and disease identification and monitoring techniques.

Threshold or action levels of acceptable pest populations may be established by the IPM scouting firm or by the growers themselves, but usually these levels are set through discussion and agreement between the grower and the firm. Once these threshold levels are reached, a treatment is recommended by the scouting firm, subject to approval by the grower. One of the direct benefits of pest scouting is that it quantifies the stages of the insects, thus permitting the grower to apply pesticide to the early instar or egg stage or to the early disease lesions. Therefore, a much lower rate of pesticide can be used and a much higher level of control will result, often eliminating the need for follow-up applications.

Climate

South Florida has a subtropical climate (Table 2). Precipitation in the Ft. Myers area, for example, averages 54 inches per year. The normal minimum temperature in January is 52°F. Parts of south Florida occasionally have freezing temperatures.

PHYSICAL AND CAPITAL RESOURCES

South Florida is characterized by flat topography and a high water table that fluctuates between 18 and 24 inches below the surface. The two generic soil types are sandy and an organic soil, muck.

Sandy Soils

Sandy soils occur on both the east coast west of Palm Beach and on the southwestern half of the state around Naples and Immokalee. The topography is flat, and the elevation is only a few feet to 10 feet above sea level. Irrigation is provided by a seepage subirrigation system. The land is laser-leveled, and a system of ditches is used to maintain the water table at the desired depth of 15 to 18 inches below the surface.

Typically, each field is rectangular, approximately 20 to 40 acres in size, and surrounded by a diked main irrigation ditch. This main ditch can be flooded with a low-lift pump to a level higher than the field. Subirrigation ditches are dug about every 80 feet parallel to the crop beds. Water flooded into these ditches seeps under the beds to wet the roots from below by raising the water table. Water reaches the plant from the perched water table by a capillary-type system. The water moves upward under the raised, plastic-covered beds, except when the fields are being drained. During excessive rainfall the water table can be lowered by reversing the system and pumping the water out of the fields.

Maintaining the water table at the 15- to 18-inch level is critical for proper root development and efficient fertilizer and water usage. A higher water table will cause excessive fertilizer leaching and pumping costs and will waste water. If the water table is too low the soil near the vegetable bed

TABLE 2 Normal Daily Temperatures and Monthly Precipitation at Ft. Myers, Florida

Month	Normal Daily Temperature (°F)			Normal Degree Days		Normal Precipitation	
	Maximum	Minimum	Monthly Average	Heating	Cooling	Inches	Days With ≥0.01 Inches
January	74.7	52.3	63.5	128	81	1.64	5
February	76.0	53.3	64.7	125	121	2.03	6
March	79.7	57.3	68.5	48	151	3.06	5
April	84.8	61.8	73.3	0	253	2.03	5
May	89.0	66.4	77.7	0	394	3.99	8
June	90.5	71.7	81.1	0	483	8.89	15
July	91.1	73.8	82.5	0	543	8.90	18
August	91.5	74.1	82.8	0	552	7.72	18
September	89.8	73.4	81.6	0	498	8.71	16
October	85.3	67.5	76.4	0	353	4.37	9
November	79.9	58.8	69.4	44	176	1.31	4
December	75.9	53.6	64.8	112	106	1.30	5
Average annual	84.0	63.7	73.9				
Average annual total				457	3,711	53.95	114

NOTE: The normal daily maximum by month is the average of each day's (midnight to midnight) high temperature for every day in that month from 1941 to 1970. The normal daily minimum by month is the average of each day's (midnight to midnight) low temperature for every day in that month from 1941 to 1970. The normal monthly temperature is the average of the normal daily maximum and minimum temperatures for that month. The normal degree days heating are the sums of the negative departures of average daily temperatures from 65°F. The normal degree days cooling are the sums of the positive departures of average daily temperatures from 65°F. To calculate the normal degree days heating or cooling, multiply the difference between 65°F and the normal monthly temperature by the number of days in the month. The normal monthly precipitation is the average of the inches of precipitation for that month from 1941 to 1970.

SOURCE: National Oceanic and Atmospheric Administration. 1980. Climates of the States, 2d ed. Detroit: Gale Research Co., Book Tower.

339

surface may become so dry under the plastic that the nutrients from the top-banded fertilizer are not dissolved and therefore cannot be absorbed by the vegetable plants.

Typical practice in this area has been to grow sugarcane or other dense plantings along the drainage ditches to minimize wind damage to the crop and also to reduce wind erosion. One grower, Ted Winsberg, grows tropical plants along each irrigation ditch, a practice that gives the fields a beautiful appearance and generates additional income.

Muck Soils

The muck soils are located in the central part of south Florida on the east and south sides of Lake Okeechobee. Fields composed of such soils are typically flooded in the off-season to control diseases and minimize soil oxidation and subsidence, soil-borne insects, and some weed problems. The irrigation system for muck soils is basically the same as that for sandy soils: fields are divided into 20- to 40-acre rectangles, and the perimeter is surrounded by dikes and irrigation ditches. The fields are flooded in 20-day cycles during the growing season. Water is left on the ground for 10 days; the fields are then drained and dried for 10 days. This process is repeated two or three times, depending on the available time between crops, and it appears to reduce significantly the populations of soil-borne pathogens, weeds, and insects. The need for pesticides is also reduced or eliminated.

The intensive cultivation of muck soils causes soil subsidence. Madeline Mellinger and John Hundley reported that up to 1 inch of muck soil is being lost each year, primarily through oxidation. Measurements at the Everglades Research and Education Center in Belle Glade indicate that, over a 40-year period, nearly 4 feet of muck soil has been lost. Until 10 to 15 years ago, the oxidation of these soils was of little concern to many farmers because the organic soil appeared very deep. In recent years, however, some muck soil areas have become too shallow to grow certain crops, and concern about conserving the remaining soil is great. Local extension personnel have observed that the only major crop that could be produced without a major loss of soil is paddy rice.

MANAGEMENT FEATURES

Pesticide Use

Because consumers, with few exceptions, demand blemish-free fresh vegetables with cosmetic appeal, commercial-scale vegetable growers produce fruit and vegetables free of insect or disease damage. Consequently, vegetable growers spend a great deal of time and money protecting their crops to ensure this cosmetic appeal.

During an interview in 1986, the extension agent in Palm Beach County said that because of the climate in this part of the state, he doubted if it would be feasible to produce vegetables on commercial-scale farms in this area, considering today's technology, without chemical pesticides. He also observed that sugarcane may be the only major crop grown on a large scale in this area that can be produced without the substantial use of chemical pesticides.

According to H. C. Mellinger (correspondence, 1987), the cane fields have been infested with the imported fire ant that feeds on cane borers, the principal insect pest of sugarcane in this area. As a result, spraying for borers has been significantly reduced; some fields have not been sprayed in 10 or more years. Sugarcane is replanted each third to fifth year, and, for the benefit of the harvesting crews, the fields are burned each year before harvesting to suppress the fire ants and leaf debris. About half of the 400,000 acres of sugarcane are hand-harvested in Florida.

Unlike sugarcane, however, vegetables require pesticide application. Still, the extension agent reports that the extensive use of IPM programs such as that offered by GCC has greatly reduced traditional pesticide usage. But he adds that because of the tropical growing conditions, even with IPM scouting, the levels of chemical usage in south Florida are still greater than those in most farming areas of the United States.

Except in parts of interior Florida, most vegetable crops are grown using raised beds covered with plastic mulch sheeting. This type of mulching system, which has been used extensively in Florida for about 15 years, has helped minimize wind erosion and the plant nutrient leaching caused by heavy rains. The use of chemicals, however, is an important part of this system. Most of the plant beds are fumigated each year with chloropicrin and methyl bromide just before the plastic is laid down. Soil fumigation and plastic mulch suppress nematodes, soil-borne diseases, and insects and obviate the use of herbicides.

Because of the area's topography and porous soils, drinking water supplies may become contaminated by agricultural pesticides. These substances need only sink 15 to 18 inches to reach the water table. The possibility of public policies banning the use of widely used pesticides is a matter of some concern to Florida vegetable growers. The extension agent also indicated that vegetable growers are also worried about federal price supports. Some growers maintain that if the price support for U.S. sugar is dropped, the cane fields will be brought into vegetable production and flood the vegetable market, suppressing prices and causing substantially reduced farm income. This view is not held by all experts in the field, however, as noted by H. C. Mellinger (correspondence, 1987).

IPM Features

The preliminary results of a 1986 survey of 40 tomato farms conducted by the University of Florida (K. Pohronezny, interview, 1986) indicate that farmers using IPM programs have been able to reduce their insecticide inputs by about 21 percent. Sixty-two percent of the growers hiring commercial scout firms reported that their net returns increased (by an average of $121.00 per acre) as a result of their participation in the IPM scouting program. The other 38 percent of the growers reported no change in net returns: scouting costs equaled their savings from reduced sprays. Among growers who monitored their own fields or relied on minimal scouting by chemical company representatives, 54 percent reported a net savings averaging $62.00 per acre (K. Pohronezny, interview, 1987). Scouting tends to reduce insecticide costs and levels of application but causes no reduction in the use of fumigants, fungicides, or bactericides.

Madeline Mellinger, president, and H. Charles Mellinger, technical director, of Glades Crop Care, Inc., maintain that their crop scouting and consulting service has a significant impact on the amount of pesticides used in the south Florida farming community (excluding the Homestead area, which is outside of their territory). They estimate that their company serves approximately one-third of the vegetable acreage in this area; another one-third of the acreage is operated by former GCC clients who now employ in-house IPM scouts. Thus, the Mellingers estimate that approximately two-thirds of the total vegetable acreage in south Florida is managed with an IPM program. In addition, there are at least three other IPM scouting companies in south Florida, ranging from a single owner-scout operation to one employing five scouts (K. Pohronezny, interview, 1986).

As part of the service provided by GCC, the customer's fields and plants are monitored twice weekly, and the grower is told what insect and mite populations are present, their instar or stage, their locations on the plants, their in-field distribution, and the size of the population. The scouts identify the diseases present, quantify their severity, and pinpoint new activity or spread. An important aspect of GCC's disease control service is a system of field management in which GCC works closely with the grower to eliminate introductory sources and reservoirs of disease in and around the fields and to eliminate or reduce the spread of a disease in the fields once the plants have become infected.

A grower who relies on the observations of a pest scout applies less insecticide than non-IPM growers for two reasons: (1) pesticides are applied only for those pests present in the field, and (2) lower rates of pesticide can be applied because the scout reports the eggs and early larval instars rather than waiting until populations of larger insects have reached critical levels. The scout also helps to identify and refine routine prophylactic and remedial insect and disease control practices used by the grower.

According to H. C. Mellinger, some growers also use the *Bacillus thuringiensis* products extensively for larval control; other more specifically targeted insecticides are also used to take advantage of the beneficial insects that may control more harmful species. K. Pohronezny has observed that this practice became quite popular in the late 1970s but has since been largely replaced by applications of a new class of insecticides, the synthetic pyrethroids.

Regarding the direct costs of a pest control program with and without the

IPM scouting, H. C. Mellinger reported that, for a fresh-market tomato crop, an average routine pesticide program applied preventatively every 2 to 5 days (without scouting) will cost the grower between $450.00 and $700.00 per crop acre for control products alone. Using IPM, a grower's direct pest control costs range from $200.00 to $300.00 per crop acre for average insect stress years. Much of this cost reduction results from the proper timing of insecticide use, which often eliminates the need for repeat applications; reduced rates of use because insecticide is applied to the early instars and stages; and the application of products only when necessary, that is, for those insects present at economic threshold levels. Another major benefit of IPM is reduced stress on the environment. Finally, there are the other benefits of reduced pesticide use, including less exposure for workers, less demand for and wear of spray rigs, fewer empty pesticide containers to dispose of, and fewer supervisory hours.

For the bell pepper crop the costs are similar to those for tomatoes; the crop growing season is longer, but the insecticide usage is slightly less intense than in growing tomatoes. The same principles apply: using biological control materials along with the other IPM tools. In fact, the pest spectrum of bell peppers makes then more amenable than the tomato crop to a greater use of biological and more specifically targeted insecticides.

Sweet corn is another widely planted vegetable crop with major insect and disease problems. The Mellingers estimate that scouting has had a substantial impact on both insecticide and fungicide usage in the sweet corn industry. Of the tens of thousands of sweet corn acres in south Florida, about 80 to 90 percent operate under an IPM program (Tables 3 and 4). Most of the insect problems in sweet corn involve larvae feeding in the stalk or ear. Methomyl (in liquid or granular form) is most commonly used for larval control, and mancozeb or chlorothalonil is commonly used for blight diseases. According to H. C. Mellinger, IPM scouting can now reduce sweet corn pesticide applications by up to 50 percent for insects and 25 percent for diseases. IPM practices on other vegetable crops have produced similar results.

Glades Crop Care finds its largest task to be one of educating growers about the life cycles of pests, disease dissemination principles, and modes of action of pesticides and their spectra. Once this educational process is completed, the grower's progress toward an effective IPM program is often swift and sure. The four farmers profiled in this case study are good examples.

John Hundley of *Hundley Farms* has been a GCC client for the past 14 years. He employs the company to scout all of the vegetable fields, which are mainly composed of peat soils. Based on GCC's findings, Hundley decides what pest population levels can be tolerated before spraying his crop. He relies primarily on flooding and cultivation for weed control; some herbicides are used, but few herbicides are registered for use on minor crops. Before hiring GCC, Hundley reported that he followed a prophylactic or regularly scheduled pesticide spray program for each crop, spraying every other day or so. If a pest build-up problem occurred, he increased the rate of pesticide application. For the past 12 years, however, GCC has monitored each field, and Hundley now sprays only when necessary to prevent an economic level of damage (the value of the crop loss exceeds the remedial treatment cost).

In his 1986 sugarcane crop, Hundley sprayed for sugarcane borers for the first time in 3 years. Normally, high populations of fire ants control the borers satisfactorily. He thinks that the reason he had to spray was because he had planted sweet corn next to the sugarcane fields, and drift from the spraying for sweet corn pests killed the fire ants in the cane.

Ted Winsberg has been growing peppers continuously on the same 350 acres for 30 years. For the past 12 years, he has been using the raised-bed

TABLE 3   Per Acre Pesticide Application for Fall Sweet Corn Under IPM in the Everglades Agricultural Area, 1980

| | Pesticide | | | | |
Date	Methomyl (Insecticide)	Toxaphene[a] (Insecticide)	Mancozeb (Fungicide)	Manganese (Fertilizer)	Cost/Active Ingredient
9/30	1 pint	1 pint	1 pound	1 pound	$ 6.40
10/6	1 pint	1 pint		1 pound	4.31
10/11	1 pint	1 pint			4.31
10/14	1 pint	1 pint			4.31
10/17	1 pint	1 pint	1 pound	1 pound	6.40
11/1	1 pint				3.25
11/2	1 pint				3.25
11/3	1 pint				3.25
11/5	1 pint				3.25
11/7	1 pint				3.25
11/9	1 pint				3.25
11/11	1 pint				3.25
11/14	1 pint				3.25
11/17	1 pint				3.25
11/19	1 pint				3.25

Costs

Insecticide, fungicide, and manganese				
$48.75	$5.30	$3.80	$0.38	$ 58.23

Application (15 applications at $2.00 each)   30.00

Herbicide (2 pounds atrazine + 1 quart
  11-E oil postemergence)   4.44

Scouting   7.50

Total $100.17

[a]The Environmental Protection Agency has cancelled toxaphene for all agricultural uses except as a livestock dip for parasites.

SOURCE: K. Shuler, Extension Service, U.S. Department of Agriculture, Palm Beach County, Florida, correspondence, 1986.

TABLE 4 Typical Per Acre Pesticide Application for Fall Sweet Corn Not Using IPM in the Everglades Agricultural Area, 1980

| | Pesticide | | | | |
Date	Methomyl (Insecticide)	Toxaphene[a] (Insecticide)	Mancozeb (Fungicide)	Manganese (Fertilizer)	Cost/Active Ingredient
9/30		1 quart			$ 2.13
10/2	¼ pound	1 pint	1 pound		6.26
10/5	¼ pound		1 pound		5.20
10/7	¼ pound		1 pound	1 pound	5.34
10/11	¼ pound		1 pound	1 pound	5.34
10/14	¼ pound		1 pound		5.20
10/17	¼ pound		1 pound	1 pound	5.34
10/20	¼ pound		1 pound	1 pound	5.34
10/24	¼ pound		1 pound	1 pound	5.34
10/27	¼ pound		1 pound		5.20
10/30	¼ pound				3.25
11/2	¼ pound				3.25
11/5	¼ pound		1 pound		5.20
11/7	¼ pound				3.25
11/9	¼ pound				3.25
11/12	¼ pound		1 pound		5.20
11/14	¼ pound				3.25
11/16	¼ pound				3.25
11/18	¼ pound				3.25
11/20	¼ pound				3.25
11/22	¼ pound				3.25
11/25	¼ pound				3.25
11/27	¼ pound				3.25

		Costs			
Insecticide, fungicide, and manganese					
$71.50	$3.19	$21.45	$0.70		$ 96.84
Application (23 applications at $2.00 each)					46.00
Herbicide (1½ pounds atrazine postemergence)					2.48
Total					$145.32

[a]The Environmental Protection Agency has cancelled toxaphene for all agricultural uses except as a livestock dip for parasites.

SOURCE: K.Shuler, Extension Service, U.S. Department of Agriculture, Palm Beach County, Florida, correspondence, 1986.

plastic mulch cultural practice. Winsberg has used GCC pest scouting for 10 years.

Peppers are planted in August and September through a layer of plastic, the top surface of which has been colored white to reflect the heat. This material costs $300.00 per acre. Later plantings (after September) are planted on black plastic, which costs $200.00 per acre. Although the use of plastic with the bed system has doubled his yields, Winsberg said that his costs have more than tripled. The ground is fumigated with methyl bromide at a cost of approximately $124.00 per acre, plus labor, equipment, and plastic.

To make the plastic mulch system work, all fertilizer must be applied before the plastic is spread over the field. Over $300.00 worth of fertilizer is applied prior to planting, including 300 pounds of nitrogen per acre. One hundred pounds of nitrogen in the form of sulfur-coated urea is broadcast before the beds are made. (The shaping of the beds helps incorporate the fertilizer into the soil.) Then 200 pounds of nitrogen, in a 16–0–23 nitrogen, phosphorus, and potassium formulation, are applied as a band fertilizer approximately 10 inches from the plants. The plastic helps eliminate the leaching of nutrients by rain.

The soil is tested each year, but in practice, all fields receive about the same application rate of fertilizer. Weeds and nematodes are controlled through the use of fumigation. The plastic mulch also controls most other weeds except in the area between the beds, which is typically sprayed once or twice with paraquat and glyphosate (K. Pohronezny, interview, 1986).

Ted Winsberg believes strongly in using pest scouting to determine the minimum frequency and dosage of pesticide application. Yet, he also said that, because of past experience, he is afraid not to spray. He reported that 12 years ago he eliminated chemical sprays in his pepper crop for 2 years because of health concerns. In addition, based on extensive readings of biological pest control literature, he released many beneficial insects to control pests. But a severe outbreak of pepper weevils caused major financial losses.

Ted Winsberg is still very much interested in using less chemical pesticide on his crops, but because of the huge investment involved in each acre of peppers (up to $3,000 in operating costs before harvesting) (Table 5), he believes that he cannot afford to not spray. He hires GCC to look for various pest problems, particularly insect pests and diseases, and an IPM scout is in the field looking for pests every second or third day. Although pesticide applications are made every second or third day, a much lower rate of insecticide is now applied as a result of recommendations from the scouting service.

Winsberg also reported that IPM scouting is saving him up to $200.00 per acre in pesticides. For example, he now sprays methomyl for worms twice per week at 1 ounce per acre; before IPM scouting, he was spraying twice a week at 1 to 2 pounds per acre. During the growing season of peppers (160 days), insecticide and fungicide sprays will cost a total of $200.00 to $300.00 per acre and involve 40 to 80 applications. This does not include the cost of fumigating, which is generally more than $100.00 per acre not including labor, equipment, and plastic to seal in the fumigant.

During the past 7 years, bacterial spot in peppers has become more and more of a problem. To control the disease, copper-containing fungicide in combination with maneb is sprayed on the plants every third day. Up to 60 pounds of bactericide is applied annually to control the spot. Winsberg expressed concern that excessive copper in the soil from the fungicide may become an increasing problem.

TABLE 5   Representative Costs for Bell Pepper Production in Palm Beach
County, 1984 (in dollars)

Category	Average/Acre
Operating costs	
Cultural labor	1,089.56
Fertilizer	314.14
Gas, oil, grease	128.86
Interest (4-month operating cost)	127.54
Machine hire	60.26
Miscellaneous	148.20
Pesticides	373.02
Plastic	268.60
Repair and maintenance	238.75
Seed and transplants	166.72
Sterilants and herbicides	110.23
Total operating costs	3,025.88
Fixed costs	
Depreciation	182.40
Insurance and licenses	94.88
Land rent	124.46
Total fixed costs	401.74
Harvesting and marketing costs	
Containers	391.68
Hauling	87.04
Picking and packing	832.32
Selling fees	174.08
Total harvesting and marketing costs	1,485.12
Total costs	4,912.74
Total receipts	4,373.76
Net return	(−538.98)
Yield (bushels)	544

SOURCE: K. Shuler, Extension Service, U.S. Department of Agriculture, Palm Beach County,
Florida, correspondence, 1986.

John Garguillio reported that he uses a three-tiered system of pest monitoring for his 1,300 acres of fresh-market tomatoes: he has employed GCC
for 6 years to provide a full-time professional crop monitoring service; he
has trained in-house scouts, who examine the fields daily; and he also uses
another private crop consultant. Based on the findings of these three
sources, and using certain threshold levels, Garguillo decides which pesticides to apply and when to spray.

The grower refused to discuss his spraying program and action threshold
levels, calling them proprietary and confidential and indicating that he
considers this to be an area in which he may have a competitive edge. He
did say that by using IPM, he has been able to cut his pesticide costs almost
in half over the past 5 years from over $500.00 per acre to $250.00 to $260.00
per acre.

Garguillo did report that he directs his field managers to apply 350 pounds of nitrogen per acre, which is broadcast and worked in with a rotary hoe prior to the shaping of the beds. The soil pH is adjusted to 5.5. He estimates that the salts in the fertilizer reduce the pH by one point. Potassium (K_2O) is applied at 1.5 to 2.0 times the amount of nitrogen. From 50 to 100 pounds of phosphorus (P_2O_5) plus 1,000 pounds of calcium are applied per acre. Based on soil tests, boron, manganese, zinc, and sulfur may also be added. As much as 60 pounds of copper-containing bactericide per acre are applied each year to control bacterial spot. After the beds are shaped, the soil is fumigated with methyl bromide and plastic is spread over a smooth seed-bed.

Fred Barfield relies exclusively on the pest scouting services of GCC and has used the company for 3 years. Barfield maintains that today's farmer cannot afford to be out looking for insects and other pests 4 to 5 days per week, which is what it takes to grow the quality and quantity of produce needed to stay in business. He therefore relies on GCC to fulfill his pest scouting requirements. He said that by spraying only when necessary, he has saved from $200.00 to $400.00 per acre in pesticide costs. Barfield fumigates his fields with 180 pounds of methyl bromide per acre prior to spreading the plastic mulch.

Before 1970, however, Barfield followed a different course. He had large areas of virgin soil, and rather than fumigate soil that had become infested with pathogens and pests, he would bring new land into production, farm it for a few years, and then convert it to cattle pasture after soil pests became too much of a problem. Today, the costs of bringing new land into production are increased by legal requirements for engineers, water-use consultants, and environmental impact studies. Consequently, he now relies on soil fumigation.

PERFORMANCE INDICATORS

Ted Winsberg markets all of his peppers through a vegetable exchange. He reported that since he began using the plastic-covered bed system to produce peppers 12 years ago, his yield has doubled to its current rate of 500 to 600 cartons per acre. (A carton is approximately 1.1 bushels.) He observed that almost every pepper grower is using the same cultural system. Winsberg begins planting peppers in early August and continues until October, and he markets his peppers from September until May. In southern Florida, according to Winsberg, producers can plant year-round, but the marketing of peppers by states further north eliminates the southern Florida producers' market during the months of June, July, and August. Buyers, and therefore trucks, will not come as far south as southern Florida if they can get the supply that they need further north, closer to northern population centers.

The price received for peppers fluctuates widely depending on weekly supply and demand. Winsberg recalls prices as high as $38.00 per carton after a large freeze and as low as $2.00 per carton. Typically, the price varies from $4.00 to $20.00 per carton. Winsberg reported that his break-even price is $5.00 per carton. To retain his labor force year-round, he will continue to harvest even if the price falls to $2.00 per carton, which is basically the cost of harvesting.

The grower indicated that one of his worst pest conditions is market related. Whenever there is a surplus of peppers on the market and the prices drop to below harvest costs, neighboring fields are often abandoned. Growers are typically reluctant to plow the peppers under because they hope for a price rise in future weeks. Yet, in order to minimize their losses, they typically discontinue their spraying programs, and pest problems tend to multiply.

John Garguillo markets all of his fresh-market tomatoes (which beginning in 1986 carried the firm's brand name, Naples Fruit and Vegetables, Inc.) through his own packing and shipping plant. He harvests nearly 75 million pounds of tomatoes per year, all of which are harvested green. Nearly 20 percent are culled at the plant and given to a local farmer for animal feed; only blemish-free tomatoes of a uniform size are marketed. Because the packing house is integrated into the business of production, Garguillo will continue to pick tomatoes as long as the packing house makes money.

Florida tomatoes are sold under a marketing order, on consignment, and are owned by the farmer all the way up to the retail level. If they deteriorate or do not sell, the farmer is not paid.

Hundley Farms' operation is vertically integrated; everything grown on the farm is marketed through a cooperative. The cooperative Hundley uses consists of five area farmers, and it is currently trying to develop brand-name recognition. In addition to the superior appearance of their products, the growers are seeking to develop a reputation for the excellent taste of their products. To this end, Hundley has changed the varieties planted on his farm and the way certain vegetables are packed, stored, and marketed. The high-sugar hybrid sweet corn is a good example: Hundley said that he has been able to market more of this corn.

The fruit and vegetable business is extremely competitive. None of the owners of the four farms visited were willing to disclose details of their spraying programs or the pest threshold levels they used to determine when they sprayed. Consequently, specific information regarding cost savings on these farms is not available. It is apparent, however, that the use of IPM by these vegetable producers has improved the monetary and environmental performances of these farms; cost savings of as much as $400.00 per acre were reported. Another benefit is that the amounts of some pesticide applications are reduced through the use of IPM scouting, through the avoidance of unnecessary insecticide spraying, by the selection of different pesticides, and by the use of lower rates of pest control materials. Soil sterilization and the application of bactericides and fungicides have not diminished, however, and the consequences of their continued use for water pollution and chemical residues on foods are unknown at this time.

CASE STUDY

8

Fresh Grapes in California and Arizona: Stephen Pavich & Sons

I N 1986, THE STEPHEN PAVICH & SONS operation included 1,432 acres, of which 823 acres are in the Harquahala Valley, Maricopa County, Arizona, west of Phoenix; 467 acres are in the Delano, California, area; and another 142 acres are in Kern County, California, near Bakersfield (Table 1).

GENERAL DATA

Grapes are by far the most important crop grown by the Paviches, accounting for 95.7 percent of their gross sales revenue in 1985. That year, they harvested 1,105 acres of grapes. By 1986, this acreage had increased to 1,125 acres; the purchase of another 160 acres in 1987 brought their total acreage to 1,285 acres of fresh grapes. On their land in Arizona (134 acres in 1985 and 307 acres in 1986), the Paviches also grow other crops (Table 2).

Climate

Kern County lies at the south end of the San Joaquin Valley of California, with most of the farmland just above 500 feet in elevation. This area is naturally a desert, and irrigation is necessary to make crop production of most crops feasible. Mean annual precipitation in Bakersfield is 5.7 inches, with 89 percent falling between November and April (Table 3). The heaviest average monthly rainfall, about 1.03 inches, occurs in February. This area typically is very hot in the summer; maximum temperatures exceed 90°F an average of 110 days per year, and minimum temperatures fall below 32°F only 11 days per year.

A similar climate is found in Tulare County, which is located near the center of the San Joaquin Valley. The Pavich grape operation in Tulare

TABLE 1   Summary of Enterprise Data for the Stephen Pavich & Sons Farm

Category	Description
Farm size	1,432 acres of cropland in 1986
Labor and management practices	Stephen Pavich, Sr., works on production and marketing; Tom Pavich and wife, Tonya, manage marketing; Steve Pavich, Jr., manages field operations. Pest scouting is done by hired entomologists on contract, plus the brothers. Twenty-five permanent hired workers do pruning, vine dressing, and other specialized duties. Approximately 350 seasonal workers are hired for harvest.
Marketing strategies	The farm is a major marketing operation selling to 19 of the 20 top retail chains in the United States plus several foreign countries. Cold storage of a part of the crop brings seasonally high prices. No premium price is asked as a result of alternative farming methods except for about 3 percent of the crop, which is certified and labeled as organic and sold through health food stores.
Weed control practices	The Paviches use no-tillage methods with a perennial rye grass and native weed cover crop, chopped periodically. Hand weeding between grape vines is also used. No herbicides are applied.
Insect and nematode control practices	A natural parasite (the *Anagrus* wasp) is important for leafhopper control but is not deliberately released. Nematodes are controlled by fumigation and a 2- to 3-year fallow period. The Paviches use IPM scouting for insect pests and occasional insecticide spot spraying.
Disease control practices	Sulfur dust is applied to prevent fungal diseases, and soils are fumigated with methyl bromide before planting to suppress pathogens and nematodes. Grapes in storage are fumigated with sulfur dioxide gas.
Soil fertility management	About 2.75 tons of composted steer manure per acre provide about 94 pounds N, 85 pounds P_2O_5, and 138 pounds K_2O per acre.
Irrigation practices	Grape vineyards are irrigated by flooding or using a modified drip (fanjet) system. The grapes require 3 to 6 acre-feet of water, depending on vineyard location, weather, winter precipitation, and the intensity of the crop. Water for irrigation comes from wells or rivers.
Crop and livestock yields	Grape yields per acre (653 boxes) exceed University of California estimated normal yields for conventional production (522 boxes); culls are 1 percent versus 15 to 30 percent for conventional production. (Data may not be exactly comparable.)
Financial performance	Producing about 1 percent of the U.S. table grape output, the Paviches are earning a substantial net income. Their preharvest cost per box of grapes ($2.12) is virtually identical to the conventional norm. The Paviches incur somewhat higher costs for some items (for example, soil fertility) and less for others (chemicals). Profits are enhanced by higher-than-average yields, extensive storage, and a nationwide marketing system.

TABLE 2   Land Use by Stephen Pavich & Sons, 1985 and 1986

	Acres	
Land Use	1985	1986
Grapes		
Bakersfield area, California		
Thompson Seedless	142	142
Delano area, California		
Almeria	0	3
Calmeria	22	46
Emperor	156	219
Red Flame Seedless	20	20
Ribier	20	10
Thompson Seedless	229	169
Subtotal, California grapes	589	609
Harquahala Valley, Arizona		
Exotic	12	12
Perlette	160	160
Red Flame Seedless	160	160
Thompson Seedless	184	184
Subtotal, Arizona grapes	516	516
Total grapes	1,105	1,125
Other crops		
Arizona		
Chili peppers	0	16
Cotton, pima	0	45
Cotton, short staple	0	105
Mixed melons	94	94
Squash	0	32
Watermelons	40	15
Subtotal, other crops	134	307
Total, all crops	1,239	1,432
Fallow, roads, buildings		
Arizona	310	137
California	193	257
Total, all land	1,742	1,826

County is at an elevation of about 560 feet. At Porterville (15 miles north of the Pavich grape operation), the normal precipitation is 11.2 inches per year, occurring almost entirely from November to April (Table 4). The maximum temperature exceeds 90°F on an average of 112 days per year; the minimum falls below 32°F on 26 days per year.

The climate in the Harquahala Valley of Arizona is also hot and dry. At Phoenix (about 70 miles east of the Pavich ranch), the normal precipitation is 7 inches per year (Table 5). The precipitation is more evenly distributed here than in California, however, with the maximum rainfall (1.22 inches) occurring in August. The elevation of the Pavich farm is about the same as

TABLE 3 Normal Daily Temperatures and Monthly Precipitation at Bakersfield, California

Month	Normal Daily Temperature (°F)			Normal Degree Days		Normal Precipitation	
	Maximum	Minimum	Monthly Average	Heating	Cooling	Inches	Days With ≥ 0.01 Inches
January	57.5	37.4	47.5	543	0	0.96	5
February	63.3	41.4	52.4	353	0	1.03	6
March	68.6	44.5	56.6	266	6	0.83	6
April	75.5	49.9	62.7	140	71	0.85	5
May	83.6	56.0	69.8	22	171	0.19	2
June	91.5	62.3	76.9	0	362	0.06	a
July	99.1	68.7	83.9	0	586	0.02	a
August	96.5	66.6	81.6	0	515	0.01	a
September	91.1	62.1	76.6	0	348	0.08	1
October	80.5	53.3	66.9	55	114	0.26	2
November	67.8	44.2	56.0	276	6	0.69	3
December	57.4	38.4	47.9	530	0	0.74	5
Average annual	77.7	52.1	64.9	Average annual total	2,185	2,179	5.72	35

NOTE: The normal daily maximum by month is the average of each day's (midnight to midnight) high temperature for every day in that month from 1941 to 1970. The normal daily minimum by month is the average of each day's (midnight to midnight) low temperature for every day in that month from 1941 to 1970. The normal monthly temperature is the average of the normal daily maximum and minimum temperatures for that month. The normal degree days heating are the sums of the negative departures of average daily temperatures from 65°F. The normal degree days cooling are the sums of the positive departures of average daily temperatures from 65°F. To calculate the normal degree days heating or cooling, multiply the difference between 65°F and the normal monthly temperature by the number of days in the month. The normal monthly precipitation is the average of the inches of precipitation for that month from 1941 to 1970.

^aLess than one-half.

SOURCE: National Oceanic and Atmospheric Administration. 1980. Climates of the States, 2d ed. Detroit: Gale Research Co., Book Tower.

354

TABLE 4 Normal Daily Temperatures and Monthly Precipitation at Porterville, California

Month	Normal Daily Temperature (°F)			Normal Precipitation	
	Maximum	Minimum	Monthly Average	Inches	Days With ≥ 0.1 Inches
January	56.3	35.8	46.1	2.17	5
February	62.8	39.5	51.2	1.61	4
March	68.5	42.3	55.4	1.53	4
April	75.4	46.8	61.1	1.30	3
May	83.8	52.6	68.2	0.46	1
June	91.9	58.8	75.4	0.06	0
July	98.6	64.5	81.6	0.01	0
August	96.9	62.5	79.7	0.01	0
September	91.7	56.9	74.3	0.16	0
October	81.4	49.1	65.3	0.62	1
November	67.4	41.3	54.4	1.51	3
December	56.2	36.1	46.1	1.72	4
Average annual	77.6	48.9	63.2	Average annual total 11.16	25

NOTE: The normal daily maximum by month is the average of each day's (midnight to midnight) high temperature for every day in that month from 1941 to 1970. The normal daily minimum by month is the average of each day's (midnight to midnight) low temperature for every day in that month from 1941 to 1970. The normal monthly temperature is the average of the normal daily maximum and minimum temperatures for that month. The normal monthly precipitation is the average of the inches of precipitation for that month from 1941 to 1970.

SOURCE: National Oceanic and Atmospheric Administration. 1980. Climates of the States, 2d ed. Detroit: Gale Research Co., Book Tower.

TABLE 5   Normal Daily Temperatures and Monthly Precipitation at Phoenix, Arizona

Month	Normal Daily Temperature (°F)			Normal Degree Days		Normal Precipitation	
	Maximum	Minimum	Monthly Average	Heating	Cooling	Inches	Days With ≥ 0.01 Inches
January	64.8	37.6	51.2	428	0	0.71	3
February	69.8	37.6	51.2	428	0	0.71	4
March	74.5	44.8	59.7	185	21	0.76	3
April	83.6	51.8	67.7	60	141	0.32	2
May	92.9	59.6	76.3	0	355	0.14	1
June	101.5	67.7	84.6	0	588	0.12	1
July	104.8	77.5	91.2	0	812	0.75	4
August	102.2	76.0	89.1	0	747	1.22	5
September	98.4	69.1	83.8	0	564	0.69	3
October	87.6	56.8	72.2	17	240	0.46	3
November	74.7	44.8	59.8	182	26	0.46	2
December	66.4	38.5	52.5	388	0	0.82	4
Average annual	83.1	55.4	70.3	Average annual total 1,552	3,508	7.05	35

NOTE: The normal daily maximum by month is the average of each day's (midnight to midnight) high temperature for every day in that month from 1941 to 1970. The normal daily minimum by month is the average of each day's (midnight to midnight) low temperature for every day in that month from 1941 to 1970. The normal monthly temperature is the average of the normal daily maximum and minimum temperatures for that month. The normal degree days heating are the sums of the negative departures of average daily temperatures from 65°F. The normal degree days cooling are the sums of the positive departures of average daily temperatures from 65°F. To calculate the normal degree days heating or cooling, multiply the difference between 65°F and the normal monthly temperature by the number of days in the month. The normal monthly precipitation is the average of the inches of precipitation for that month from 1941 to 1970.

SOURCE: National Oceanic and Atmospheric Administration. 1980. Climates of the States, 2d ed. Detroit: Gale Research Co., Book Tower.

that of Phoenix—1,100 feet. Daily maximum temperatures exceed 90°F an average of 165 days per year, and freezing temperatures occur an average of 12 days per year. Average relative humidity in June at Phoenix is 12 percent, compared with 23 percent in Bakersfield.

PHYSICAL AND CAPITAL RESOURCES

Soil

The Pavich operation in the Delano area includes vineyards at both Richgrove and McFarland. The Richgrove vineyards have Exeter and Ducor loam soils, underlain with a hardpan at 3 feet, and they require heavy ripping before a vineyard can be established. The McFarland area vineyards have much deeper soils of alluvial silt (Hanford loam) with no hardpan. Soils in the Kern County vineyard are alluvial fan soils formed by the Kern River. The soils in the Arizona vineyard are deep loess soils with no hardpan but with an occasional rock outcropping.

Irrigation Systems

All of the Pavich grapes are grown using irrigation. Irrigation water for the California vineyards is obtained from surface water; the Paviches' Arizona vineyards are irrigated from wells.

The amount of irrigation water applied per acre varies widely from 3 to 5 acre-feet in California and from 4 to 6 acre-feet in Arizona. According to Steve Pavich, factors that tend to increase the irrigation water requirements for Arizona include lower-than-normal winter precipitation, higher-than-normal summer temperatures, and extraordinarily heavy crops. For the Tulare County vineyards, irrigation water costs $65.00 per acre-foot, compared with $3.00 per acre-foot in Kern County. The 20-fold difference between the two counties in the price of water can be attributed to the age of their respective water districts.

The construction of canals, dams, and other facilities was more heavily subsidized by the federal government in older water districts, such as that serving the Bakersfield area, as compared with more recently constructed districts, which include the district serving the Delano area. Given the average yield of grapes per acre in the Paviches' Tulare County vineyards (653 boxes per acre in 1985), this difference in water prices adds about $0.10 per 23-pound box to the cost of producing grapes, or $0.04 cents per pound of grapes.

In Arizona the Paviches irrigate their fields from four irrigation wells: one is powered by electricity (200 horsepower, 440 volts), and three are powered by 450-horsepower natural gas engines with gearhead motors. The approximate cost of this irrigation water is $75.00 per acre-foot.

All but 190 acres of the Paviches' California vineyards are irrigated by gravity-flood irrigation. Of those 190 acres, the Paviches use a conventional drip irrigation system on 80 acres (1 gallon per hour per orifice); on the other 110 acres (one vineyard), the Paviches use a fanjet system, which is a modified drip irrigation system featuring a computer that controls the flow of water to a series of hoses feeding fanjets. The Paviches report that the fanjets distribute water uniformly throughout the vineyard at a rate of up to 11 gallons per hour per orifice, with 622 orifices per acre. Liquid fertilizer or other substances can be injected into the irrigation water at rates ranging from 6 ounces per hour to 450 gallons per hour. The Paviches consider this high maximum rate an important safety factor, reducing the risk of "getting behind" in irrigating and thus suffering damage to crops.

Several advantages of the fanjet system, in comparison with flood irrigation, are cited by the Paviches: (1) fanjets produce little or no compaction of the soil; (2) they saturate the soil for a shorter time, which allows a periodic drying of the soil and reduces the incidence of root rot; (3) they provide uniform coverage of the entire vineyard; and (4) they are highly cost-effective. The Paviches plan to expand the fanjet system to another 330 acres in the near future.

Buildings and Facilities

The Paviches have cold storage capacity for approximately 200,000 boxes of grapes (half of this capacity is in Arizona and the other half is in California). They also have housing for 300 persons in Arizona and 100 persons in California. Miscellaneous sheds and warehouses are also available for the storage of equipment and materials.

Machinery

The Paviches have an extensive inventory of machinery and equipment, including 11 wheel-type tractors, 9 half-ton pickups, 1 three-quarter-ton pickup, 5 one-quarter-ton pickups, 7 2-ton trucks, 4 electric forklifts, 4 fertilizer spreader trucks, and miscellaneous implements (plows, disks, grasscutters, a ripper, etc.). Virtually all of the equipment is duplicated in the California and Arizona operations.

MANAGEMENT FEATURES

Soil Fertility

From 1966 until 1970 the Paviches applied commercial fertilizer to their vineyards. They used 16-16-16 nitrogen, phosphorus, and potassium (NPK) fertilizer or 15-5-25 NPK. Since 1971, however, they have not applied any commercial fertilizer to their vineyards because they decided to switch to an alternative fertilization method, relying on legumes (until recently) and compost as their principal sources of nutrients. Their premise is that healthy grape plants are achieved through a proper balance of nutrients in the soil,

TABLE 6   Nutrient Content of Four Batches of Compost Used in Pavich California Vineyards

Nutrient	Percentage Nutrient Composition				Median (percent)	Pounds/Acre[c] (percent)
	6/26/85[a]	7/29/85[a]	8/14/85[a]	11/11/86[b]		
Nitrogen	1.7	1.7	1.7	2.2	1.7	94
Phosphate	1.4	1.6	1.5	1.6	1.55	85
Potash	3.7	2.5	2.4	2.5	2.5	138
Calcium	3.0	2.4	2.4	4.3	2.7	148
Magnesium	1.0	1.3	1.2	1.4	1.25	69
Sodium	0.7	NA	NA	0.3	NA	—
Organic matter	32.9	30.4	32.5	NA	32.5	—
Water	16.0	34.4	27.8	NA	27.8	—

NOTE: NA indicates data were not available. A dash indicates a negligible percentage.

[a]Dellavalle Laboratory, Inc., Fresno, California.
[b]A&L Western Laboratories, Modesto, California.
[c]Based on median analysis and assuming 2.75 tons of compost applied per acre.

including various trace elements. High priority is given to providing ample calcium for plant nutrition and good water penetration.

The Paviches apply about 6,000 tons (about 2,000 tons in California and 4,000 tons in Arizona) of composted steer manure per year to their entire farm. This translates to about 2.5 to 3.0 tons per acre of grapes and provides approximately 94 pounds of nitrogen, 85 pounds of phosphorus (P_2O_5), and 138 pounds of potassium (K_2O) per acre. (These figures are based on the medians of laboratory test results; see Table 6.) The rate of nitrogen application used by the Paviches is somewhat higher than the recommended rate.[1] The compost is spread by small trucks driven between the rows of vines.

The Paviches purchase more than 2,000 tons per year of ready-made compost from a local firm in California. The compost is produced from cow manure, cotton gin trash, and an inoculant. The inoculant contains bacteria and fungi selected to be resistant to high temperatures, an addition that greatly expedites the completion of the composting process (about 45 days). Laboratory tests of four batches of compost are summarized in Table 6. Price quotations (not including hauling costs) were provided by the producers (Table 7).

TABLE 7   Price of Compost for Pavich California Vineyards

Tonnage of Compost	Dollars/Ton
<500	18.00
500–1,999	16.00
≥2,000	15.00

In Harquahala Valley, Arizona, this composting service is not available, so the Paviches purchase about 4,000 tons of bovine manure ($1.00 per ton plus $9.00 hauling cost) and compost it themselves, with improvised equipment. Their composting process takes about 90 days to complete.

For trace elements, the Paviches rely on a special preparation, an enzyme-digested mixture of fish waste materials from a cannery, plus kelp, which has an analysis of 5-1-1 NPK along with calcium and micronutrients. The fish material is applied as a foliar spray at least once each year, with extra applications when the vines are stressed by pests.

In previous years the Paviches grew a legume (fava beans) as a green manure crop between the rows of the vineyards to produce nitrogen (Madden et al., 1986). This practice has been terminated, however, because excessive nitrogen can be detrimental. Luxury consumption of nitrate can lead to the overproduction of foliage and shading of the grape berries, thereby reducing their market quality. In addition, producing such a thick foliar mass may cause pest problems to become more severe. The Paviches view the nutrient balance in plant tissue as an essential factor in determining the populations of certain pests. For example, Steve Pavich, Jr., has observed that a very high nitrate content relative to calcium in plant tissue encourages spider mite populations and various diseases. It may also cause softer fruit, thereby reducing shelf life (Albrecht, 1975). Further research is needed to test these relationships.

Over the years, Steve Pavich, Jr., has had literally hundreds of tests done to determine the nutrient needs of his vineyards, including both soil tests and plant tissue (petiole and leaf) samples. He estimates that he has spent approximately $20,000 for these tests, which he finds accurate but limited in value.

He offered the example of one tissue analysis that showed a nitrogen concentration of 200 parts per million (ppm) compared with a standard of 800 to 1,200 ppm, clearly indicating that fertilizer was necessary. Yet, 1 week later following the irrigation of the vineyard, a second tissue analysis showed the nitrogen concentration to be 1,600 ppm. No fertilizer had been added to the vineyard between these two tests. His conclusion was that the results of the laboratory tests must be interpreted very carefully and in the context of other evidence such as the appearance of the vine's foliage and the levels of nutrients already applied.

Planting and Tillage

Grapes are a perennial crop, and it is theoretically possible for them to remain in production for many years. In a commercial fresh grape operation, however, it is important that the grape vines in a vineyard be uniformly productive, providing a high yield of quality grapes of uniform size, in large bunches, and with minimal pest damage. Most fresh grape growers, including the Paviches, find it necessary to periodically replace the vines in some of their vineyards because of low productivity or lack of uniform fruit quality.

The process of removing and reestablishing a vineyard is expensive, typically exceeding $6,700 per acre (Klonsky, 1986). First, the vines are removed with heavy equipment, usually a backhoe or bulldozer, and taken out of the vineyard. The soil is then ripped deeply in both directions (north and south followed by east and west). The land is left fallow for 2 to 3 years; no weeds or crops are permitted to grow on the field during this time. Populations of nematodes and other pests are reduced by a combination of chemical fumigation and solar heating of the soil. The bare field is then fumigated with chloropicrin and methyl bromide and ripped one more time in one direction. Following this treatment, grape vines are transplanted into the vineyard. Whereas the University of California grape enterprise budget recommends planting 454 vines (rootings) per acre, using a spacing of 8-by-12 feet (Klonsky, 1986), the Paviches plant 519 vines per acre, using a spacing of 7-by-12 feet.

Weed Control

Until recently the Paviches used a French plow for weed control. This device is articulated so that it operates between the vines to scrape the top layer of soil away and return loose soil onto the berm beneath the row of grape vines. However, Tom Pavich's calculations proved French plowing to be quite costly, and the Paviches now use a nonchemical, no-tillage procedure for weed control in their vineyards, using labor in place of conventional herbicides (Flaherty et al., 1981). This method is much less expensive than the French plow.

After the vines are established in the vineyards the Paviches plant a permanent cover of perennial rye grass *(Lolium perenne)*. A berm or mound of soil is shaped around the rows of vines; a bare berm is shaped in the center between the rows of vines. In this way, the irrigation water is forced to flow closer to the vines. A specially adapted grass chopper (with the center flails set shorter in the middle to rise above the central berm) is used periodically to chop the grass and native weeds of the permanent ground cover. The Paviches also observe that the ground cover supports populations of various beneficial predators and parasites that feed on pests in the vineyards. In addition to flail-chopping the ground cover, the Paviches hire workers to hoe or pull weeds from among the vines.

Insect Control

Steve Pavich, Jr., reports that the primary insect pest in their vineyards is the grape leafhopper *(Erythroneura elegantula)*. In addition to monitoring the various vineyards himself, Steve Pavich hires entomologists on contract to weekly monitor insect populations and the various conditions that indicate trends in their numbers.

Although some successful efforts to augment or colonize populations of natural enemies to control grape insect pests have been documented (Ridgway and Vinson, 1977), no success has been observed in efforts to control grape leafhoppers. Occasionally, the Paviches have released beneficial predators or parasites grown in insectaries into their grape vines, melons, or other crops. Their current assessment, however, is that these efforts have been futile, and Steve Pavich reports that the practice has been abandoned. Naturally occurring populations of beneficial parasites and predators, if not decimated by the application of nonselective pesticides, can often control pest populations below the economic threshold of damage.

The *Anagrus* wasp (a tiny parasitic insect) occurs naturally, and when its population is in high enough concentration, it effectively controls the grape leafhopper.[2] Unfortunately, this parasite is much less effective against a close relative of the grape leafhopper, the variegated leafhopper *(Erythroneura variabilis)*, which is becoming a serious pest in fresh grapes in some parts of California. In addition, the increased application of insecticides to control the variegated leafhopper is leading to more secondary outbreaks of mite pests such as spider mite (Settle et al., 1986).[3]

During 1986, as in most years, the Paviches were able to produce their entire grape crop in Arizona without the use of insecticides. This favorable pest situation is due in part to isolation; the vineyard nearest to their Arizona operation is 55 miles away. Most conventional grape growers in California apply four to six applications of insecticides. Most years, the Paviches are able to avoid spraying insecticides on any of their vineyards in California. For example, in 1986 they were able to avoid applying any insecticide to their entire 142 acres in Kern County. In their Tulare County vineyards, it was necessary to spray a total of 142 acres once with methomyl for leafhopper control, late in the season (August). Thus, a total of 13 percent of the entire Pavich grape operation (23 percent of the California vineyard acreage) was sprayed one time, whereas 100 percent of the conventionally produced grapes are normally sprayed one or more times, both with insecticide and fungicide (Flaherty et al., 1981).

Conventional pest control practices (see Dibble, 1982) involve the application of various pesticides. Although most pesticides are poisonous (with notable exceptions, such as sulfur for controlling mildew and some biological preparations such as *Bacillus thuringiensis*), the toxicity of pesticides used in grape production varies over a wide range. The LD_{50} ratings give a general indication of the order of magnitude of acute health risk associated with each pesticide. The lower the LD_{50} rating, the more toxic the chemical.[4] Table 8 presents the technical description and LD_{50} ratings for various pesticides used in grape production in California in 1982 (Dibble, 1982).

Dibble has also summarized the relative effectiveness of the various pesticides against specific grape pests. The various pesticides incorporated in the University of California farm management enterprise budget (Klonsky, 1986) are listed in Tables 9 and 10. The levels, frequencies, and specific materials that are applied depend on the weather, the location of the farm,

TABLE 8   Pesticides Used in Grape Production

Pesticide	Active Ingredient	Chemical Classification	LD$_{50}$ (milligrams/ kilogram body weight)[a] Oral	Dermal
Azinphos-methyl	O,O-Dimethyl S-[(4-oxo-1,2,3-benzotriazin-3 (4H)-yl)methyl] phosphorodithioate	Organic phosphate	13	220
Bacillus thuringiensis	Bacillus thuringiensis (bacterial spores)	Bacteria	Exempt	
Carbaryl	1-Napthyl N-methylcarbamate	Carbamate	850	4,000
Carbophenothion	S-((p-Chlorophenylthio)methyl) O,O-diethyl phosphorodithioate	Organic phosphate	30	1,270
Cryolite	Sodium fluoaluminate	Inorganic	10,000	—
Demeton	O,O-Diethyl O-[2-(ethylthio) ethyl] phosphorothioate	Organic phosphate	6	14
Diazinon	O,O-Diethyl O-(2-isopropyl-4-methyl-6-pyrimidinyl) phosphorothiote	Organic phosphate	466	900
Dimethoate	O,O-Dimethyl S-(N-methylcarbamoylmethyl) phosphorodithioate	Organic phosphate	215	400
Dioxathion	2,3-p-Dioxanedithiol-S,S-bis-(O,O-diethyl phosphorodithioate)	Organic phosphate	43	235

[a]LD$_{50}$, or the Lethal Dose 50, is the dose of a substance that kills 50 percent of the test animals exposed to it. The lethal dose can be measured orally or dermally.

SOURCE: Dibble, J. E. 1982. Insect and Mite Control Program for Grapes. University of California Cooperative Extension Leaflet 21102. Berkeley. July.

TABLE 9   Herbicide Costs per Acre for Conventional Production of Mature Thompson Seedless Grapes, California, 1986

Date	Problem	Material	Amount/ Vineyard Acre	Cost/Acre (dollars)	Application Cost (dollars)
January–March	Weeds, preemergence, and clean-up	Diuron	0.5 pounds	1.65	8.00
		Simazine	0.5 pounds	1.65	
		Paraquat	1.0 pint	6.00	
		Surfactant	0.2 gallons	2.81	
May–June	Maintenance	Glyphosate	1.0 pint	10.00	6.00
		Surfactant	0.2 gallons	2.81	
Subtotal				24.92	14.00
Total costs				38.92	

SOURCE: Klonsky, K. 1986. Thompson Seedless Grapes for Table Use—Sample Costs to Establish a Vineyard; Sample Costs for a Mature Vineyard. Davis, Calif.: University of California Cooperative Extension.

and other management considerations. These budget items are designed to be typical of good management practices.

The Paviches use an electrostatic sprayer (which emits electrically charged droplets that adhere to the grape plants) when they apply foliar spray or the occasional application of insecticide. This machine was redesigned at their request in a local machine shop with an improved agitator and heavy-duty pumps.

The Paviches' Arizona grape operation is isolated from other (conventional) grape producers, but the California vineyards are literally surrounded by conventional grape-growing operations. Steve Pavich, Jr., observed that his neighboring California vineyards are heavily infested with spider mites. In spite of the risks of infestation from neighboring vineyards, however, the Paviches have found it necessary to spray only once in 15 years, on only 40 acres, to control the spider mites. Steve Pavich, Jr., believes that this is a result of their improved cultural practices, including (1) expert vine dressing, (2) maintaining proper soil nutrient balances, (3) maintaining a permanent ground cover to support beneficial insect populations, and (4) obtaining the advice of qualified field entomologists.

Through the advice of these experts, the Paviches avoid unnecessary applications of insecticides that would reduce the populations of natural predators and parasites, especially the beneficial western vineyard mite (*Metaseulis occidentalis*), which is a predator of the spider mite (*Tetranychus pacificus* and *Eotetrancychus willametti*), and the beneficial *Anagrus*, which is a parasite of the grape leafhopper (*Erythroneura elegantula*) (Dibble, 1982). The fact that the Pavich vineyards have a greatly reduced load of nonspecific pesticides probably accounts for much of their success in avoiding secondary infestations, such as spider mites.

Steve Pavich, Jr., also attributes the natural control of spider mites to the nutrient balance in the soil. He has observed that a deficiency of calcium in

TABLE 10 Pesticide and Growth Regulator Costs per Acre for Conventional Production of Mature Thompson Seedless Grapes, California, 1986

Date/Growth Stage	Problem	Material	Amount/ Vineyard Acre	Cost/Acre (dollars)	Application Cost (dollars)
Mid-April	Powdery mildew	Triadimefon	4 ounces	11.75	8.00
Bloom	Powdery mildew	Triadimefon	6 ounces	17.63	
	Grape leaf skeletonizer and omniverous leaf roller	Sodium fluoaluminate	8 pounds	8.80	
	Growth regulator	Gibberellic acid	12 grams	13.44	
	Bunch rot	Fungicide[a]		25.00	12.50
Set	Leafhopper	Endosulfan	2.5 pounds	10.00	
	Powdery mildew	Triadimefon[b]	6 ounces	17.63	
Second set	Growth regulator	Gibberellic acid	32 grams	35.84	12.50
	Growth regulator	Gibberellic acid	32 grams	35.84	12.50
July-August	Leafhopper	Methomyl[c]	1 pound	14.00	8.00
Preharvest	Bunch rot	Fungicide dust[a]	2 applications	40.00	10.00
Subtotal				229.93	63.50
Total costs				293.43	

[a]Materials and rates are not included because of wide variation among growers.
[b]May be combined with leafhopper spray.
[c]Some growers may put on a second leafhopper spray.

SOURCE: Klonsky, K. 1986. Thompson Seedless Grapes for Table Use—Sample Costs to Establish a Vineyard; Sample Costs for a Mature Vineyard. Davis, Calif.: University of California Cooperative Extension.

relation to nitrate in the plant tissue is often accompanied by a rapid growth in the population of spider mites. This relationship has not been scientifically established, but it is worthy of additional research.

In the fall of 1985 the Paviches began renting a 142-acre vineyard of Thompson seedless grapes in Kern County near Bakersfield. Prior to their rental of this vineyard, very few agricultural chemicals had been applied there, although a limited amount of herbicide had been used. The Paviches improved the pruning and other cultural practices in the vineyard. They found that adequate nitrogen was already available. Water penetration in the soil was very limited, however, as a result of poor drainage. Limestone (2 tons per acre) was applied to provide calcium and to improve soil drainage. The 1985 crop yielded 825 boxes per acre, an abnormally high yield that occurred in a year when the entire San Joaquin Valley had record high yields. This vineyard was sprayed once (with methomyl) in 1985 to control leafhopper. During 1986 no sprays were applied.

During a tour of the Tulare County vineyards, a 40-acre vineyard of Emperor grapes was observed. The extension entomologist who was accompanying the site visit recalled that during 1985 this vineyard had been heavily infested with leafhopper. But in 1986 he noted that the population of grape leafhoppers near harvest time was below the levels that would cause economic damage. Many of the grape leafhopper eggs found under the grape leaves were seen to be parasitized by the *Anagrus* wasp. Steve Pavich, Jr., estimated that the yield in this vineyard would be 750 to 800 boxes per acre. No insecticide was applied to this vineyard in 1986.

Disease and Nematode Control

Many diseases afflict fresh grape production in the area of California in which the Pavich vineyards are located. The Paviches apply sulfur several times a year for disease control.

Nematode infestations are a major threat to the longevity of grape vines. In addition to causing parasitic damage to the root system, nematodes are also a vector for grape fanleaf virus *(Xiphanema)*. Although he is aware that scientific evidence does not support his belief, Steve Pavich, Jr., strongly suspects that the application of compost may suppress root knot nematode *(Meloidogyne* spp.) and other soil-borne pests and pathogens in vineyards (Kerry, 1981; Van Gundy, 1985). The Paviches also use the more conventional method of controlling nematodes and other soil-borne pests: fumigating the soil prior to planting the grapes (with chloropicrin and methyl bromide) and maintaining the soil in a bare fallow condition for a period of 2 to 3 years between vineyard removal and reestablishment. Grapes held in storage for a matter of weeks or months are fumigated with sulfur dioxide gas, which is standard practice in the industry.

Labor

The management and operation of the Stephen Pavich & Sons farm requires an extensive labor force. Stephen Pavich, Sr., who is semiretired, works on all phases of the production and marketing operations; Tom Pavich (who holds an M.B.A. degree) and his wife, Tonya (B.A., communications), are in charge of marketing operations; and Steve Pavich, Jr. (B.S, viticulture), is primarily responsible for the management of field operations.

The labor force includes 25 regular hired workers. The Paviches emphasize the importance of a permanent labor force, particularly in the case of the highly skilled vine dressers. If the vines are not pruned properly, the results can be disastrous for the vineyards. Approximately 350 seasonal workers are also hired, primarily for the harvest period.

The University of California grape enterprise budget (K. Klonsky, University of California extension farm management specialist, interview, 1986) indicated that field laborers were being paid $4.85 per hour, including payroll taxes and fringe benefits. The Paviches pay $5.00 per hour as a base wage, plus fringe benefits and incentives, bringing the average employee wage to $6.00 and above per hour. They report that wages are higher in the Delano area than in other areas of the state such as Fresno County. Tom Pavich estimated that the higher labor costs add approximately $1.00 to the cost of producing each box of grapes. A compensating factor, however, is the very high yields and excellent quality obtained in this area.

PERFORMANCE INDICATORS

Environmental Impact

In view of the fact that the Paviches use very little if any insecticide, no herbicide, and only sulfur as a fungicide, their grape operation poses a greatly reduced environmental threat with regard to residues of agricultural chemicals in groundwater or on food, pesticide drift onto neighboring farms, or injury to workers on the farm.

Like conventional grape producers, the Paviches fumigate their cold-storage grapes with sulfur dioxide and their soil with the highly toxic combination of chloropicrin and methyl bromide. This soil fumigant is applied roughly 2 to 3 years prior to the first harvest of grapes and poses no toxic residue threat to consumers of the grapes. Sulfite residues on the grapes resulting from their prolonged exposure to sulfur dioxide may cause health problems for some allergy-prone consumers. However, Steve Pavich noted that these grapes are labeled with an appropriate warning (correspondence, 1987).

Economic Performance

Pavich grapes are sold throughout the United States and in several foreign countries: Canada, Guatemala, Hong Kong, Indonesia, Japan, Mexico, New Zealand, Panama, Saudi Arabia, Taiwan, and the United Kingdom. Their

U.S. buyers include 19 of the top 20 retail food chains, plus several smaller regional outlets and a few wholesalers.

Health food stores purchase 3 percent of the Pavich grapes, under the brand name "Pavich." All of the grapes sold to health food stores come from acreage on which no pesticides have been applied for at least 2 years. The Paviches charge these specialized outlets a 12 to 25 percent premium of about $1.00 to $2.00 per 23-pound box for grapes certified as produced with practices that comply with the state's law governing organic foods. Steve Pavich, Jr., reports that the higher price is necessary to cover the additional costs of certification and special handling and storage. The vast majority of the grapes produced by the Paviches would qualify as organic according to the existing legislation in California; and, in fact, all of the Paviches' California and Arizona vineyards are in the process of being certified as organic, pending at least a 2-year period in which forbidden chemicals are not used.

Tom Pavich reported in a 1986 interview that his family's 1985 production of fresh grapes was 704,360 boxes (23 pounds per box) or 8,100 tons. This was about 1 percent of the total U.S. production of fresh grapes.[5] Tom also said that the Paviches produced 26.4 percent of the fresh grapes grown in Arizona, including 61.9 percent of the Flame seedless grapes. (This information was corroborated by M. Shine of the U.S. Department of Agriculture Market News Service in Phoenix.) The total cash value of the Pavich grapes in 1985 was $5.7 million or 1.7 percent of the total U.S. fresh grape sales.[6]

In 1986 the Paviches produced a substantially higher sales volume than in 1985, the result of increased production and higher prices for fresh grapes. Their production exceeded 800,000 boxes, which was due to higher yields compared with 1985 plus 20 additional acres in production. With an additional 160 acres purchased and new leases in 1987, the Paviches expected to exceed 1 million boxes.

Grape production involves substantial outlays of capital. Table 11 lists the expenses incurred in establishing an acre of Thompson seedless fresh grapes, according to the University of California (UC) farm management extension enterprise budget (Klonsky, 1986). The Paviches use similar procedures for establishing a vineyard, with a few exceptions, and incur somewhat higher capital costs. In the first 2 years after the vines are planted, no grapes are harvested; in the third year there is a light harvest (435 boxes). The accumulated cost over the 3-year establishment phase is $6,711 per acre.

The Paviches plant vines using a spacing of 7-by-12 feet, as contrasted with the UC assumption of spacing of 8-by-12 feet. Decreasing the size of the spacing increases the number of plants per acre from 454 to 519 (although both of these spacings are considered standard [Klonsky, 1986]). The Paviches also keep their land fallow for 3 years rather than the usual 2, which both increases operating costs and delays the beginning of production 1 more year, as compared with the UC budget.

This delay in replanting is intended to further reduce pathogen populations (especially nematodes), thereby enhancing the longevity of the vine-

TABLE 11   Sample Costs per Acre to Establish a Conventional Thompson
Seedless Table Grape Vineyard in California, 1986

Item	Costs/Acre (dollars)		
	1st Year	2nd Year	3rd Year
Cultural costs			
Fumigation	460		
Land preparation			
4 hours chisel and labor	60		
4 hours disk, float, and labor	72		
Rootings: 454 at 37¢ (20 rootings, 2nd year)	168	7	
Trim and store	36	2	
Machine planting			
(2 hours labor, 2nd year)	50	10	
Stakes (treated): 454		454	
End posts (treated): 11		50	
Stake and set end posts		131	
Wire		119	
String four wires and staple		100	
Attach crossarms and braces: 32 hours labor			155
Crossarms			177
Training and suckering: 24 hours labor, 2nd year;			
18 hours labor, 3rd year		116	87
Prune and tie: 5 hours labor, 2nd year; 18 hours			
labor, 3rd year		24	87
Rabbit control	15	8	19
Irrigation: 5 hours labor each year	29	29	29
Water power and/or district tax: 60 feet pumping 1,			
2, 3.5 acre-feet at $32.70	33	65	114
Cultivation and irrigation preparation	60	72	72
Fertilizer: 30¢/pound (30 pounds N), 2nd year; 50			
pounds N, 3rd year; $5/acre for application		14	20
Pest management and disease control, includes			
mildew	17	27	293
Herbicides: materials and application	35	32	32
Miscellaneous labor, materials	34	29	34
Total cultural costs	1,069	1,288	1,120
Harvest costs			
Contract at $45/ton, pick and haul			225
Total harvest costs, custom			225
Overhead costs			
County taxes	39	39	39
Office and business costs	30	30	30
Total overhead costs	69	69	69
Total cash costs	1,138	1,357	1,413
Accumulated cash costs	1,138	2,495	3,908
Depreciation			
Building, equipment, and irrigation	65	65	65
Interest on investment at 12.5 percent			
Building, equipment, and irrigation	55	55	55
Land ($3,000/acre)	375	375	375
Interest in accumulated cash cost	142	312	489
Total interest on investment	572	742	918
Total cash and fixed costs for the year	1,775	2,164	2,397

TABLE 11 *(Continued)*

Item	Costs/Acre (dollars)		
	1st Year	2nd Year	3rd Year
Credit for production at $75/ton for juice			−375
Net cost for the year	1,775	2,164	2,772
Accumulated net cost	1,775	3,939	6,711
Yield (tons/acre)	—	—	5

Pesticide Costs During Establishment

1st Year:	Sodium fluoaluminate 8.0 pounds (for grape leaf skeletonizer and omniverous leaf roller)	Cost = $8.80 + $8.00 application cost
	Methyl bromide (fumigant)	Cost = $460
2nd Year:	Sodium fluoaluminate 8.0 pounds (for grape leaf skeletonizer and omniverous leaf roller)	Cost = $8.80 + $8.00 application cost
	Endosulfan 2.5 pounds (for leafhopper)	Cost = $10.00 + $8.00 application cost
3rd Year:	Same as pesticide production cost for that particular variety	

NOTE: Totals may not be exact because of rounding. Costs are based on a 120-acre unit, vines spaced at 8-by-12 feet, and yield of 5 tons. Wages include Social Security, Workmen's Compensation, and insurance. Skilled supervisory labor, $5.70/hour; unskilled labor, $4.85/hour; tractor $6.48/hour.

SOURCE: Klonsky, K. 1986. Thompson Seedless Grapes for Table Use—Sample Costs to Establish a Vineyard; Sample Costs for a Mature Vineyard. Davis, Calif.: University of California Cooperative Extension.

yards. The Paviches expect their vines to continue in peak production for 30 to 40 years or longer, depending on soil quality, compared with a typical longevity of about 20 years.

Table 12 lists selected operating expenses from the UC and Pavich budgets for a mature vineyard of Thompson seedless grapes (Klonsky, 1986). UC budgets are available for several other varieties of grapes. Although the Paviches provided detailed accounting data, the categories they use to report their cost data are not comparable with several items in the UC farm management extension budgets. Soil fertility costs were somewhat higher on the Pavich operation than in the UC budget, but on the other hand the Paviches incurred no expenses for herbicides. The differences are trivial, however, as a proportion of total cash costs. The Pavich preharvest cost per box ($2.20) is about the same as the UC enterprise budget cost ($2.14).

Tom Pavich reports that labor accounts for about 55 percent of their grape production costs. The UC budget contains several items that are custom hired; the Paviches do all their own work except an occasional insecticide application, which is generally done by a commercial applicator. The spe-

TABLE 12   Comparative Per Acre Costs of Pavich McFarland Vineyard and UC Enterprise Budget: Selected Practices Used in Mature Thompson Seedless Grape Vineyard, California, 1986

Item	University of California Enterprise Budget (dollars)	Pavich Farm, McFarland Vineyard (dollars)
Field labor wage rate, includes Social Security, Workmen's Compensation, insurance	4.85/hour	6.00/hour
Selected preharvest cash costs		
Irrigation preparation and cultivation	48.72	c
Irrigation labor	29.10	44.00
Irrigation water	130.80	280.00
Chemical fertilizer (60 pounds N)	23.00	0
Alternative fertilizers (compost, foliar sprays)[a]	0	100.00
Soil amendments	0	30.00
Herbicide	38.92	0
Ground cover maintenance	0	22.00
Growth regulator (gibberellic acid)	85.12	85.12
Disease and pest control materials	141.81	220.00
Application of pest control material (gibberellic acid)	63.50	c
Biological pest control adviser	c	3.00
Pruning	158.90	214.00
Training	c	18.00
Suckering	c	18.00
Tying	35.00	44.00
Thinning	198.85	143.00
Girdling	67.90	42.00
Pull leaves	c	107.00
Brush disposal	10.00	c
Total preharvest cash costs[b], excluding interest on operating capital	1,118.30	1,436.00
Preharvest cash cost per box[b]	2.14	2.20
Vines/acre	454	519
Vine spacing	8 x 12 feet	7 x 12 feet
Percentage of culls[d]	15–30 percent	1.0 percent
Yield, 23-pound boxes per acre (net of culls)	522	653

[a]Foliar spray materials cost $20.00 plus $80.00 for compost ($15/ton × 5.33 tons/acre).

[b]Totals and average costs per box may not be comparable due to differences in accounting procedures. Totals may not be exact because certain data were not available or comparable.

[c]Comparable category of data not available.

[d]Data may not be exactly comparable.

SOURCES: Tom and Steve Pavich, Jr., interviews, 1986; Klonsky, K. 1986. Thompson Seedless Grapes for Table Use—Sample Costs to Establish a Vineyard; Sample Costs for a Mature Vineyard. Davis, Calif.: University of California Cooperative Extension.

cific herbicides and other pesticides assumed in the UC budget are listed in Tables 9 and 10.

In interpreting the cost data, it is important to recognize that the nature of the Pavich production system is such that it is virtually impossible to place several expense items in a single category. For example, compost is applied in part for its nutrient value as a fertilizer, but it is also intended to have a beneficial effect on the biological balance in the soil, enhancing populations of pathogens, predators, and other antagonists that may help to control certain pests (Cook, 1986). Consequently, the Paviches view compost not just as a fertilizer but also as a major part of their pest control strategy.

Detailed accounting data from the Pavich operation have been examined, but they are not reported here because of issues of confidentiality. The income of the Pavich operation is clearly enhanced by the fact that the family has an extensive marketing and storage system that enables it to receive a higher price for its grapes than if it were selling them through another marketing firm. By holding about one-eighth of their total grape production in cold storage for 2 months (until just before the Chilean grapes enter the U.S. markets in December), the Paviches are able to earn a significantly higher price than growers who sell all of their grapes at harvest.

Without revealing confidential information regarding the profits of the Pavich grape operation, it can be stated unequivocally that their management and marketing practices are succeeding financially, in terms of cash flow, market share (1 percent of the U.S. total crop of fresh grapes), and a favorable debt-to-asset ratio.

ENDNOTES

1. The University of California enterprise budget for fresh (table) grapes calls for 60 pounds of nitrogen per year. The extension recommendation for this area ranges from 30 to 40 pounds for mature grapes in good health to 80 pounds for weak or unhealthy vines. In some locations, irrigation water pumped from a depth of 400 to 500 feet contains 15 to 20 ppm of nitrate, which provides the equivalent of 40 to 50 pounds of nitrogen with normal levels of irrigation (information provided by W. Peacock, extension viticulture farm adviser, Tulare County, California, 1987).

2. Flaherty et al. (1981) describe the *Anagrus* wasp's predation of the grape leafhopper as follows:

> Egg parasite of grape leafhopper. The most important natural enemy of the grape leafhopper in commercial vineyards is a tiny, almost microscopic wasp called *Anagrus epos* (Girault). Its progeny develop within the egg of the grape leafhopper, resulting in its death. Its size is about 0.3 mm (1/100 inch).
>
> These parasitic wasps are particularly valuable because of their amazing ability to locate and attack grape leafhopper eggs. Also, their short life cycle permits them to increase far more rapidly than do leafhopper

populations. Their nine to ten generations during grape growing season make them capable of parasitizing 90 to 95 percent of all leafhopper eggs deposited after July.

This parasite overwinters on wild blackberries, *Rubus* spp., on which it parasitizes the eggs of a non-economic, harmless leafhopper, *Dikrella* spp. These overwintering wasp populations tend to be along rivers that have an overstory of trees sheltering both wild grapes and wild blackberries. When the blackberries leaf out in February, the lush, new foliage apparently stimulates heavy oviposition by the *Dikrella* leafhoppers. The *Anagrus* parasites increase enormously on these eggs so that by late March and early April there is widespread dispersal of the newly produced *Anagrus* adult females. Fortunately, their dispersal occurs at the same time that grape leafhopper females begin to lay eggs. Vineyards located within a five- to ten-mile range will usually benefit immediately from the immigrant parasites. Vineyards distant from actual refuges may not show *Anagrus* activity until midsummer or later. (pp. 100–103)

3. University of California researchers D. Gonzales and T. Wilson are currently conducting experiments with the introduction of several alternative strains of *Anagrus* wasp from Mexico and Colorado in search of an effective parasite of the variegated leafhopper. This research is supported by the University of California Agricultural Experiment Station with funds provided by the California Wine Growers Association, the California Table Grape Commission, and the Raisin Advisory Board (D. Gonzales, Division of Biological Control, University of California, Riverside, telephone interview, 1987).

4. The toxicity of a pesticide is expressed by the terms oral and dermal LD_{50}. LD_{50} means the minimum single dosage needed to kill 50 percent of a group of test animals, usually rats or rabbits; the lethal dosage is of the pure compound and is given in so many milligrams of pesticide per kilogram of the animal's body weight.

Oral LD_{50} is a measure of the toxicity of the pesticide when administered internally to the test animals.

Dermal LD_{50} is a measure of toxicity when the pure compound is applied to the skin of the test animals. Generally, the oral application is more toxic than the dermal.

5. The total volume of fresh (table) grapes produced in the United States in 1985 was 781,090 tons (U.S. Department of Agriculture, 1987).

6. The total value of sales of fresh (table) grapes produced in the United States in 1985 was reported as $225 million (U.S. Department of Agriculture, 1987). Presumably, this total is for the grapes only, excluding the cost of boxes (about $1.50 per box). When the Pavich sales volume is adjusted to exclude the cost of boxes, their sales volume is about $4.6 million, or 1.7 percent of the total U.S. sales of fresh table grapes.

REFERENCES

Albrecht, W. A. 1975. The Albrecht Papers, Charles Waters, ed. Kansas City, Mo.: Acres U.S.A.

Cook, R. J. 1986. Interrelationships of plant health and the sustainability of agriculture, with special references to plant diseases. Amer. J. of Alternative Agriculture 1(1):19–29.

Dibble, J. E. 1982. Insect and Mite Control Program for Grapes. University of California Cooperative Extension Leaflet 21102. Berkeley, Calif.: University of California, Cooperative Extension.

English-Loeb, G. M., D. L. Flaherty, L. T. Wilson, W. Barnett, G. M. Leavitt, and W. H. Settle. 1986. Pest management affects spider mites in vineyards. California Agriculture 40(3&4):28–30.

Flaherty, D. L., F. L. Jensen, A. N. Kasimatis, H. Kido, and W. J. Moller. 1982. Grape Pest Management. Agricultural Sciences Publication No. 1405. Berkeley, Calif.:

Kerry, B. R. 1981. Progress in use of biological agents for control of nematodes. Pp. 79–90 in Biological Control in Crop Production, Beltsville Symposium in Agricultural Research, No. 5, G. C. Papavizas, B. Y. Endo, D. L. Klingman, L. V. Knutson, R. D. Lumsden, and J. L. Vaugh, eds. Totowa, N.J.: Allanheld, Osmun.

Klonsky, K. 1986. Thompson Seedless Grapes For Table Use—Sample Costs to Establish a Vineyard; Sample Costs for a Mature Vineyard. Davis, Calif.: University of California, Cooperative Extension.

Madden, P., S. Dabbert, and J. Domanico. 1986. Regenerative Agriculture: Concepts and Selected Case Studies. Staff Paper No. 111. University Park, Pa.: Department of Agricultural Economics and Rural Sociology, The Pennsylvania State University.

Ridgeway, R. L., and S. B. Vinson, eds. 1977. Biological Control by Augmentation of Natural Enemies. New York: Plenum Press.

Settle, W. H., L. T. Wilson, D. L. Flaherty, and G. M. English-Loeb. 1986. The variegated leafhopper, an increasing pest of grapes. California Agriculture 40(7&8):30–32.

Shine, M. 1986. USDA Market News Service in Phoenix. Telephone interview.

Teriotdale, B. L. and W. D. Gubler. 1984. Disease Control Guidelines for Grapes. Berkeley, Calif.: University of California, Division of Agricultural Sciences.

U.S. Department of Agriculture. 1987. Agriculture Statistics 1987. Washington, D.C.

Van Gundy, S. D. 1985. Biological control of nematodes: Status and prospects in agricultural IPM systems. Pp. 467–478 in Biological Control in Agricultural IPM Systems, Marjorie A. Hoy and Donald C. Herzog, eds. New York: Academic Press.

9

Integrated Pest Management in Processing Tomatoes in California: The Kitamura Farm

T HE KITAMURA FARM is located in Colusa County, north of Sacramento, California, on the border with Sutter County. The farm, a total of 305 acres, is on the west bank of the Sacramento River (Table 1). The Kitamuras own 40 of these acres, the site of their walnut orchard; they rent the balance of the land from a large estate.

GENERAL DATA

The Kitamuras currently produce 160 acres of processing tomatoes, using a modification of the integrated pest management (IPM) program developed by the University of California. The farm also includes about 70 acres of vine seeds (including cucumbers, squash, and watermelons) and 30 acres of beans. The Kitamura Farm is a family operation, run by David and Diann Kitamura in partnership with David Kitamura's brother. Both David and Diann Kitamura are trained in IPM pest scouting, and they have been participating in the University of California IPM program for tomatoes since 1984.

Climate

The climate in the area surrounding the Kitamura Farm is hot and dry during the summer with cool nights, ideal for the production of tomatoes. Normal daily temperatures reach a maximum in excess of 85 degrees from June through September (Table 2). Annual precipitation is approximately 17 inches, falling mostly between mid-October and April. Normally, less than 1 inch of precipitation occurs between May and October, during the bulk of the tomato-growing season. Low precipitation and low humidity are impor-

TABLE 1 Summary of Enterprise Data for the Kitamura Farm

Category	Description
Farm size	305 acres, of which 160 are planted with processing tomatoes
Labor and management practices	All management is provided by the farm operators (David and Diann Kitamura), including pest scouting. All labor is provided by the operators, David Kitamura's brother (and partner), and one hired worker; eight workers are hired for harvest.
Marketing strategies	Tomatoes are sold under contract with a major processor. An increased market share is awarded because of the grower's very low percentage of rot and insect damage.
Weed control practices	Preemergence herbicides (napropamide and pebulate) are used. If nightshade occurs postemergence, pebulate is applied; otherwise, trifluralin is used.
Insect and nematode control practices	Crop rotations reduce insect pest problems. Sulfur dust controls russet mites. IPM scouting enables a minimal use of insecticides.
Disease control practices	Tomatoes are grown in a rotation of no more than 1 year. Mold is controlled by the early termination of irrigation.
Soil fertility management	Starter fertilizer (13 pounds N/acre) is used, plus 100–120 pounds N/acre side-dressed.
Irrigation practices	Flood irrigation is interrupted 40 days prior to harvest to prevent mildew (30 days is the usual practice). Configuration of furrows is reshaped at the final or penultimate cultivation, from 30- to 60-inch centers.
Tomato yields	The yield was 35.5 tons/acre in 1986, which was above the county average (29.2 tons/acre).
Financial performance	The farm had an estimated cost savings of $7,297 in 1986 through reduced pesticide use from IPM pest scouting done by the farmer. Innovative irrigation scheduling reduces crop loss as a result of mold. The farm is solvent, with a debt-to-asset ratio of less than 10 percent.

tant for disease control in tomatoes because moist atmospheric conditions lead to the development of mold and other diseases.

PHYSICAL AND CAPITAL RESOURCES

Soils

The soil in this area is an alluvial sandy loam, which is typically deep, well drained, and highly fertile.

Buildings and Facilities

The buildings and facilities on the Kitamura Farm are minimal. The farm has a small machine shop in which the Kitamura brothers repair and overhaul their machinery.

TABLE 2 Normal Daily Temperatures and Monthly Precipitation at Sacramento, California

Month	Normal Daily Temperature (°F)		Monthly Average	Normal Degree Days		Normal Precipitation	
	Maximum	Minimum		Heating	Cooling	Inches	Days With ≥0.01 Inches
January	53.0	37.1	45.1	617	0	3.73	10
February	59.1	40.4	49.8	426	0	2.68	9
March	64.1	41.9	53.0	372	0	2.17	8
April	71.3	45.3	58.3	227	26	1.54	6
May	78.8	49.8	64.3	120	98	0.51	3
June	86.6	54.6	70.5	20	185	0.10	1
July	92.9	57.5	75.2	0	316	0.01	[a]
August	91.3	56.9	74.1	0	286	0.05	[a]
September	87.7	55.3	71.5	5	200	0.19	1
October	77.1	49.5	63.3	101	48	0.99	3
November	63.6	42.4	53.0	360	0	2.13	7
December	53.3	38.3	45.8	595	0	3.12	9
Average annual	73.2	47.4	60.3	Average annual total	2,843	1,159	17.22	57

NOTE: The normal daily maximum by month is the average of each day's (midnight to midnight) high temperature for every day in that month from 1941 to 1970. The normal daily minimum by month is the average of each day's (midnight to midnight) low temperature for every day in that month from 1941 to 1970. The normal monthly temperature is the average of the normal daily maximum and minimum temperatures for that month. The normal degree days heating are the sums of the negative departures of average daily temperatures from 65°F. The normal degree days cooling are the sums of the positive departures of average daily temperatures from 65°F. To calculate the normal degree days heating or cooling, multiply the difference between 65°F and the normal monthly temperature by the number of days in the month. The normal monthly precipitation is the average of the inches of precipitation for that month from 1941 to 1970.

[a]Less than one-half.

SOURCE: National Oceanic and Atmospheric Administration. 1980. Climates of the States, 2d ed. Detroit: Gale Research Co., Book Tower.

Machinery

The Kitamura Farm machinery inventory includes one self-propelled to-mato harvester (1976), two crawler tractors, two 115-horsepower wheel-type tractors, three smaller wheel-type tractors, a vacuum planter, a power take-off-driven sprayer, four pickup trucks, one flat-bed truck (1952), and miscel-laneous other equipment.

The vacuum planter is used for seeding the tomatoes directly into the soil; transplanting is no longer used in the production of processing toma-toes in this area. The Kitamura Farm has a power takeoff-driven pump sprayer for ground application of various sprays. When the stage of devel-opment of the crop or the soil condition prevents the use of the ground spray rig, however, the Kitamuras rely on aerial application at a cost of $4.50 to $5.00 per acre for each application. About 30 percent of the insecticides used on the farm are applied aerially; virtually all herbicides and fungicides are applied with ground sprayers.

The Kitamura brothers do all the tractor work on the farm. In addition to repairing and overhauling their machinery, they replace worn bearings and other expendable parts of the tomato harvester each year in preparation for the coming season. They hire a service firm to replace all of the various conveyor belts on the tomato harvester. The tomato harvester machine has been modified by the Kitamuras to make it operate more efficiently. Ordi-narily, they harvest all of their tomatoes with their own machine. During one year when rain was forecast, however, they hired a custom operator to assist with the harvest to prevent the loss of the crop.

MANAGEMENT FEATURES

Processing tomatoes are planted in stages, organized in cooperation with the processor with whom the grower has a marketing contract. The first planting (45 acres) was made on April 3–4, 1986; the second (55 acres) on April 17–19; and the third (60 acres) on May 7–10. Planting is staggered to achieve an orderly harvesting schedule with only a certain proportion of all the tomatoes coming ripe at the same time. This is an advantage to both the grower and the processing plant.

Soil Fertility

The Kitamuras apply about 20 gallons of starter fertilizer (8-24-0) per acre and side-dress 32 percent nitrogen solution at about 120 pounds of nitrogen per acre for the first of three plantings; they apply about 100 pounds of nitrogen per acre for the second and third plantings. The Kitamuras avoid the use of aqua ammonia because they say that it is reported to cause softening of the fruit. No scientific documentation of this claim has been located. The fertilizer they apply is ammonium nitrate (33.5-0-0) together with a slower release source of nitrogen.

Tillage, Irrigation, and Crop Rotation

Tomato production requires a clean seedbed. Consequently, tillage is necessary to dispose of the residue from the previous crop. The tomatoes are seeded with a planter set for 30-inch single rows; the alternate rows are left unseeded, thereby achieving 60-inch rows. The blank row is maintained to channel the irrigation water closer to the tomato rows when the plants are young. At the time of the last or next-to-last cultivation, the blank middle row is split with a disk cultivator, and the soil is pushed toward the tomato rows on either side, leaving a central furrow on 60-inch centers. Subsequent irrigations occur down this furrow, located 30 inches from the tomato plants. Gravity furrow irrigation is used throughout. Irrigation water is provided as part of the land rental agreement. The water is pumped from the Sacramento River, which is adjacent to the farm.

The Kitamuras rent their tomato land from a large estate. Each year a certain area of the estate is set aside for tomatoes. Typically, a field that will be used for tomatoes has been out of tomato production for at least 6 or 7 years, producing crops such as dry beans, safflower, wheat, or other field crops. The common practice in the area is a 2- or 3-year planting of other crops before planting tomatoes in a field (W. L. Sims, correspondence, 1987). The Kitamuras do not practice any deliberate crop rotation other than this extended wait between tomato plantings on a certain field, a consequence of their renting and not owning the land they farm. Decisions as to which specific fields will be in tomatoes in the next year are out of the control of the growers.

The University of California
IPM Program for Tomatoes

In 1984 the University of California introduced an IPM program to reduce damage to processing tomatoes by two prominent pests: the fruitworm and the beet army worm. The University of California (1985) IPM manual for tomatoes contains color illustrations of all the prominent insect pests and the various tomato diseases. It also discusses management guidelines, preferred pesticides, natural parasites and predators, analysis of their life cycles, biological controls, monitoring procedures, and other essential information.

In its initial stage this IPM program was tested on about 2,000 acres of processing tomatoes in the Sacramento Valley in northern California (L. T. Wilson, interview, 1986; F. G. Zalom, interview, 1986). Participating growers received special training in IPM from University of California extension specialists. Data were collected from 82 farms producing processing tomatos, 22 of which were in the University of California IPM program (Antle and Park, 1986; Grieshop et al., 1986). Of those farmers who initially participated in the IPM program, 71 percent said that they planned to continue in the program. The possibility of financial gain was cited as the primary motivating factor in joining the program.

Processing tomatoes are harvested mechanically. The ripe fruit is conveyed into 12.5-ton gondola tanks that are carried by trucks. Two of these gondola tanks (a total of 25 tons) are carried by each truck from the field to an inspection station and then to the processing plant. At the inspection station, sample tomatoes are taken from each gondola tank. There is a strong monetary incentive for growers to minimize worm damage in the tomatoes. At inspection stations tomatoes are examined for mold, green fruit, worms, and materials other than tomatoes in the load. If more than 2 percent worm damage is found in the load, it must either be resorted by the grower at considerable expense or discarded. The percentage of defective fruit is subtracted from the gross weight of the load of tomatoes in determining payment for the grower. Insect damage can also cause the fruit to drop off the plant before it is harvested, thereby potentially reducing yields and income. The amount of yield reduction depends on when the fruit is dropped; the more mature the dropped fruit, the greater the loss of yield (Zalom et al., 1983).

IPM for processing tomatoes involves three interrelated components: cultural practices, monitoring, and treatment. IPM emphasizes preventive methods that produce economical, long-term solutions to pest problems while minimizing hazards to human health and the environment (University of California, 1985). The prominent insect pests in tomatoes in California include cutworms, flea beetles, green peach aphids, potato aphids, tomato russet mites, cabbage loopers, vegetable leafminers, tomato fruitworms, beet army worms, tomato pinworms, and stink bugs. The most frequent diseases encountered in processing tomatoes in this area are damping-off, phytophthora root rot, fusarium wilt, verticillium wilt, buckeye rot, pythium ripe fruit, bacterial speck, black mold, grey mold, tobacco mosaic, and curly top.

Good crop management practices, including weed and other pest control, irrigation, and fertilization are essential to a successful IPM program. All cultural practices are interrelated within the growing system. Important factors include the selection of the appropriate field, preferably one with deep, uniform soil (with 4 or more feet of root zone) to avoid various disease problems. The land must be properly prepared to minimize weed problems and to provide the appropriately shaped seedbed. Other essential cultural practices include the placement of the seeds at the appropriate depth, spacing, and correct timing with regard to soil temperature, the stage of the season, and the intended date of harvest. The selection of a cultivar is important for avoiding various diseases.

Proper irrigation practices are critical for IPM and for the successful production of processing tomatoes. Either too much or too little water can be disastrous. Normally, tomatoes require 3 to 4 acre-feet of water per growing season. Ideally, the soil should be essentially depleted of water by harvest time. In this way, mold damage caused by dew formed from water evaporated from moist soil will be minimized.

Another essential aspect of IPM is the maintenance of healthy tomato plants through proper fertilization, cultivation, and irrigation practices. Sanitation is also important in obtaining a clean source of tomato seed (free of various disease pathogens and weeds), using soil that has either been fumigated (at a very high cost per acre) or in a rotation to reduce the populations of various pests. Weed control is an essential part of general pest control. Keeping weed populations low along field borders helps prevent infestations of pests including weeds, insects, and vertebrates (University of California, 1985).

The second essential component of the IPM program is monitoring for pest populations. The University of California has developed a systematic method of scouting tomato fields (Wilson et al., 1983; Zalom et al., 1983), which is presented to growers in 2-day training sessions. The instruction has been summarized on a videotape by the agricultural extension service and experiment station personnel. The person scouting the field must become proficient at both collecting samples of tomato fruit and leaves, usually 100 of each, and counting the incidence of insect eggs or other pest problems. Scouting may be done by professional pest control advisers, by the farmer, or by hired workers.

The third component of an IPM program is treatment. When it is determined through monitoring that pest populations have reached the level at which they will cause economic damage, the grower is advised to use an appropriate control measure (Zalom et al., 1983). An economic level of damage is estimated on the basis of the value of the predicted fruit damage versus the cost of treatment. Presumably, when crop values are extremely high and treatment cost is inexpensive, the threshold of economic damage is at a rather low level of pest population. Conversely, if the price of the crop is relatively low and the cost of treatment high, it is appropriate to permit a higher level of pest damage before initiating treatment. This is one of the fundamental concepts of IPM.

The appropriate treatment, once a pest has reached a damaging level, usually includes the application of an insecticide. In some instances, predators such as parasitic wasps may be released by the grower or by the pest control adviser, but often these biocontrol practices are not effective quickly enough to bring a rapidly growing population of pests under control. Growers are advised to contact their local extension farm adviser to determine the appropriate pesticide and level of application.

The effects of IPM on processing tomatoes fall into four categories: (1) changes in the cash cost of production, particularly for pesticides; (2) changes in crop yields and revenue; (3) risks associated with growing tomatoes, such as public health hazards or the development of resistance to pesticides by pests; and (4) environmental impacts associated with the use of pesticides. When an IPM system is used, the number of pesticide applications tends to decline, and different kinds of spray material are used. In some cases a much smaller amount of a more selective pesticide (causing less damage to natural predators and parasites) is used. This is generally not the case with processing tomatoes, however (L. T. Wilson, telephone interview and correspondence, 1987).

The monetary impacts associated with IPM include the cost of IPM pest scouting and possible changes in yields or in the price received for the product as a function of quality. Yields can be altered when the incidence of cull fruit or the tonnage of harvested fruit per acre changes. Revenue can also be influenced by changes in prices received by the grower as a result of differences in fruit quality (percentage of insect damage, mold, and other quality factors).

A comprehensive study of the results of adopting the University of California tomato IPM program was undertaken by Antle and Park (1986). Their results show that, on average, the use of IPM in processing tomatoes will both increase income and reduce the risk of crop damage and loss. Fields in the IPM program had 39.5 percent lower average worm damage (significant at the 1 percent level) resulting in a higher net value of about $7.70 per acre. More importantly, a field of tomatoes using IPM has a 25 percent chance of having more than 1 percent insect damage, compared with an 80 percent chance for fields not in the IPM program. Tomatoes grown under the IPM program have an almost zero likelihood of being rejected for damage, whereas non-IPM fields have a 5.6 percent risk of rejection (University of California, 1985).

The Kitamura Farm Insect Control Program

The Kitamuras have modified the University of California IPM program to meet their own preferences and needs. The university recommends an economic threshold of five to seven eggs in a sample of 100 leaves. The Kitamuras followed this guideline during 1984 and 1985. During 1985 none of their loads of tomatoes was rejected for excessive worm damage (none of the loads exceeded 2 percent worm damage, the state inspection limit). Some of the loads were found to have almost 1 percent worm damage, however, and although this level of damage was not sufficient to reject the load, it was high enough to be unacceptable to the Kitamuras. They decided to apply a more stringent threshold of three to four eggs in a sample of 100 leaves, rather than the five to seven eggs recommended by the University of California. Even with this threshold, they have needed fewer insecticide applications than most growers who follow conventional spraying recommendations.

In addition to being concerned about the risk of having tomatoes rejected at the inspection station, the Kitamuras also indicated that it was in their financial interest to keep worm damage very low, well below the legal limit, in the hope that the packer might grant a larger contract in future years.

In 1986 the Kitamuras' entire 160 acres of tomatoes were treated with sulfur dust to control tomato russet mite (*Aculops lycopersici*). No other insecticide was applied on the first or second plantings, a total of 100 acres of tomatoes. In the third planting of 60 acres, however, the Kitamuras discovered that the number of eggs of the tomato fruitworm (*Heliothis zea*) had exceeded the critical level, indicating the need for treatment. Consequently, a single aerial application of methomyl was made on the 60 acres.

Early in the season the Kitamuras scout their fields once each week. They place pheromone traps in the fields to detect moths, and once moths are detected, the frequency of scouting is increased to every 3 days. The acreage is rectangular, and the scouting takes less than an hour per visit to cover the 160 acres. Typically, the scouting is done for about 1 month, with intensive (a 3-day schedule) scouting for about 2 to 3 weeks during the growing season.

Weed Control

The Kitamuras use a preemergence herbicide, a combination of napropamide and pebulate, on all their tomato fields. A postemergence herbicide is applied to control nightshade if this weed becomes a problem.

Disease Control

The Kitamura approach to disease control includes three major components: (1) the selection of disease-resistant tomatoes; (2) growing tomatoes in soil in which crops other than tomatoes have been grown for several years, thereby reducing the populations of nematodes and other pests specific to tomatoes; and (3) their innovative irrigation program.

The standard irrigation recommendation for tomatoes is to terminate irrigation 30 days prior to harvest. In this way the ground surface dries, and very little dew, if any, forms on the tomato plants, thus keeping the incidence of mold quite low. The Kitamuras decided to extend this dry period to 40 days in the hope of further reducing mold damage while maintaining high yields. Their plan was successful: even though the tomato plants appeared to be stressed by lack of moisture during a field visit at harvest time, the yield was the highest the Kitamuras have had since they began producing tomatoes in 1970. The effect of moisture stress depends on a number of factors such as soil type, season length, cultivar of the crop grown, and ambient temperatures.

In 1986 the Kitamuras' yield was so high that they were able to meet their contract obligation with only 120 acres out of their 160. No mold damage was found during inspection of their tomatoes until after a late-season, 1-inch rainfall (an amount of rain that normally results in major losses due to mold). Diann Kitamura reported that their percentage of mold damage was 2.5 percent. Rain at harvest time often causes total loss of the crop (L. T. Wilson, interview, 1987), and even in normal years, with no rain at harvest time, an average of 1.1 percent of the tomatoes have mold damage (W. L. Sims, correspondence, 1987).

Labor

The farm is operated with a labor force of five family workers, one full-time hired worker, and eight seasonal hired workers. The individuals inter-

TABLE 3   Acres and Yields of Processing Tomatoes on Kitamura Farm Compared With Colusa County, 1970–1986

Year	Kitamura Farm		Colusa County	
	Acres	Average Yields (tons/acre)	Acres	Average Yields (tons/acre)
1970	140	21.4	3,300	24.3
1971	140	20.0	4,530	25.2
1972	140	27.8	4,720	25.6
1973	150	27.2	6,060	22.0
1974	202	19.8	9,220	21.6
1975	220	31.1	9,530	22.4
1976	226	22.3	8,000	21.0
1977	210	22.6	10,100	24.3
1978	150	26.9	8,300	22.2
1979	206	23.3	8,440	24.9
1980	175	32.0	6,060	26.7
1981	156	19.1	8,190	22.1
1982	160	28.0	10,650	28.1
1983	135	23.0	11,900	25.0
1984	—	18.5	13,400	24.0
1985	—	25.0	12,100	28.5
1986	160	35.5	11,300	29.2

NOTE: From 1970 to 1983, the Kitamura Farm was not under integrated pest management (IPM); from 1984 to 1986, the farm was under IPM.

SOURCE: Colusa County data from Colusa County Cooperative Extension Service. 1970–1986. Agricultural Crop Report, County of Colusa, California.

viewed for this case study were two of the principal operators, David and Diann Kitamura. David Kitamura has been farming with his father since he was a child. His wife, Diann, is a licensed pest control adviser. David Kitamura and his brother do most of their own mechanical work on the farm equipment, usually during slack time in the winter.

PERFORMANCE INDICATORS

Tomato Yields

Table 3 presents the Kitamura Farm average yields compared with the Colusa County averages. The farm's 1986 average yield—35.5 tons per acre versus the county average of 29.2 tons per acre—was characterized as outstanding. In the previous 2 years, the farm's yield was less than the county average. It is not possible to draw valid conclusions about the yield effects of IPM on the basis of these very limited data, however. A far larger sample with controls for soil quality and other factors would be required to draw such inferences.

Finanical Performance

The Kitamuras reported that their accountant describes their business as "quite solvent." The land they own, the 40-acre walnut orchard, is not mortgaged. They owe some money on their newer equipment, particularly the tractors, and on the operating capital required for producing the crop of tomatoes. Yet they report having a debt-to-asset ratio of substantially less than 40 percent, the level frequently used as a critical point indicating a possible risk of financial vulnerability.

Diann Kitamura has estimated that without IPM, the farm would have spent approximately $8,800 on various insecticides for the tomatoes during the 1986 season. This estimate is comparable to the county average per acre cost of pest control and is based on following a conventional pesticide-based control program using manufacturers' recommended application rates. The Kitamuras reported spending a total of $1,482 on pest control, with a savings of $7,318—an average of $45.73 per acre. It has not been possible, however, to document this result. Diann and David Kitamura do their own scouting, so there is no direct cost to them in running their IPM program. The savings of $45.73 per acre reported by the Kitamuras is substantially above the $7.70 average savings estimated by Antle and Park (1986). The difference no doubt reflects the success of the Kitamuras in eliminating virtually all insecticide applications through scouting.

Environmental Impacts

The reduction of insecticide use associated with the IPM program in tomatoes varied according to the stage of the season when the tomatoes were planted. Fields that were planted at midseason had a 12 percent reduction in the amount of insecticide applied compared with a 40 percent reduction on late-season fields. One of the major changes effected by the IPM program, however, was a change in the type of pesticide material applied (Antle and Park, 1986).

In general, in the results reported by Antle and Park (1986), pesticide sprays were applied slightly less frequently on IPM farms (1.5 times versus 1.7 times), and the number of pounds of insecticide applied to the tomato fields was reduced. Yet because a more expensive material was used on average, the IPM and non-IPM fields in the study had virtually identical pesticide costs. Whether the reduced quantity of insecticide (a 22 percent average decrease in pounds applied through the season) and the different toxicological and ecological properties of the pesticide material applied on the IPM farms constitute a benefit to the environment could be determined only through further analysis. The relative effects of the various kinds of materials used and their possible additive effects would need to be carefully examined before a determination of ecological impact could be made (Antle and Park, 1986).

The material most frequently applied to tomatoes in California was the

TABLE 4   Pesticide Application Reported on Tomatoes in California, 1984

Chemical	Number of Application(s)	Pounds Applied	Acres or Units Treated	Type
Anilazine	105	3,502.00	3,274.50	A
Azinphos-methyl	169	6,263.70	8,150.45	A
Bacillus thuringiensis	82	603.84	4,591.84	A
Benomyl	39	531.50	1,924.96	A
Benzoic acid	1	1.29	35.00	A
Capsicum oleoresin	3	450.00	245.00	A
Captafol	171	23,203.56	12,551.50	A
Captan	6	320.56	102.40	A
Carbaryl	770	69,118.91	39,681.19	A
Carbolic acid	2	2.77	102.00	A
Chloramben, ammonium salt	1	3.88	3.00	A
Chlorine	6	1,795.00	1,809.00	T
Chloropicrin	31	13,488.44	815.81	A
Chloropicrin	1	0.08	0.28	T
Chlorothalonil	1,057	126,160.23	69,908.76	A
Chlorpropham	1	201.85	140.00	A
Chlorthal-dimethyl	7	2,091.75	302.00	A
Copper	14	276.55	520.00	A
Copper hydroxide	87	7,715.55	3,828.10	A
Copper-zinc sulfate complex	3	45.00	9.50	A
Demeton	20	392.46	1,442.30	A
Diazinon	130	2,832.50	6,726.49	A
Di-capryl sodium sulfosuccinate	1	8.48	26.00	A
Dichlorobenzalkonium chloride	3	328.50	2,758.00	T
Dichlorophen	3	23,206.08	476.00	A
1, 2-Dichloropropane, 1, 3-Dichloropropene, and related C-3 compounds	321	1,565,872.87	25,875.66	A
1, 3-Dichloropropene	9	33,610.53	589.80	A
Dicofol	69	4,706.05	4,191.50	A
Dimethoate	156	3,221.23	7,846.30	A
Dinoseb	23	3,252.77	1,904.00	A
Diphenamid	102	10,605.20	4,212.50	A
Disulfoton	71	6,031.38	4,087.40	A
Endosulfan	802	64,667.66	59,925.50	A
Ethephon	590	16,570.87	28,914.30	A
Ethion	13	2,967.05	1,360.00	A
Ethylene dibromide	4	5,288.42	178.00	A
Fenbutatin-oxide	5	0.69	1.03	A
Fensulfothion	30	10,350.74	1,590.00	A
Fenvalerate	1,804	24,864.80	125,008.40	A
Fonofos	142	11,286.11	7,304.40	A
Garlic	3	180.00	245.00	A
Glyphosate, isopropylamine salt	13	594.63	1,085.00	A
Lindane	20	335.25	796.50	A
Malathion	15	828.52	496.67	A
Mancozeb	887	46,214.30	35,588.86	A
Maneb	113	17,516.93	9,611.45	A
Metalaxyl	764	4,368.42	25,387.91	A

(Continued on page 386)

TABLE 4   *(Continued)*

Chemical	Number of Application(s)	Pounds Applied	Acres or Units	
			Treated	Type
Metaldehyde	4	7.20	15.00	A
Methamidophos	677	32,398.13	37,020.80	A
Methiocarb	2	0.76	94.30	A
Methomyl	2,236	83,589.99	136,767.02	A
Methoxychlor	17	486.70	768.00	A
Methyl bromide	40	44,901.33	629.18	A
Methyl bromide	1	3.92	0.28	T
Methyl isothiocyanate	1	1,386.39	24.00	A
Methyl parathion	320	18,698.83	24,832.00	A
Metribuzin	78	2,750.12	6,390.50	A
Mevinphos	98	1,776.91	5,940.50	A
Naled	24	742.58	946.00	A
Napropamide	345	24,466.79	17,752.40	A
Oxamyl	124	4,179.77	7,062.50	A
Paraquat dichloride	243	8,647.30	15,373.19	A
Parathion	308	14,825.11	21,771.50	A
Pebulate	165	21,523.59	6,378.50	A
Permethrin	4	43.75	276.38	A
Phosphamidon	100	2,579.54	4,977.00	A
Piperonyl butoxide	6	16.69	412.00	A
Pyrethrins	6	1.80	412.00	A
Sodium fluoaluminate	9	715.10	332.00	A
Strychnine	15	91.84	22,728.00	A
Sulfur	1,543	3,451,374.20	133,187.55	A
Toxaphene	6	276.37	186.00	A
Triadimefon	617	5,996.68	52,910.15	A
Trichlorfon	10	393.60	421.00	A
Trifluralin	114	8,822.36	10,466.90	A
Xylene	518	33,175.17	36,595.99	A
Xylene range aromatic solvent	613	89,424.53	48,072.88	A
Zinc	2	6.28	185.00	A
Zinc phosphide	3	67.72	135.00	A
Zinc sulfate	24	209.34	1,108.00	A
Compound 1080	1	2.20	275.00	A
Commodity total	16,943	5,969,461.06		

NOTE: In the Acres or Units column, Type A means acres treated; Type T means tons of tomatoes postharvest.

SOURCE: State of California, Department of Food and Agriculture. 1985. U.S. Annual Pesticide Use Report by Commodity, January Through December, 1984. Sacramento, Calif.

insecticide methomyl; 2,236 applications of this substance were made in 1984 to 136,767 acres of tomatoes. A total of 83,590 pounds of methomyl were applied, or about 1.6 pounds of active ingredient per acre. The second most prevalent pesticide was fenvalerate; 1,804 applications were made, for a total of 24,865 pounds on 125,000 acres (Table 4).

The success of the Kitamuras' IPM practices in avoiding crop losses, cutting costs, and maintaining high-quality harvests and top prices are significantly influenced by two factors: the predictability of weather patterns in the central valley of California and the Kitamuras' management skills. When summer or fall rains are unexpectedly late, the incidence of plant disease, need for fungicide treatment, and risk of crop losses rise greatly. In most years, however, the combination of soils, climate, rotational patterns, and IPM on this farm are successful in sustaining high levels of production with minimal adverse environmental effects.

REFERENCES

Antle, J. M., and S. K. Park. 1986. The economics of IPM in processing tomatoes. California Agriculture 40(3&4):31–32.

Grieshop, J. I., F. Zalom, and G. Miyao. 1986. Exploratory Study on the Adoption of the IPM Tomato Worm Monitoring Program by Tomato Growers in Yolo County—Descriptive Statistics. IPM Implementation Group, Division of Agriculture and Natural Resources, University of California, Davis. September.

University of California. 1985. IPM for Tomatoes. Publication No. 3274. Davis, Calif.

Wilson, L. T., F. G. Zalom, R. Smith, and M. P. Hoffmann. 1983. Monitoring for fruit damage in processing tomatoes: Use of a dynamic sequential sampling plan. Environmental Entomology 12(3):835–839.

Zalom, F. G., L. T. Wilson, and R. Smith. 1983. Oviposition patterns by several lepidopterous pests on processing tomatoes in California. Environmental Entomology 12(4):1133–1137.

10

Livestock Farming in Colorado: Coleman Natural Beef

T HE MAIN HEADQUARTERS OF COLEMAN NATURAL BEEF (CNB) is located 1 mile west of Saguache, Colorado, but the Colemans operate their owned and leased pastures and irrigated meadows over a 50-mile distance. They have been ranching in the area since the 1970s and own 13,000 acres, 4,500 of which is near Saguache. They lease another 13,000 acres (some from family members and relatives) in the same area. The Colemans' base ranch is located in the very arid northwest section of the San Luis Valley, a great, dry lake bed, approximately 8,000 feet in elevation. The summer ranch is 50 miles west of Saguache on Highway 114 along the Continental Divide, in the high meadows and mountain lands known as the Cochetopa. The meadowland (known as parkland in the West) is at an elevation of approximately 10,000 feet. The summer ranch consists of 8,500 owned acres with grazing permits on 250,000 acres of Forest Service and Bureau of Land Management summer lease land.

GENERAL DATA

The CNB operation is a beef ranching business with an extensive marketing component selling to a specialized clientele for premium prices. The Colemans' own beef cattle herd totals 2,500 cow-calf units; they contract with local ranchers to provide an additional 12,500 head to meet the demands of their clientele (Table 1).

Climate

The climate and topography on both the main ranch at Saguache and the Cochetopa play a key factor in the way this traditional cattle ranch is run.

TABLE 1 Summary of Enterprise Data for Coleman Natural Beef

Category	Description
Size	21,500 acres owned, 13,000 leased, 250,000 made available by grazing permits; 2,500 beef cattle
Labor and management practices	Two brothers provide all the management: one focuses primarily on production; the other handles marketing and public relations. Four full-time hired men work the ranch, each receiving about $1,000 per month plus housing, paid utilities and insurance, and use of a vehicle. The man on the summer ranch receives similar pay and benefits. Jim Coleman's wife works off-farm.
Livestock management practices	All replacement cattle are produced on the ranch; 300 cows are artificially inseminated each year, and breeding bulls are selected from their offspring. After calving, half the cattle and calves are herded 50 miles to summer pasture; half are pastured at the base ranch. Feeder stock is wintered on rented wheat pasture in eastern Colorado and western Kansas. Breeding stock fed at the base ranch are given hay produced without pesticides or fertilizers, plus supplemental cottonseed cake. In spring, feeders are shipped to feedlots and fed under contract with specific feeds that do not contain growth-promoting chemicals. The death rate is 2 percent at Coleman Natural Beef (CNB) compared with the industry average of 6 percent, although these data are not precisely comparable.
Marketing strategies	CNB has approximately $15 million in annual sales, including on-farm production plus cattle from selected neighbors, totaling 15,000 head. The 25 percent premium above regular carcass prices is key to the profitability of the operation.
Weed control practices	All plant material is harvested as hay, including weeds.
Insect and nematode control practices	The dry climate and high elevation greatly reduce the incidence of pests. Cattle receive an injection of ivermectin to eliminate scabies and lice; no other pest control is used.
Disease control practices	Three-way inoculations are used, compared with the conventional seven-way vaccine. There are some problems (minor) with lumpjaw and pinkeye (due to feeding of foxtail and barley). Sick animals are segregated, given medication, and removed from the natural program.
Soil fertility management	No fertilizer or lime is used.
Irrigation practices	All hay acreage and some pasture are flood-irrigated from mountain streams.
Crop and livestock yields	Feeder cattle are estimated to be 25 to 50 pounds smaller than those of conventional producers who apply subtherapeutic doses of antibiotics or other growth-stimulating drugs. Daily rates of gain are slightly less than those of conventional producers.
Financial performance	Net returns to the ranching operation (producing feeder cattle with minimal inputs of chemicals) are reported to be less than hired labor wages. The finishing and beef packing and sales operations, using CNB's own beef plus that of neighbors, appear to be financially stable because of the 25 percent premium received.

Annual precipitation in Saguache is 8.7 inches, falling mostly in the summer (Table 2). Snowfall averages 31 inches per year. Annual rainfall at the summer ranch is 10 inches, and average snowfall is 2 feet. The snowfall may vary from less than 7.5 inches on the meadowlands to many feet in the mountains.

Soil

The Colemans describe their soils in the San Luis Valley area as alkaline desert soils. They describe their Cochetopa mountain meadow soils as acid. Data on the pH of the soils are not available.

Buildings and Facilities

Barns are provided for the horses but not for the cattle. Seven-foot-high wooden windbreaks are provided in places where trees are too sparse to provide protection. The corrals are constructed of lodge-pole pine logs. One squeeze chute, sorting pens, and a loading ramp are provided at each of the several areas in which the cattle are grazed. Instead of a barn for sick animals, a 500-acre pasture is provided for ill or, occasionally, snake-bit cattle that require special observation, increased feeding, or medication.

Irrigation and Haying

Because of the limited rainfall in the area, all of the hay ground and some of the pastureland (more than 6,000 acres) on the farm are gravity-irrigated from mountain streams, which are primarily fed by snow melt runoff. Two hired men manage the floodgates during the summer. Native hay is cut once, both on the Cochetopa and at the base ranch, yielding about 1 ton per acre. The grass is basically native fescue, timothy, and wheat grasses, except for 150 acres of alfalfa from which two cuttings per year are taken yielding about 3 tons per acre. The Colemans apply no fertilizer or lime. It is not known whether the productivity of the cropland can be maintained indefinitely without the application of fertilizer or lime.

Labor

In addition to Jim and Mel Coleman (brothers), the Coleman ranch operation has four full-time hired men, who are paid about $1,000 per month. They also receive housing, paid utilities and insurance, and use of a vehicle. The man on the Cochetopa ranch receives similar pay and benefits. Because the cattle ranch has proven only marginally profitable, Jim Coleman's wife works as a school nurse in Saguache to help cover family living expenses.

TABLE 2 Normal Daily Temperatures and Monthly Precipitation at Saguache, Colorado

Month	Normal Daily Temperature (°F) Maximum	Minimum	Monthly Average	Normal Precipitation (inches) Rain	Snow
January	36.6	5.6	21.1	0.24	3.9
February	41.4	11.5	26.4	0.23	3.6
March	49.0	17.5	33.3	0.36	4.5
April	58.8	25.0	41.9	0.57	3.8
May	68.4	33.9	51.2	0.75	0.8
June	77.5	41.3	59.4	0.61	0
July	81.6	47.6	64.6	1.78	0
August	79.2	48.0	62.6	1.67	0
September	73.5	37.6	55.6	0.83	0.2
October	63.4	28.7	46.1	0.88	4.2
November	47.7	16.4	32.1	0.43	4.5
December	37.6	7.5	22.6	0.39	5.5
Average annual	59.6	26.6	43.1		
Average annual total				8.74	31.0

NOTE: The normal daily maximum by month is the average of each day's (midnight to midnight) high temperature for every day in that month from 1941 to 1970. The normal daily minimum by month is the average of each day's (midnight to midnight) low temperature for every day in that month from 1941 to 1970. The normal monthly temperature is the average of the normal daily maximum and minimum temperatures for that month. The normal monthly precipitation is the average of the inches of rain or the liquid equivalent of snow for that month from 1941 to 1970.

SOURCE: National Oceanic and Atmospheric Administration. 1980. Climates of the States, 2d ed. Detroit: Gale Research Co., Book Tower.

Machinery

Jim Coleman, his son Tim, and one hired hand harvest approximately half the hay (2,000 tons) using their own equipment, most of which is at least 6 years old. The farm has one self-propelled swather, which cuts, conditions, and places the hay into windrows; a square baler; and a round baler, all purchased in about 1980. The Colemans harvest approximately half of their hay in round bales and the other half in square bales in Saguache. They have two tractors and self-loading hay wagons so that square bales can be handled with relatively little labor. Haying starts in July and lasts 5 to 6 weeks. Other equipment used by the Colemans are front-end loaders on the tractors to set out round bales for feed and a 10-year-old dozer that is used to make winter trails to feed bunks and for irrigation ditch work.

MANAGEMENT FEATURES

During a site visit in July 1986, the Coleman beef herd appeared to be in excellent condition. The cows are primarily offspring of a foundation herd of Herefords purchased 25 years ago. Since then, Red Angus, Limousin, Gelbvieh, and other breeds have been introduced into the herd by artificial insemination. Black Angus have been used to breed first-calf heifers for ease of calving. Between 300 and 400 heifers each year are selected as herd replacements.

Breeding

The Colemans have not purchased any cattle since 1959. Instead, they artificially inseminate 300 cows per year (and have been doing so for more than 25 years) to upgrade their foundation herd. According to D. Lamm, assistant director for agriculture and research at Colorado State University, this practice is generally carried out by only the top 3 to 5 percent of the beef cattle operators in the state (correspondence, 1987).

The Colemans watch the bull calves from these cows closely and select the best as herd replacement bulls, running 5 bulls for every 100 cows. Before the cattle are taken to the summer pastures, they go through one heat cycle at the ranch, during which they are carefully watched. During this heat cycle, approximately two out of three cows will become pregnant.

The Colemans report that, because of the high altitude and a genetic predisposition, animals have a tendency to get brisket disease. This disease is associated with congestive right heart failure as a result of the stress of high altitudes. Animals born at high altitudes, however, tend to adjust more easily than imported stock. Neighboring ranchers who have purchased animals from lower elevations have lost many cattle because of this problem; yet, the Colemans' death loss for all of their cattle from all causes is typically under 2 percent, compared with an industry average of 6 percent.*

---

*Exactly comparable data have not been located. Ensminger (1983), however, estimates

In November of each year, all 2,500 cows are tested for pregnancy. No open (nonpregnant) cows are kept. Age is not a selection factor, and some pregnant cows are 10 years old or older. Jim Coleman reported that they can pregnancy-check approximately 500 cows per day (D. Lamm, correspondence, 1987). The ranch has excellent working corrals at different locations, and the herd is subdivided into groups of 300 cow-calf units for ease of management.

Much of the ranch work is done on horseback. The Coleman ranch owns 30 horses, and some of the hired men ride their own horses. The cattle are observed twice daily, once in the morning and again in the evening.

Feeding

Approximately half of the cows are kept in Saguache year-round, and the other half are herded to the Cochetopa summer pastures approximately 50 miles away after calving each spring. The calves are weaned off their mothers in the Cochetopa in later October and November, and by the first of December the cattle are being herded back down to Saguache for the winter.

About 4,000 tons of hay per year are harvested from the ranch. Half of this amount is harvested in the upper range, usually under contract for $27.00 per ton. The other half is baled by Jim Coleman, his son, and one hired man. At least 1 year's hay reserve is always kept for emergencies. If there is any excess above that amount, it may be sold.

During the winter the younger cows get approximately 1 pound of cottonseed cake each day and all the hay that they can eat without wasting it. The old cows are fed cottonseed cake every other day with free choice of hay. Replacement heifers are fed 1.5 to 2.0 pounds of ground barley or wheat, or both, to promote growth.

Animal Health

The beef cattle are inoculated with a three-way vaccine for the common diseases of brucellosis, blackleg, and malignant edema; they are also injected with ivermectin to eliminate scabies and lice. No medicines or growth hormones other than these materials are given prophylactically. The primary cattle health problems on the ranch other than big brisket have been lumpjaw and pinkeye. Jim Coleman has said that these diseases are related to feeding foxtail and barley, which have barbs on the seed. Mel Coleman also indicated that climate makes a big difference in animal health problems. In this area the weather is dry with a great deal of sunshine. Believing that animal stress causes illness, which may involve costs associated with

---

that "calf losses from birth to weaning average 6 percent. About 1.5 million head of cattle die in feedlots each year." Based on an estimated 25.8 million head of fed steers and heifers slaughtered in 1983, feedlot deaths also average about 6 percent (U.S. Department of Agriculture, 1986).

medicines and veterinary services, the Colemans have based their herd health program on minimizing stress. Because the grazing area per animal is large, diseases associated with confinement are eliminated. Animals requiring drugs for illness or chemicals for external parasites are red-tagged and do not enter the natural beef feeding program. Instead, these animals are sold through regular marketing outlets.

The Colemans report that shipping fever and other related weaning stresses have been kept extremely low by conditioning the calves on hay at the ranch prior to trucking them to winter wheat pastures in eastern Colorado and western Kansas.

The cows are bred to calve in late February through May. Of those cows who calve early, the best are selected for artificial insemination. Bull calves are castrated at 2 to 4 weeks of age, except for those sired by artificial insemination; these bulls are retained for possible selection as breeding stock. Strong, healthy calves weighing 450 to 500 pounds are weaned in the fall without the use of hormone implants. According to Colorado State University livestock specialists, these weaning weights compare favorably with those of calves that do receive growth stimulant implants (D. Lamm, correspondence, 1987). The Colemans report that their use of artificial insemination has improved the herd's weaning weight and average daily gains from weaning to market. Other economically important factors taken into account in herd improvement include carcass grade and area of the rib eye. Also, by selecting for ease of calving among first-calf heifers, the Colemans report that they have significantly reduced calving problems and parturition death loss.

Marketing

The Colemans have developed a unique system for marketing their beef. In 1979 they were being squeezed by rising land payments, higher costs of production, and falling cattle prices; profit margins were nonexistent. They knew that they had to try something different to prevent failure of the firm.

The Colemans stopped using most of the drugs, fertilizers, and pesticides commonly used in their operation during the 1960s. They have never used any growth hormones or implants. According to R. E. Taylor, a professor of animal sciences at Colorado State University, feed additives, such as subtherapeutic doses of antibiotics, are routinely applied by only about 5 percent of Colorado beef producers. A much more common practice is the use of growth stimulant implants in the animal's ear, which increases weaning weights by 10 to 25 pounds (R. E. Taylor, correspondence, 1987). Before they started CNB, the Colemans sold their calves to other beef feedlot firms that fed antibiotics and growth stimulants. Although their feeder calves brought top prices, the prices were still too low to cover all their expenses and land payments.

Mel Coleman decided to devise a marketing scheme in which he would control the way their cattle were being fed, and then sell the beef to health-conscious consumers at a premium price, thereby covering their higher costs of production. He felt that the growing health food market was ready for natural beef and devised a way of managing his cattle throughout the feeding process so he could be assured that the cattle were free from chemicals (antibiotics, implants of growth hormones, and feed residues of pesticides). As business grew, Mel Coleman began to specialize in marketing, and Jim operated the ranch.

The current demand for Coleman Natural Beef has outgrown the Colemans' own herd's production capacity. They have therefore devised the Coleman Certified Rancher Program, entering into agreements with certain neighbors to raise calves using methods compatible with CNB policies. Today, CNB sales exceed 15,000 head per year, including cattle from other ranches in the Coleman Certified Rancher Program. Such a vertically integrated system, from cow conception to the supermarket, took several years to develop. In the first year, 1979, only one beef carcass was sold. In 1980 sales were up to three per week. As debts mounted, Mel Coleman looked outside Colorado and entered the expanding health food markets in California and then in Massachusetts. In 1985 Coleman Natural Beef and other natural meat product sales totaled almost $15 million. Mel Coleman reported that as the CNB volume has increased, so have the logistical problems. The Colemans' major challenges have included making sure their calves were fed properly and then butchered, packed, and shipped on time.

All of the weaned and conditioned calves are trucked from the Saguache base ranch in late November to rented wheat pastureland in eastern Colorado and western Kansas. They arrive on wheat pasture weighing 450 to 500 pounds and are grazed there for 90 to 100 days. One man from the Coleman ranch operation goes with the cattle to take care of them.

Because of the low rainfall and minimal weed and insect problems in eastern Colorado and western Kansas, it is not difficult for the Colemans to find wheat fields that have had little or no application of chemical herbicides or fertilizers. In nonirrigated wheat pastures, summer fallowing of the cropland every second or third year with tillage usually controls the weed population without herbicides.

According to a telephone interview with Mel Coleman (1986), the Colemans' feeder cattle gain an average of 2.25 pounds per day while on wheat pasture. D. Lamm, a Colorado State University livestock specialist, observed that a normal rate of gain may be about 2 pounds per day while the cattle are on range, but depending on such factors as weather and the genetic potential of the beef animals, an average rate of 2.25 pounds per day is attainable (correspondence, 1987). During adverse weather conditions, however, the animals sometimes gain no weight at all.

When the Coleman feeder cattle reach 700 pounds, they are shipped to a custom feedlot where they are fed until they reach a weight of 1,000 to 1,100 pounds. In a 1987 telephone interview, Mel Coleman reported that his cattle gain an average of 2.75 pounds per day while in the feedlot. D. Lamm, however, states that the normal rule of thumb is that heifers will gain 2.8 pounds per day and steers will gain 3.2 pounds per day while in a feedlot, implying that the Coleman rate of weight gain is slightly below the expected range for feedlot operations using conventional practices.

While in the feedlot the cattle must be fed according to Coleman specifications, thus requiring that they be fed separately from other cattle because their special ration contains no growth hormones or chemical feed additives. This practice requires additional labor expenditure by the feedlot operator and results in a somewhat slower average daily weight gain, for which the Colemans have to pay an extra fee to the feedlot operator.

The Colemans strive to purchase feed grains that have been produced without chemical pesticides or fertilizers, and toward that end, chemical tests are done by commercial laboratories to check for pesticide residue in the feed. The effectiveness of this approach in assuring residue-free feed grains is difficult to verify. It is unlikely that the analytical methods used by commercial laboratories could detect at very low levels (in the parts per billion range) all of the pesticides used on feed grain crops. Nor has this committee found any data indicating that natural beef is different from conventionally produced beef in composition or nutritive value. The implications of alternative management practices, such as those used by the Colemans, on meat product quality and safety deserve further study, however. Carefully controlled protocols and sensitive analytical methods are needed.

Once the cattle reach about 1,100 pounds, they are slaughtered at a Denver packing plant. Again, keeping their cattle separate requires more logistical attention from the Colemans and added cost. Currently, to prevent their cattle from being mixed with other beef, the Colemans have managed to ensure that theirs are the first batch of cattle to be slaughtered on a given day (normally on Fridays). The carcasses are hung in cold storage for 1 week, and then the beef is transported to the Coleman sales and distribution office, where it is packaged and shipped. Some beef is cut into prime cuts and vacuum-packed in plastic; some is cut into portion control cuts. Some is shipped as carcass beef, depending on what the retailers order.

The Colemans receive a price for their beef that is approximately 25 percent higher than the regular carcass price. The cattle are bred and fed to reach a low Choice grade, with a yield score of most of their beef in the number 2 grade. The Colemans guarantee their beef; if a consumer or retailer is dissatisfied, the beef is replaced. Their goal is to have all their cattle grade number 2 and still make the Choice grade without having to trim off excess fat.

PERFORMANCE INDICATORS

Management factors tend to reduce the Colemans' cash operating costs, as compared with typical midwestern crop-livestock operators and feedlots. The ranch's death loss from calving problems and disease has typically been very low: 2 percent versus an industry average of about 6 percent. For example, in the year ending July 1986, out of the Colemans' 2,500 cow-calf units on the summer ranches, only 8 animals died (6 calves and 2 cows), all reportedly struck by lightning. Good management practices (not unique to the Coleman ranch) are used to reduce stress on the beef animals. Crowding is avoided through the use of ample acreage per animal unit. Infections from outside herds are avoided by not introducing new animals to the herd. Improved genetic material is introduced through artificial insemination rather than by the purchase of breeding stock. Feeder calves are conditioned prior to shipping.

Fertilizers and pesticides are not used in the production of crops or pasture on the ranch. The effects of this method of production on the quality of the beef (in terms of food safety and nutritional value) have not been scientifically established. The Colemans have observed that significantly higher yields of hay could be obtained by fertilizing but only if they introduced plant cultivars other than native species. They have also noted that these pasture improvement practices increase costs and the risk of more sizable crop failure during low-rainfall years. Native species are reputed to be more winter-hardy and drought-resistant than other hay cultivars that the Colemans or their neighbors have tried.

The ranch uses a minimum of buildings and machinery and provides natural and manmade windbreaks rather than barns to protect the cattle from severe winter weather. Haying equipment and other machinery are considered minimal for an operation of this size.

Although their costs appear to be relatively low, the Colemans' production may be reduced as a result of some of their management practices—not using fertilizer or growth hormones, for example. Yet, these management practices are apparently essential to receive premium prices. The Colemans are obtaining a premium price of up to 25 percent or more for raising and marketing their cattle as natural beef, catering to people willing to pay a higher price to obtain foods produced with minimal or no chemical inputs. Nonetheless, the owners report that the Coleman ranch is currently earning a return on labor and management that is less than the wages they pay their hired personnel. The rate of return on their investment thus is extremely low—if not actually negative—at present. Clearly, the marketing operation is supporting the ranch: more than 12,000 additional cattle are marketed under agreements with neighbors who produce their beef to CNB specifications.

There does not appear to be any technical efficiency incentive for beef producers to adopt the Coleman technology; the premium price appears to be necessary to keep this beef operation going. Today CNB is one of several suppliers of natural beef. If the total market production of natural beef increases more rapidly than demand, the 25 percent premium price now obtained could be greatly reduced, thereby substantially reducing the profits of the operation.

REFERENCES

Ensminger, M. E. 1983. The Stockman's Handbook. Danville, Ill.: Interstate Press.

U.S. Department of Agriculture. 1986. Livestock and Poultry—Situation and Outlook Report. Report No. LPS-22. Economic Research Service. Washington, D.C.

11

Rice Production in California: The Lundberg Family Farms

THE LUNDBERG FAMILY FARMS is located in northern California in Richvale, Butte County, about 30 miles southeast of Chico. A family partnership owned by four brothers, the farm consists of 3,100 acres (Table 1). The Lundbergs produce about 1,900 acres of rice each year using largely conventional methods that include the use of chemical fertilizers and pesticides, but on the Lundberg Farms the level of pesticides used is somewhat less than the recommended amounts. Unlike many producers the Lundbergs' conventional production practices also involve disposal of rice straw by decomposition in the soil rather than burning. Besides their conventionally managed acreage, the Lundbergs also produce about 100 acres of rice without pesticides or chemical fertilizers as an experiment. They refer to this 100 acres as their organic rice because the methods that they use comply with the California organic farming law. They have been experimenting with the production of organic rice for 18 years.

Rice is the only cash crop grown on the Lundberg Family Farms. Purple vetch *(Vicia benghalensis)* is also grown as a green manure crop and nitrogen source on the experimental acreage. On the Lundberg Farms, as throughout northern California, rice production, both conventional and organic, is on flooded land.

GENERAL DATA

The unusual features of this farm are the extensive field experiment in a continuing effort to develop economically viable methods of producing rice without chemical pesticides and fertilizers; the incorporation of rice straw into the soil in lieu of burning, which is practiced in both conventional and organic production; and the extensive marketing system.

TABLE 1 Summary of Enterprise Data for the Lundberg Family Farms

Category	Description
Farm size	3,100 acres (100 acres experimental)
Labor and management practices	Four Lundberg brothers operate the farm plus a large marketing and processing operation. Extensive farm management input is provided by a salaried production manager, who also does all pest scouting for the rice acreage. The rice production operation employs 6.5 person-years of regular year-round labor. Seasonal workers are hired for 8 weeks in the spring and 6 weeks in the fall. Labor requirements are higher for the alternative rice operation than for the conventional one because of repeated irrigation and surface tillage practices used in fallow operations.
Marketing strategies	Rice from the 100 experimental acres is sold at a premium price (about 50 percent) through the farm's extensive marketing and processing operation (along with the output of several other growers) as organically grown. Most of the rice produced on the farm is processed or sold raw as ordinary rice.
Weed control practices	A 2-year rotation is used for experimental organic rice: the rice crop in year 1 is followed by fall-sown vetch; year 2 has summer fallow and fall vetch. Repeated flooding and shallow tillage is used for weed control in the fallow year. Reduced rates of herbicide are applied on conventional rice fields.
Insect and nematode control practices	Tadpole shrimp (a crustacean) is controlled by irrigation on the experimental acres, alternating wet and dry fields. Nematodes are not a problem in inundated fields, and other pests are less problematic.
Disease control practices	The rice straw is rolled down, decomposing sclerotia of stem rot pathogens. The farmer says stem rot is not a serious problem. There is no other major rice disease.
Soil fertility management	The farm uses a 2-year rotation, rice and vetch-fallow-vetch on the experimental acres. The nitrogen supply is considered inadequate, reducing yields. No other fertilizer is applied. A 3-year rotation (rice, rice, and vetch-fallow-vetch) and commercial fertilizer are used on the other acreage.
Irrigation practices	Rice fields are alternately flooded and drained to control tadpole shrimp until the rice stand is established. The depth of the inundation depends on the growth stage of the rice. Five acre-feet of water are used.
Crop and livestock yields	The experimental (nonchemical) rice yields 44 hundredweight versus the Lundbergs' 74 hundredweight/acre conventional average, or the 110 hundredweight/acre on the most productive farms in the county.
Financial performance	Experimental nonchemical rice is generally less profitable than conventionally produced rice despite premium price, due to insufficient nitrogen and lower yield. Premium prices for yields in organic rice would dissipate if production increased significantly.

The marketing enterprise of the Lundberg Family Farms is extensive, including a modern milling and processing plant employing up to 70 people. In addition to their own organically grown rice, the Lundbergs contract with 10 other growers in the local area who use methods in compliance with the state law governing organic farming. The Lundbergs are well known for their marketing and processing system through which they market not only organically produced rice but also conventionally grown rice from their farm and others in the area.

Climate

Normal precipitation at Orland, 30 miles northwest of the Lundberg farm, is about 20 inches per year (Table 2). Two or more inches of precipitation fall each month during November through February; less than 1 inch of precipitation falls per month during May through September. Climatic conditions at Orland are a good approximation of those prevailing on the Lundberg Farms. The elevation at the farm is approximately 200 feet above sea level.

PHYSICAL AND CAPITAL RESOURCES

Soil

The Lundberg Farms' soil is vertisol, largely of Stockton clay adobe, with some 40 to 45 percent of the area underlain by a calcareous cemented hardpan (D. Mikkelsen, interview, 1986). The topsoil is natural, self-generating soil that is high in phosphorus, potassium, calcium, and other nutrients. The land is quite flat; most fields required only moderate leveling prior to the advent of rice production and now need only a minimal finish leveling every few years.

Buildings and Facilities

Except for a machine shop, the main buildings and facilities on the farm—an extensive, modern milling and rice cake processing plant—are associated with the postharvest operations. The milling and processing facilities include rice storage bins, in which high levels of carbon dioxide can be maintained to prevent insect damage during storage; a rice drier; a cleaning mill; various sorting machines; a packaging plant; rice cake production machinery; and two warehouses.

Farm Machinery

The machine inventory on the farm, other than the processing and marketing equipment, consists of nine crawler tractors; four wheel tractors; three 90-horsepower wheel tractors; two 150-horsepower wheel tractors; three 60-foot land-planes; three disks with 30-inch blades, 20 feet long; two

TABLE 2 Normal Daily Temperatures and Monthly Precipitation at Orland, California

Month	Normal Daily Temperature (°F)			Normal Precipitation	
	Maximum	Minimum	Monthly Average	Inches	Days With ≥ 0.1 Inches
January	53.0	35.3	44.2	4.37	7
February	59.6	38.9	49.3	2.95	5
March	64.7	41.0	52.8	1.84	5
April	72.2	45.4	58.8	1.32	3
May	81.1	51.9	66.5	0.53	2
June	89.5	58.6	74.1	0.37	1
July	96.4	61.4	78.9	0.11	0
August	94.6	59.4	77.0	0.20	0
September	89.7	55.7	72.8	0.35	1
October	78.5	48.9	63.7	1.13	2
November	63.8	41.4	52.6	3.18	5
December	53.8	36.3	45.0	3.61	6
Average annual	74.7	47.9	61.3	Average annual total 19.96	37

NOTE: The normal daily maximum by month is the average of each day's (midnight to midnight) high temperature for every day in that month from 1941 to 1970. The normal daily minimum by month is the average of each day's (midnight to midnight) low temperature for every day in that month from 1941 to 1970. The normal monthly temperature is the average of the normal daily maximum and minimum temperatures for that month. The normal monthly precipitation is the average of the inches of the precipitation for that month from 1941 to 1970.

SOURCE: National Oceanic and Atmospheric Administration. 1980. Climates of the States, 2d ed. Detroit: Gale Research Co., Book Tower.

32-foot chisel plows; a 22-foot cage roller; a 22-foot rubber wheel roller; a 15-foot tiller; six pickups; four 1.5-ton flatbed trucks; two diesel tractor-trailer rigs; and miscellaneous implements. Aerial planting and some harvesting operations are custom hired, and a no-tillage grain drill is rented, which accounts for the absence of such equipment in the machinery inventory.

MANAGEMENT FEATURES

Rotations and Cultural Practices

Rice is conventionally grown in northern California as a more or less continuous cash crop, with as many as 2 to 5 years of continuous rice in a given field, followed by 1 year of fallow for releveling (Wick, 1975). Some California rice fields have produced a rice crop in each of the past 30 to 50 years (D. Mikkelsen, interview, 1986). Many farmers are now using a 3-year rotation of 2 years rice followed by 1 year fallow (Wick, interview, 1986), however, because of the federal price-support program requirement that 35 percent of a farm's rice allotment be idle.

The experimental method currently used by the Lundbergs on the 100 acres grown without pesticides or chemical fertilizers is a 2-year rotation that alternates rice with purple vetch and fallow. Following the rice harvest in October and November, the rice straw is spread and rolled onto the soil to expedite its decomposition. (The greater the contact of the straw with the soil, the more rapidly it will decompose.) One of two kinds of roller is used, depending on soil conditions. If the soil is compacted, the field is first chiseled; then a rubber-wheeled roller is used to mash down the straw. A steel cage roller is used, drawn by a crawler tractor, in cases in which the soil is soft enough so that the straw may be incorporated into the soil without prior chiseling. After the straw is rolled down in the fall, purple vetch seed is sown by airplane.

In the Lundbergs' experimental system, unsprouted rice seed is planted, in contrast to conventional planting methods in which rice seed is soaked, partially sprouted for 24 hours, and then drained prior to seeding by air into flooded fields. Dry, unsprouted rice seed is used when drilling directly into dry soil because the tender growing points of the partially sprouted rice seed would be damaged by the mechanical action of a drill and germination rates would be low. Using this method, the rice seed is drilled directly into the soil until the appropriate moisture is available for sprouting (based on moisture, temperature, and soil contact), at which point the field is "flushed" (rapidly and briefly irrigated).

Following germination, and until the rice reaches a height of 2 to 4 inches, the Lundbergs allow the soil to become rather dry. When the rice begins to show stress from a lack of moisture, the field is flushed again. After the rice plants have become fully established (3 to 5 inches tall), the fields are kept flooded until they are drained in preparation for harvest (3 to 4 weeks earlier) so that the soil dries out enough to support the harvest machinery.

No crop is harvested from a field in a fallow year. Instead, purple vetch is planted in the fall following the rice harvest and again in the fall of the fallow year. The vetch normally grows rather slowly during the fall and becomes dormant during cold temperatures in winter, but by April or May it has usually produced abundant foliage that makes an excellent green manure crop or mulch. In the spring of the fallow year, the vetch is flail-chopped and disked under, along with the largely decomposed rice straw. The field is then laser-leveled and alternately flushed and shallow-tilled with an implement to control weeds. In some years, depending on weed populations, a fallow field may be treated with as many as three cycles of flooding and tillage.

In the spring of the year in which rice is to be planted, the leguminous foliage is flail-chopped, along with the largely decomposed rice straw, leaving a mulch on the soil. A heavy no-tillage drill is then used to plant rice seed into this mulch. The drill leaves the soil bare above the narrow rows (about 8 inches apart) in which the rice seed is planted. The areas between the rice rows remain covered with the mulch, which helps control weeds.

The rationale for these management practices is based on weed and pest control and improved soil fertility. The mulch is thought to inhibit weed seed germination and thus compensate for the disadvantage of dry seeding (the delayed emergence of the rice crop) as compared with the conventional practice. Seeding into mulch, followed by intermittent flooding in the early stages of rice growth and development, also breaks the life cycles of water pests, such as the seed midge, tadpole shrimp, and rice water weevil, which need continuous flooding to survive.

The Lundbergs estimate that the vetch supplies about 120 to 130 pounds of nitrogen per acre. A University of California soil scientist, D. Mikkelsen, has estimated that the nitrogen supplied may actually be in the range of 60 to 120 pounds (interview, 1986).

Mikkelsen has also observed that the flail-chopped mulch of vetch "tends to float and is blown by the wind to the nearby levees" (correspondence, 1987). This tendency may be an impediment to widespread adoption of this procedure.

The alternative methods used by the Lundbergs have been evolving from year to year as their experimentation followed an orderly sequence of objectives. Until 1986, their objective had been to find an economical method of controlling weeds without the use of herbicides. Having attained this goal to their own satisfaction, they now recognize that the next important objective is to enhance available nitrogen in the soil by methods other than the use of chemical fertilizers forbidden by the state's law on organic farming.

Except for the 100 acres on which they produce rice without chemical pesticides and fertilizers, the Lundbergs use methods similar to those of conventional growers in their area, with two exceptions. One is that they have not burned rice straw since 1960; all of their straw is rolled down each

TABLE 3   Rice Yields on the Lundberg Family Farms'
Experimental Organic Fields Compared With Other Sources
(hundredweight/acre)

Source	1985	1986
Lundberg organic	44.0	27.0
California statewide	73.5	76.0[a]
Butte County Rice Growers Association	74.0	80.0[a]

[a]These figures are estimates.

fall, and the largely decomposed residue is disked under in the spring prior
to planting. The Lundbergs are not alone in this practice; rice straw decomposition is gradually becoming a more common cultural method in the
Sacramento Valley rice production area.

The second exception is that the Lundbergs seek to minimize their application of herbicides. They ordinarily apply 3 pounds of molinate per acre,
compared with the recommended rate of 4 pounds per acre (Wick and
Klonsky, 1984), or the common practice of 5 or more pounds per acre (G.
Brewster, interview, 1986). The legal limit is 9 pounds per acre (J. Hill,
correspondence, 1987). The Lundbergs fallow each rice field once every 3 to
5 years. During the first year after fallow, they find that they can sometimes
omit the herbicide application entirely with no appreciable weed damage to
the rice crop. The success of the fallow method of weed control varies from
field to field, however, and with different weather conditions. In some years
the weed populations are rather high, forcing a choice between reduced
yields and herbicide application. On their experimental fields, the Lundbergs take the lower yields; on their other fields, they apply a reduced rate
of molinate and take yields comparable to those of other growers.

The experimental method of rice production currently practiced by the
Lundbergs has the advantages of breaking the reproductive cycle of various
weeds and other pests and pathogens and dramatically reducing (to nil) the
use of pesticides. It has the disadvantage of significantly lowering yields
and economic returns, even in comparison with statewide averages that
have been adjusted for the impact of rotation (Table 3).

Labor

The Lundberg farming operation employs the equivalent of 6.5 year-
round, full-time people, as well as 7 seasonal workers for 8 weeks in the
spring and 6 weeks in the fall. The labor required for the alternative rice
operation is somewhat greater than that required for conventional rice
growing because of the repeated irrigation and surface tillage to control
weeds during the fallow year. As stated earlier the number of cycles of
irrigation and surface tillage varies from time to time and from one field to
another, depending on weather conditions and weed populations.

Soil Fertility

Until recent years the Lundbergs applied chicken manure to their experimental fields. To reduce costs the Lundbergs now rely on legumes as the source of nitrogen. The field operations manager (G. Brewster, interview, 1986) attributes the current low yield in the experimental field to a lack of nitrogen. Alternative fertility management practices, including the use of a combination of purple vetch (*Vicia benghalensis*) and bell beans (*Vicia faba*), are being explored.

In the 1940s, 1950s, and 1960s, vetch was used extensively in Butte County as a green manure crop. The use of vetch was discontinued because of the availability of inexpensive inorganic fertilizers and because, in wet winters, patches of the vetch would drown out, causing irregularities in the uniformity of field nitrogen distribution. Areas lacking in nitrogen had to be spot-treated, causing lodging (and poor yields) in areas in which overlaps occurred and poor yields in areas that did not get spot-treated. Today, with laser leveling, semidwarf varieties of rice, and improved drainage, former problems with using legumes as a nitrogen source might be more easily overcome (J. Hill, correspondence, 1987).

Insect Control

The development of a rice crop proceeds through four phases: (1) the seedling stage, from germination until the initiation of tillering; (2) the vegetative stage, from the onset of tillering until the beginning of panicle formation; (3) the flowering stage, from panicle initiation through fertilization of the rice flowers; and (4) the ripening stage, from flower fertilization until the rice is mature and ready for harvest. The duration of these phases and the severity of the pest problems that may accompany them depend on the choice of cultivar; the temperature of the soil, air, and irrigation water; the length of the growing day; and other environmental conditions and cultural practices (Flint, 1983).

Gordon Brewster, the manager of the Lundberg Family Farms field operations, carefully scouts all of the farm's fields on a continuous basis throughout the growing season. Before working for the Lundbergs, Brewster was a researcher with Occidental Petroleum, where he was in charge of developing agricultural chemicals for rice production. He uses the latest chemical technology for pest control on the conventional acreage, but he uses only those methods officially approved by state law as organic on the experimental fields.

Disease Control

The major disease afflicting rice in northern California is stem rot, a fungal disease. The causal organism, *Magnaporthe salvinii*, is best known in its sclerotial stage as *Sclerotium oryzae* (Webster et al., 1981).

All cultivars of rice currently being grown in California are susceptible to stem rot fungus, although some cultivars exhibit some degree of tolerance. The first sign of the disease in the field is the appearance of small, dark lesions on the rice stem (or culm) at the water level. The lesions expand as the season progresses, eventually destroying the sheaths. The adverse effects of the disease are a reduction in the size of the panicle (the number of rice seeds per panicle), a reduction in grain quality, and increased incidence of lodging (rice plants bending horizontally rather than standing straight, making harvesting difficult and causing the loss of grain).

The inoculum of stem rot is carried over from one year to the next in sclerotia (compacted masses of fungus mycelium that serve as the dormant stage of the fungus), which are associated with rice straw from the previous year. The principal method of controlling stem rot is burning the rice straw following harvest to achieve total removal of the crop residue and any stem rot sclerotia. D. Mikkelsen (interview, 1986) estimates that rice straw is burned on approximately 95 percent of all California rice acreage, either in the fall following harvest or in the spring. This practice causes severe air pollution and is currently controlled by law in California.

Straw burning is the recommended disposal and stem rot control method, provided that it is done only during designated times and by prescribed methods (Flint, 1983). Incorporation of rice straw is not recommended for managing stem rot (Flint, 1983). Mikkelsen, however, says that the practice now recommended by the University of California Cooperative Extension as an alternative to burning is to chop the straw (either with an attachment to the harvest combine or as a separate operation with a flail chopper) so as to maximize the contact of the rice with the soil and moisture, thereby expediting the decomposition of the straw and the sclerotia (interview, 1986).

The Lundbergs maintain that stem rot is not a severe problem in their fields because of the methods they use to expedite the decomposition of the straw and because they subsequently incorporate it into the soil. However, this claim has not been tested experimentally.

The incidence of stem rot is affected by a number of cultural practices, most notably the destruction of sclerotia by burning or decomposition. Stem rot is more serious in dense stands than in more sparse stands of rice. Consequently, high seeding rates and excessive nitrogen application (which promotes more extensive growth of foliage) tend to increase stem rot damage. Improperly timed applications of herbicides—in particular, MCPA—late in the season and at high rates of application tend to injure and stress the rice plants, predisposing them to stem rot disease. The application of MCPA is recommended no later than the first 55 days after planting to provide the best control of weeds and to reduce the risk of phytotoxicity or chemical damage to the rice plant (Flint, 1983).

Webster et al. (1981) conducted experiments in Butte and Yolo counties on alternative methods of managing rice straw residue, depending on the severity of stem rot disease. The treatments included burning the straw in the fall, followed by disking, and five management practices that did not involve burning. These five methods involved various tillage practices intended to incorporate all or part of the straw into the soil. The results of the experiment indicated that burning the straw in the fall, followed by disking, was the most effective method of controlling stem rot. This method resulted in significantly lower numbers of viable sclerotia per gram of soil, lower disease severity ratings, and somewhat higher yields of rice. The treatments included in this research did not, however, include the methods currently employed by the Lundbergs.

Other fungal pathogens that cause diseases afflicting rice in this area include *Achyla klebsiana* and *Pythium* (causing seed rot), the *Rhizoctonia* species (causing sheath blight), and *Helminthosporium oryzae* (producing brown leafspot). These diseases appear to be less of a problem in California than in other regions of the United States and in humid areas of other countries; they cause less damage in California than does stem rot (Flint, 1983).

Control of Tadpole Shrimp and Insects

The tadpole shrimp, a hardy crustacean, reaches a maximum length of 3 inches. It is able to survive the dormant stage for many years in dry soil and to revive quickly as soon as the soil is irrigated. About 9 days after hatching, tadpole shrimp begin their reproductive phase by digging into the soil, uprooting new rice seedlings, or cutting off new leaves. The muddy water caused by the digging reduces light penetration and slows the emergence of rice seedlings. Although low populations of tadpole shrimp do not cause economic damage, high populations have been known to greatly diminish rice stands and reduce yields.

Conventional practice is to control tadpole shrimp by irrigation management or application of pesticides. Growers are advised to flood the field as fast as possible, and seed as soon as possible after flooding has been initiated. During the seedling stage, when populations of tadpole shrimp are found to be above economic damage thresholds (30 or more dislodged seedlings per square foot), a chemical treatment is needed (Flint, 1983). Parathion can be applied at 0.1 pint per acre at a cost of $2.17 per acre (Table 4). Alternatively, if algae are becoming a problem, both tadpole shrimp and algae can be controlled simultaneously by application of finely ground copper sulfate (5 pounds per acre) at a cost of $4.02 per acre. These costs include $1.90 per acre for aerial application (Wick and Klonsky, 1984).

In their experimental field the Lundbergs prevent damage by tadpole shrimp through intermittent irrigation during the early stages of rice growth, a process that delays the anaerobic stage of irrigation (perpetual flooding) until after the rice plants have reached a height of 6 to 8 inches. At this stage, the tadpole shrimp do not cause injury to the rice (Flint, 1983).

Rice water weevil larvae feed on the roots of rice plants, causing loss of yield by inhibiting growth, tillering, and plant vigor (Flint, 1983). The wee-

TABLE 4 Preharvest Cultural Cost of Producing an Acre of Rice Using Conventional Methods Versus Lundberg Family Farms' 1985 Experimental Alternative Methods

Operation	Labor/Acre Hours	Labor/Acre Dollars	Fuel and Repairs (dollars)	Materials	Dollars/Acre	Hypothetical Conventional Rice Farm (percent acres treated)[a]	Methods (dollars) Conventional	Organic
Rice seedbed preparation								
Chisel	0.16	1.15	4.79			100	5.94	5.94
Stubble disk plow	0.16	1.15	5.65			100	6.80	0
Disk harrow (2 times)	0.32	2.30	11.30			100	13.60	0
Flail-chop vetch						0	0	2.00
Subtotal							26.34	7.94
Fertility management								
Preplant fertilizer				16-20-0 (NPK): 250 pounds at $200.00/ton	25.00	100	25.00	0
				Custom air: $2.10/hundredweight	5.25	100	5.25	0
				Nitrogen aqua ammonia: 100 pounds at $70.00/ton	17.50	100	17.50	0
				Custom ground: $10.00/acre	10.00	100	10.00	0
				Ammonium sulfate: 500 pounds at $97.00/ton	24.25			
				Custom air: $2.10/hundredweight	10.50	0	0	0
				Urea: 220 pounds at $200.00/ton	21.70	0	0	0
				Custom air: $2.70/hundredweight	5.94	0	0	0
				Zinc sulfate: 50 pounds at $200.00/ton[a]	5.00	40	2.00	0
				Custom air: $3.00/hundredweight	1.50	40	0.60	0
Plant purple vetch				Seed: 100 pounds/acre at $11.00/hundredweight	11.00	0	0	11.00
				Aerial planting	3.00	0	0	3.00
Topdress postplant fertilizer				Ammonium sulfate: 150 pounds at $97.00/ton[b]	7.28	60	4.37	0
				Custom air: $2.10/hundredweight	3.15	60	1.89	0
Subtotal							66.61	14.00

	Conventional			Organic
Pest control				
Rice water weevil				
Carbofuran 5G: 10 pounds at $33.53/50 pounds	6.71	60	4.02	0
Custom air: $3.00/acre	3.00	60	1.80	0
Rice leafminer and/or tadpole shrimp				
Parathion: 0.1 pint at $108.00/5 gallons	0.27	100	0.27	0
Custom air: $1.90/acre	1.90	100	1.90	0
Barnyard grass (watergrass)				
Molinate 10 G: 40 pounds at $26.00/50 pounds	20.80	90	18.72	0
Custom air: $3.60/acre	3.60	90	3.24	0
Broadleaf weed				
MCPA: 14 ounces active ingredient at $69.90/5 gallons formulated product	3.05	85	2.59	0
Custom air: $4.20/acre (10 gallons of water)	4.20	85	3.57	0
Blackbirds, muskrats, waterfowl, and Norway rats	5.00	100	5.00	0
Subtotal			41.11	0
Other preharvest costs				
Close and maintain levees/boxes	3.08	100	3.08	3.08
Flood	17.22	100	23.84	20.97
Plant				
Seed: 135 pounds at $11.00/hundredweight	14.85	100	14.85	14.85
Custom treat and soak: $2.00/hundredweight	2.70	100	2.70	0
Custom haul seed: $0.22/hundredweight	0.30	100	0.30	0.30
Custom air: $4.15/hundredweight	5.60	100	5.60	0
Custom no-tillage drill	10.00	0	0	10.00
Subtotal			50.37	49.20
Total, preharvest cultural costs per acre of rice			184.43	71.14

[Levee detail values shown in the source alongside "Close and maintain levees/boxes": 0.10, 0.72, 2.36; 0.75, 5.40, 1.22]

[a]Because zinc is now adequate in most soils, it is not applied every year.

[b]Not all growers topdress. Some use 20 to 30 pounds more of aqua ammonia fertilizer before planting.

[c]With good water management, as little as 4 acre-feet are used for continuous flooding of rice.

[d]Practices vary greatly among conventional growers. Percentages *will change* from grower to grower.

SOURCES: Conventional rice inputs and costs from C. M. Wick and K. Klonsky, Sample Costs of Rice Production, Butte County (Davis, Calif.: University of California, 1984); organic rice inputs and costs from Gordon Brewster, field manager, Lundberg Family Farms, interview and correspondence, August 1986.

vil is more prevalent in fields where rice is grown continuously (without interruption of a fallow year or a different crop) and in areas where rice is the prevalent crop. Crop rotation, therefore, is a recommended practice for preventing weevil damage. Where weevil populations are high, preflooding application of a pesticide (10 pounds of carbofuran at $9.71 per acre, including aerial application cost) is recommended (Wick and Klonsky, 1984).

Rice leafminer larvae burrow into rice leaves located under or near the surface of water, thereby reducing vegetative development and, consequently, yields. Preventive measures include shallow irrigation in the germination stage, especially when temperatures are low and plant growth is slow. Parasitic wasps and high temperatures (promoting rapid growth above the water line) are natural control factors. The recommended insecticide is parathion (0.1 pint at $2.17 per acre), which also controls tadpole shrimp (Wick and Klonsky, 1984).

Leafhopper, armyworm, and grasshopper ordinarily do not cause significant damage to rice crops in the area of California in which the Lundberg's farm is located.

Mosquitoes are a nuisance to human populations near rice fields and sometimes are vectors of disease. Local mosquito abatement officials, who spray heavily infested areas with ethyl parathion, reported that they refrain from applying pesticides near fields where the farmer is producing crops without pesticides, including the Lundbergs' 100-acre experimental field. Mosquito Abatement Districts in many areas now use *Gambusia* (mosquito fish) in irrigation canals as a biological control measure, which is highly effective in open water where predator fish are not prevalent and water quality is favorable.

Weed Control

The control of weeds is a major concern for rice growers. Watergrass (*Echinochloa phyllopogon* and *E. oryzoides*) can be controlled largely through continuous flooding to a depth of 3 to 4 inches for 21 to 28 days after planting. Although continuous flooding provides good control of most terrestrial weeds, it tends to encourage various aquatic plants such as certain grasses, sedges, bulrushes, arrowheads, waterhyssop, water plaintain, pondweeds, and algae. The Lundbergs rely largely on crop rotations (rice-vetch and fallow or rice-rice-vetch and fallow) for weed control. Crop rotation is recommended for cases of severe infestations (Flint, 1983).

Watergrass becomes a particularly severe problem where the water depth is maintained at less than 3 to 4 inches. For this reason, it is important that the rice field be leveled every few years (sometimes annually), because sometimes the soil settles in some areas or is rutted by harvesting machines or the wind when the soil is fall-tilled and left bare.

Application of herbicides is the conventional method of weed control. The Lundbergs use molinate, one of the most widely used herbicides in the area, as necessary to control weeds in their conventional rice. During the site visit to their farm, one field of 320 acres was observed in which no herbicide had been applied. Although rice plant and weed plant populations were not counted, it appeared that the stand of rice was quite heavy and uniform; only an occasional weed was evident. The Lundbergs attributed effective weed control to their rotation sequence and cultural practices during the third (fallow) year of the rotation used in their conventional rice. They applied only 32 pounds of nitrogen per acre to this field, which yielded approximately 70 hundredweight per acre in 1986, compared with a county average of 74 hundredweight in 1985. County average yield data for 1986 were not available when this report was prepared.

Irrigation

The conventional method of producing rice includes flooding the rice field continuously from before planting (May 15 to June 1) until the rice is mature. The soil is then drained and dried enough to support harvesting equipment prior to harvest in October or November. The direct seeding and intermittent flooding (flushing) that the Lundbergs practiced during stand establishment of their organically produced rice can increase the time between planting and harvest by 7 to 14 days (J. Hill, communication, 1986). Rice varieties that came into use after 1981 have somewhat shortened growing season requirements, which are compatible with the Lundbergs' practices.

The water must be relatively warm, preferably about 70 to 75°F. Water temperatures above 90°F or below 60°F are detrimental to rice growth. Cold water during the growth stage seriously retards seedling and stand development. Slow-growing seedlings are vulnerable to various pests; weeds are also more problematic when the rice stand is sparse. Where water is cold, rice growers sometimes allow the irrigation water to stand in warming basins before it flows into the rice fields in order to prevent a reduction (by as much as 45 percent) in rice yields (Miller et al., 1980; Flint, 1983). Well water temperatures usually fall in the 66 to 76°F range. Water diverted from the nearby Oroville Dam on the Feather River to the Lundberg farm is below 55°F until May 15 and below 63°F in midsummer (Miller et al., 1980).

The ideal water depth depends on the developmental stage of rice. Shallow water, 1 to 4 inches, favors stand establishment and tiller development, particularly when the short statured varieties of rice are grown. As the rice grows taller, deep water is preferred for controlling various terrestrial weeds (most notably watergrass, the most serious weed in rice production in California) and discouraging growth in rodent populations.

From the completion of tillering (about 60 days after planting) until 3 weeks before heading (development of panicles), water depth has little effect on rice plant development. However, from 3 weeks before heading until the panicles are developed, water depth is very important, particularly in areas subject to cool night temperatures. Empty florets increase when rice plants are subjected to cool temperatures, and rice yields are greatly reduced. When irrigation water is kept relatively deep at this time, panicle development is protected from the low ambient air temperatures that are most likely to occur at night (Miller et al., 1980).

Typically, between 5 and 9 acre-feet of water are delivered to a rice field. Most of this water moves through the field and is reused in the network of rice irrigation districts, eventually being returned to public waterways. The crop requires about 3 acre-feet, including what is lost through evapotranspiration. Rice uses about 10 percent more water than alfalfa (D. Mikkelsen, interview, 1986). The Lundbergs apply an average of 5 acre-feet of water.

The Lundbergs' intermittent flush-irrigation practices can delay harvest by 7 to 14 days (J. Hill, correspondence, 1986). However, with currently used rice cultivars, this delay is problematic.

Flooded rice fields where green manure crops or straw have been incorporated occasionally exhibit a buildup of organic acids (lactic, butyric, acetic, and propionic) in the soil. These acids later break down into carbon dioxide, which can (if present in excessive quantities) inhibit plant respiration and uptake of water and nutrients (D. Mikkelsen, interview, 1986). This problem tends to occur when large quantities of straw or other plant materials are buried deeply in the soil and subjected to anaerobic decomposition. The toxic gases usually develop during the first 20 days. When this problem occurs, the fields may have to be drained and dried out to aerate the soil to deactivate the production of phytotoxic hydrogen sulfide (Miller et al., 1980). Toxic gas production by rice fields is not considered an environmental threat to air quality (D. Mikkelsen, interview, 1986).

PERFORMANCE INDICATORS

Rice Yields

The Lundbergs have continued to experiment with various nonchemical approaches to rice production. Their experimental method of production has changed substantially each year. During the 1986 case study site visit, they reported that the experimental rice enterprise became profitable for the first time in 1985, with a yield of 44 hundredweight per acre. However, in 1986 the yield dropped to 27 hundredweight (see Table 3), and the experimental crop sustained a financial loss. Furthermore, the yield of the experimental rice is obtained only every other year because of the 2-year rotation (rice and legume-fallow-legume); the annual average yield, therefore, is one-half the measured yield in a given year.*

Most conventional rice growers use a rotation with 1 year of fallow and 2 years of rice production because of federal price-support program rules. A

---

*Prior to price-support program changes in 1981, a more intensive rotation (5 in 6 years) was common. On clay soils, no alternative crop is ordinarily grown in the non-rice year. On lighter soils, rice is rotated with a cash crop (wheat, safflower, and others) (J. Hill, correspondence, 1987).

grower receives a de facto yield of two-thirds the average production per acre harvested. The average yield of rice grown in Butte County during 1985 was 74 hundredweight per acre harvested (see Table 3). The season average price was $7.90 per hundredweight. In contrast, the Lundbergs paid an average of $11.75 per hundredweight to their 10 contract growers using nonchemical methods (G. Brewster, interview, 1986).

Yield data for these farms are not available. However, during the 1986 site visit interview, the Lundbergs indicated that the yields and net returns from their experimental fields and those of their contract farmers vary considerably. For example, they reported a yield of 69 hundredweight from one of their organic contract farms that uses a 5-year rotation: 1 year of no-tillage rice followed by 1 year of legume-fallow and leveling followed by 3 years of oats and vetch harvested as either hay or seed (depending on prices).

Financial Performance

Over a period of years the average annual yield of rice grown under the Lundbergs' alternative system is substantially less than that of conventional rice. The question remains, however, whether the reduced yield is more than offset by the higher price received for certified organic rice and the lower production costs that appear to be possible, at least in some years. The Lundbergs follow a budgeting approach based on the University of California Farm Management Extension enterprise budgets (Wick and Klonsky, 1984) in analyzing the economic approach of their farm operations. Brewster was asked to examine the 1985 University of California rice budget and to indicate the comparable costs incurred on their experimental 100-acre field in 1985 (Tables 4 and 5).

Because the Lundbergs plant the rice seed by no tillage into the flail-chopped mulch of purple vetch, they have a cash cost of seedbed preparation only one-third that of the conventional rice growers—$7.94 per acre compared with $26.34 per acre. The soil fertility management program on the Lundbergs' experimental acreage is substantially less expensive than that on their conventional acreage because of the nitrogen and organic matter supplied by the vetch. The entire fertility management cost is $16.00 per acre, the cost of planting and flail-chopping the purple vetch. By comparison, the conventional rice fertility management program cost is $66.61.

However, as previously noted, the Lundbergs view their soil fertility on the experimental acreage as deficient in nitrogen. They are modifying their experimental method to meet more adequately the nitrogen requirements of the rice crop. Lack of nitrogen is clearly indicated as the primary factor limiting their experimental rice yields.

Another major difference between conventional and alternative rice occurs in the cost of pest control: no direct cost incurred by the Lundbergs versus $41.11 per acre for the conventional pest control program. The conventional approach includes a per acre application of 10 pounds of carbofuran for control of rice water weevil, 0.1 pints of parathion for control of

TABLE 5   Summary of Costs and Returns/Acre for Conventional Production
Versus Lundberg Family Farms' Experimental Organic Rice Production, 1985

| | Dollars/Acre | |
Item	Conventional	Organic
Direct cash costs		
Preharvest cultural costs	185.34	71.14
Harvest costs		
Drain and open levees	3.26	3.26
Custom harvest, haul, and dry: $1.81/hundredweight	134.24	79.82
Postharvest costs		
Mow levees, clean around boxes	2.01	2.01
Burning rice straw	2.45	0
Rolling rice straw or chisel	0	15.00
Total, direct cash costs	327.30	171.23
Revenue during crop years		
Conventional rice:		
74 hundredweight at $7.90/hundredweight	584.60	—
Lundberg experimental rice:		
44 hundredweight at $11.75/hundredweight	—	517.00
Net return over cash costs	257.30	345.77
Fallow year costs		
Triplane (including move crawler)	5.49	0
Roto spike: 3 times at $10.00/acre	0	30.00
Landplane (including move crawler)	0	20.48
Flush-irrigate: 3 times at $3.50/acre	0	10.50
Laser level (custom hire)	60.00	60.00
Plant purple vetch	0	14.00
Total, fallow year costs	65.49	134.98
Net return over cash cost/acre/year	149.70	105.40
Cash cost/hundredweight rice	4.86	6.96

NOTE: Conventional rotation is 2 years of rice followed by 1 year of fallow. The 1985 Lundberg
Family Farms' experimental rotation was 1 year of rice followed by 1 year of legume-fallow.

SOURCES: Conventional rice yield from C. M. Wick and K. Klonsky, Sample Costs of Rice
Production, Butte County (Davis, Calif.: University of California, 1984); organic rice yield and
price from Gordon Brewster, field manager, Lundberg Family Farms, interview and
correspondence, August 1986; state average price from California Crop and Livestock Reporting
Service (California Field Crop Review 7[2]:1).

rice leafminer and tadpole shrimp, 40 pounds of molinate for control of
barnyard grass, and 14 ounces of MCPA for control of broadleaf weeds. The
costs of these and other options are listed in Table 4 (Wick and Klonsky,
1984).

Other preharvest costs are similar, with two exceptions. First, the Lundbergs use 5 acre-feet rather than 6 acre-feet of water; the difference is
attributed to more careful management (G. Brewster, interview, 1986). Second, they plant the experimental rice by no tillage, using rice seed that has
not been soaked or treated, at a cost of $10.00 per acre for planting, in
addition to the cost of the seed. The conventional method is to soak,
partially sprout, and treat (with a fungicide such as captan) the rice seed by
aerial spraying prior to planting, at a cost of $8.30 per acre.

The overall preharvest costs total $184.43 per acre for conventional rice; for the alternative rice produced by the Lundbergs in 1985, the costs were $71.14 per acre. The conventional rice budget, however, is for a 3-year rotation (rice-rice-fallow), while the Lundberg budget is for a 2-year rotation (rice-legume and fallow legume). Consequently, when net returns are calculated on the basis of these budgets, it is necessary to transform the costs and returns per acre harvested into average figures per acre per year based on the crops in the rotation. In making these calculations, it is assumed that the conventional rice yield is 74 hundredweight per acre (county average), compared with 44 hundredweight obtained by the Lundbergs in 1985 from their experimental rice. Consequently, harvest costs (roughly proportional to yields) are substantially less for the alternative than for the conventional operation (see Table 5).

The Lundbergs do not pay to burn rice straw. They use tillage practices that expedite decomposition of the straw and incorporate it into the soil. The cost of the Lundberg approach is higher than the cost of burning rice straw—$15.00 versus $2.45 per acre harvested. The total monetary value of the nutrients retained in the field and the improved organic matter in the soil associated with decomposing rather than burning the straw is unknown. However, D. Mikkelsen (correspondence, 1987) estimates the decomposed straw reduces nitrogen fertilizer needs by about 20 percent, a potential savings of about $9.00 per acre (based on Klonsky and Wick data; see Table 4). A method to measure the additional value of organic matter and nutrients other than nitrogen has not been developed.

The Lundbergs' total direct cash costs per acre for organic rice are roughly one-half those of the average conventional producer in the area ($171.23 versus $327.30). The organic rice yield is lower than that of conventional rice, but this is offset by the higher price for organic rice. The values of the conventional and organic rice crop per acre harvested were similar ($584.60 versus $517.00). The net return over direct cash operating costs per acre of rice harvested was $257.30 for conventional rice and $345.77 for the Lundbergs' experimental alternative rice. When these net returns are adjusted for the rotation and costs of the fallow year are taken into account, however, the results are reversed: $149.70 per acre of rotation per year for the conventional rice and $105.40 for the Lundberg experimental crop in 1985. In other years, net returns were lower for the Lundbergs' organic production and higher for conventional rice, further widening the difference between the two types of rice.

Overhead and indirect costs, such as interest on operating expense, bookkeeping, depreciation, insurance, taxes, and other necessary expenses are not taken into account in the calculations for producing conventional and alternative rice. Most of these indirect costs would be approximately the same for experimental and conventional rice producers, so the per acre differences in net returns would not be significantly affected by this omission.

The Lundbergs are aware that what they call the organic rice market is rather fragile; the yield and acreage of organic rice have been increasing. By reducing the production quota for each of the contract growers, they hope to avoid a catastrophic decline in prices. For many years the Lundbergs have been able to maintain a substantial premium price for organically grown rice. For example, as of January 1986, the price of all rice in California (including an approximately $4.00 per hundredweight government program payment) averaged $7.90 per hundredweight, compared with $11.75 per hundredweight for rice certified as organic (California Crop and Livestock Reporting Service, 1986; Lundbergs, interview, 1986).

The Lundbergs reported that they were subsidizing their experimental rice production by approximately $50,000 per year in 1982 (Madden et al., 1986). The revenue from the sale of their 100 experimental acres of rice was well below expenses. At that time, they indicated their willingness to continue subsidizing their experimental enterprise because they hoped that it would become profitable. They were willing to make this sacrifice because they were concerned about the health implications of pesticide use. The Lundbergs say that they are committed to developing profitable cultural practices that minimize environmental damage and residues of agricultural chemicals on the food they produce and market.

Environmental Impacts

Production of rice by conventional, recommended practices gives rise to several environmental concerns—notably, water pollution caused by pesticides and air pollution created by burning rice straw. The Lundbergs do not burn rice straw on any of their acreage.

According to the University of California manual for IPM for rice (1983), some of the pesticides used in rice are hazardous to people. The person most at risk is the applicator; other people who spend time in the fields (field workers and irrigators) may also be exposed. People in surrounding areas may suffer pesticide poisoning when sprays drift from the field into populated locales.

It is also important to consider the hazard pesticides may have for fish, wildfowl, and domestic animals, especially sheep grazing on levees. Migrating waterfowl may die if they are in the fields during application of various insecticides. Fish may die if pesticide-contaminated water from paddies or soak water drains into streams and bodies of water flowing into streams or rivers.

Cumulative levels of certain pesticides draining from Sacramento Valley rice fields into the Sacramento River have caused concern about drinking water quality and taste and health. In the Sacramento area and other locations, agricultural pesticide concentrations in water are high enough (in the parts per billion range) for short periods of time to cause an offensive taste; however, the health implications are unclear. These problems can be mitigated to some extent by water recirculation systems now in common use that allow for decomposition of the pesticides before water is let out of the fields.

An emerging problem in conventional rice production is the development of resistance to pesticides among strains of various pests. In some areas, tadpole shrimp are resistant to parathion, and mosquitoes that breed in rice fields have been found to be resistant to particular insecticides.

Using pesticides can also induce emergence of secondary pests (National Research Council, 1986). When substantial numbers of the natural predators and parasites are killed because of pesticide use, certain secondary pest populations may begin to rise.

REFERENCES

California Crop and Livestock Reporting Service. 1986. California Field Crop Review 7(February 26):1.

Flint, M. L. 1983. Integrated Pest Management for Rice. Publication No. 3280. Berkeley, Calif.: Division of Agricultural Sciences, University of California.

Madden, P., S. Dabbert, and J. Domanico. 1986. Regenerative Agriculture: Concepts and Selected Case Studies. Staff Paper No. 111. University Park, Pa.: Department of Agricultural Economics and Rural Sociology, The Pennsylvania State University.

Miller, M. D., D. W. Henderson, M. L. Peterson, D. M. Brandon, C. M. Wick, and L. J. Booher. 1980. Rice Irrigation. Leaflet 21175. Berkeley, Calif.: Division of Agricultural Sciences, University of California.

National Research Council. 1986. Pesticide Resistance: Strategies and Tactics for Management. Washington, D.C.: National Academy Press.

Webster, R. K., C. M. Wick, D. M. Brandon, D. H. Hall, and J. Bolstad. 1981. Epidemiology of stem rot disease of rice: Effects of burning versus soil incorporation of rice residue. Agriculture and Natural Resource Publications, University of California. Hilgardia 49(3)February:1–2.

Wick, C. M. 1975. Cash Costs of Total Rice Residue Incorporation into the Soil. Rice Review. Oroville, Calif.: Butte County Agricultural Extension Service.

Wick, C. M., and K. Klonsky. 1984. Sample Costs of Rice Production, Butte County. Davis, Calif.: Cooperative Extension Service, University of California.

Glossary

Acre-foot The volume of irrigation water that would cover one acre to a depth of one foot.

Agricultural resource base The soil, water, climate, and other natural resources necessary to produce a crop.

Agricultural Stabilization and Conservation Service (ASCS) A U.S. Department of Agriculture (USDA) agency responsible for administering farm price and income support programs as well as some conservation and forestry cost-sharing programs; local offices are maintained in nearly all farming counties.

Allelopathy The suppression of the growth of one plant species by another.

Band application A method of applying fertilizer in bands near plant rows where the fertilizer will be more efficiently used rather than applying it in an application to the entire soil surface.

Base acres The acres on a farm that are eligible for federal program payments. Base acres for each year are calculated as the average number of acres enrolled in a specific commodity program during the previous 5 years.

Biomass Matter of biological origin; for example, the living and decaying matter in soil as opposed to the inorganic mineral components.

Bureau of Reclamation A federal agency responsible for building dams and canals and providing water to local water districts. The districts then sell the water to agricultural producers.

Cash grains Grains commonly produced for sale, such as corn, oats, and wheat, as opposed to hay and other grains that are grown principally as feed for animals or seed.

Center pivot irrigation   An irrigation system that pumps groundwater from a well in the center of a field through a long pipe, elevated on wheels, that pivots around the well and irrigates the field in a large circular pattern.

Commodity Credit Corporation (CCC)   A wholly owned government corporation created to stabilize, support, and protect farm income and commodity prices. The CCC and the ASCS administer the federal farm programs.

Commodity price and income support programs   Federal programs designed to support crop prices and farm income. These programs include all commodity-specific programs (such as the corn program) under which commodity price support levels are established, set asides are determined, direct payments and nonrecourse crop loans are made to farmers, and agricultural land is diverted from production through paid land diversions and other provisions.

Conservation Reserve Program (CRP)   A program authorized under the Food Security Act of 1985 that allows up to 45 million acres of highly erodible land to be placed into a 10-year reserve. Land in the reserve must be under grass or tree cover to protect it from erosion. It is not allowed to be used for hay production or livestock grazing.

Cover crop   A crop grown for its value as ground cover to reduce soil erosion, retain soil moisture, provide nitrogen for subsequent crops, control pests, improve soil texture, increase organic matter, or comply with erosion control requirements of federal commodity programs. Commonly used cover crops include the clovers, vetch, alfalfa, and rye.

Crop residues   The remains of crop plants after harvest. Residues are frequently left in fields to supply organic matter to the soil and help cover the soil surface, which reduces erosion losses.

Crop rotation   The successive planting of different crops in the same field over a period of years. Farmers using rotations typically plant a part of their land to each crop in the rotation. A common 4-year rotation is corn-soybeans-oats-alfalfa.

Crop yield   The amount of a crop harvested, commonly expressed in bushels or other units per acre.

Cross-compliance   A provision of the Food Security Act of 1985 designed to control the expansion of a farmer's base acres and limit federal payments for the production of program crops. In general, cross-compliance stipulates that to receive any benefits from an established crop acreage base, the farmer may not exceed his or her acreage base for any other program crop.

Cultivation   To mechanically loosen or break up soil between the rows of growing crops, uproot weeds, and aerate the soil. Soil around crops is generally cultivated one to three times per season, depending on soil type, weather, weed pressure, and herbicide use.

Cultural pest control   Pest control practices that generally refer to physical or mechanical changes in an agricultural method. These may include clearing crop residue soon after harvest, crop rotations, clearing weeds from the field borders, changes in irrigation, or altering the timing or way of planting.

Deficiency payment The per unit of production (bushel or pound) payment that is made directly to producers enrolled in the commodity programs. It is usually calculated as the difference between the target price and the loan rate or market price, whichever is higher.

Denitrification The bacterial reduction of nitrate to nitrogen gas (NO_2), nitrous oxide (N_2O), and nitric oxide (NO). Denitrification occurs under anaerobic conditions and results in loss of available nitrogen from the soil.

Direct payments Payments made by the federal government to agricultural producers enrolled in commodity programs. A deficiency payment is the most common form of a direct payment. Deficiency payments can be made in cash or in certificates entitling the producer to receive an equivalent cash value of crops from the CCC based on the current loan rate.

Diversion payments A per acre payment available in certain years as an option to producers enrolled in commodity programs who divert land from production of a program crop in addition to the acreage required by the set-aside provisions of a specific commodity program.

Eutrophication The process by which a body of water becomes rich in nutrients. This can happen naturally or by human activity, usually in the form of industrial or municipal wastewater or agricultural runoff.

Farmers Home Administration (FmHA) The USDA agency that makes loans to farmers and homeowners. The FmHA is generally the farmer's lender of last resort.

Federal crop insurance A federally subsidized crop insurance program available to farmers to protect them against unavoidable crop losses caused by drought, fire, hail, floods, and other natural disasters.

Feed grains Grains such as corn, barley, oats, and sorghum that are commonly fed to animals. Many feed grains are also consumed by people.

Fixed costs Costs of production that generally do not change as a result of the volume or type of crop produced. Fixed costs include insurance, rent or land mortgage payments, interest, and machinery depreciation.

Gene transfer The process of moving a gene from one organism to another. Current biotechnology methods permit the identification, isolation, and transfer of individual genes as a molecule of DNA. These methods make it possible to transfer genes between organisms that would not normally be able to exchange them.

Government farm program outlays Total costs associated with federal commodity price and income support, storage, disaster, and related programs. In the case of storage payments, outlays involve payments to producers and grain handlers.

Green manure The use of leguminous crops as a source of nitrogen when they are plowed into a field.

Highly erodible land   Land that has an erodibility index of greater than 8. This index is based on a field's inherent tendency to erode from rain or wind in the absence of cover crop. The erodibility index is based on the Universal Soil Loss Equation (USLE) and the Wind Erosion Equation (WEE), along with a soil's T-value, which is a measure of the amount of erosion in tons per year that a soil can tolerate without losing productivity. For most cropland soils, T values fall in the range of 3 to 5 tons per acre.

Horizontal resistance   A plant's ability to uniformly resist all strains of a pathogen. Many different physiological and morphological traits that act independently or jointly to block the virulence of a pest determine such resistance. These traits are nonspecific and polygenic, in contrast to vertical or specific gene resistances that may be overcome by a mutated strain of the pathogen.

Inputs   Items purchased to carry out a farm's operation. Such items include fertilizers, pesticides, seed, fuel, and animal feeds and drugs.

Integrated pest management (IPM)   A pest control strategy based on the determination of an economic threshold that indicates when a pest population is approaching the level at which control measures are necessary to prevent a decline in net returns. In principle, IPM is an ecologically based strategy that relies on natural mortality factors, such as natural enemies, weather, and crop management, and seeks control tactics that disrupt these factors as little as possible.

Intercropping   The planting of one crop into another crop, either between the rows or into the stubble of a previous crop.

Living mulch   An understory of vegetation that helps reduce soil erosion and adds organic matter to the soil, but which does not compete heavily with the crop for water and nutrients.

Loan rate   The commodity-specific dollar amount per unit of production (bushel or pound) that the CCC uses in making nonrecourse loans to producers. The loan rate is also known as the price support level. A major change in the Food Security Act of 1985 was to adjust the loan rates for each commodity between 75 and 85 percent of the average prices received by farmers for the previous 5 years, excluding the high and the low years. When the market price falls below the loan rate, producers who have taken out nonrecourse loans may turn over their crop to the CCC as repayment of the loan. The government accumulates stocks of commodities in this way.

Low intensity animal production   Systems of animal rearing for food products that strive to use less capital, energy, and fewer purchased inputs than conventional confinement systems. An example of a low intensity system is a pasture and hutch swine production system.

Marketing order   A voluntary agreement among a majority of the producers of a commodity that must be approved by the USDA and follow certain guidelines. It generally applies to fruit and vegetable producers and is primarily designed to control supply and price by setting acreage limits, marketing quotas, and grading standards.

Method A systematic way to accomplish a specific farming objective by integrating a number of practices. Examples include weed control, tillage, or soil fertility methods that generally entail the integration of a number of practices such as rotary hoeing, cultivation, manure spreading, and crop rotations.

Multiline Crop seed composed of a mixture of several breeding lines of the same variety, each containing a different resistance gene to a specific pest. A single breeding line is vulnerable to crop failure when a pathogen mutates and regains virulence over the single resistance gene. In contrast, the mixing of different resistance traits in a multiline greatly reduces the probability of crop damage.

Net farm income The sum of all income minus expenses from the farm operation, which includes maintenance and depreciation of all buildings, machinery, and dwellings located on the farm. To derive this figure, gross income (income before expenses) is adjusted to account for net quantity changes in inventory and year-to-year carryover.

Nitrogen fixation The chemical transformation of atmospheric nitrogen (N_2) into forms available to plants for growth. Certain species of symbiotic and free-living bacteria can accomplish nitrogen fixation. The more efficient forms are symbiotic with plants, where a food supply and a protected environment are provided to the bacteria within root nodules. The bacteria in turn supply fixed nitrogen to the plant. Strains of the genus *Rhizobium* are the symbiotic, nitrogen-fixing bacteria that associate with leguminous crops such as beans, clover, and alfalfa. Seeds of leguminous crops are often inoculated with a slurry of *Rhizobium* spores to promote nitrogen fixation by the crop.

Nonpoint water pollution Pollution of water that does not enter waterways from a specific "point" source, such as a pipe. Nonpoint pollutants are often carried from dispersed, diverse sources into water channels by rain-induced runoff. Runoff from streets, open pit and strip mines, and agricultural fields are prominent examples.

Nonrecourse loan Participants in federal commodity programs may obtain loans from the CCC by pledging planted or stored crops as collateral. These loans enable producers to pay for planting costs or to store crops for later sale. The producer can settle the loan by paying it back with interest or by turning the stored crop over to the CCC when the loan period ends. Loans are generally paid off when market prices rise above loan rates. Crops are frequently forfeited to the CCC at the end of the loan period when market prices are below loan rates.

Organic matter Living biota present in the soil or the decaying or decayed remains of animals or plants. The living organic matter in the soil decomposes the dead organic matter. Organic matter in soil increases moisture and soluble nutrient retention, cation exchange, and water infiltration and can reduce soil erosion.

Output A marketable product of a farming operation, such as cash crops, livestock products, or breeding stock.

Pesticides Chemicals used by farmers to control pests such as weeds (herbicides), insects (insecticides), plant diseases (fungicides), nematodes (nematicides); to regulate plant growth; or to simplify harvest (dessicants).

Polyculture The growing of many crops at once in the same field.

Practice A way of carrying out a discrete farming task such as a tillage operation, particular pesticide application technology, or single conservation practice. Most important farming operations—preparing a seedbed, controlling weeds and erosion, or maintaining fertility—require a combination of practices, or a method. Most farming operations can be carried out by different methods, each of which is a unique combination of different practices.

Recharge The replenishment of an aquifer with water from the land's surface.

Ridge tillage A type of soil-conserving tillage where the soil is formed into ridges, and seeds are planted on the tops of the ridges. The soil and crop residues between the rows remain largely undisturbed during planting.

Rotary hoe A tool pulled behind a tractor, designed to control weeds by dislodging weed seedlings at a very early stage of growth from the soil.

Row crops Crops that require planting each year and are grown in rows, such as corn, soybeans, and sorghum.

Scouting The inspection of a field for pests (insects, weeds, or pathogens). Scouting is a basic component of IPM systems. It is used to determine whether pest populations have reached levels that warrant intervention for control and to help determine the appropriate method of control.

Set aside The percentage of a commodity program acreage base that must be idled in a given year. The purpose is to help reduce commodity supplies and limit the cost of farm programs. This idled land must meet federal requirements for weed and erosion control. If these requirements are not met, the farmer loses eligibility for program payments and loans.

Small grains Crops with small kernels, such as wheat, barley, oats, rice, and rye.

Soil depth profile A vertical profile of distinct zones within a soil, called soil horizons. The top, or A horizon, is the zone of leaching (eluviation) and is most abundant in biomass composed of roots, bacteria, fungi, worms, and microscopic animals. The second, or B horizon, is the zone of accumulation (illuviation); it contains little living matter and is often richer in clays, iron, and aluminum oxides that have percolated down and accumulated from the A horizon. The C horizon is composed of the weathered rock and true parent material of the soil.

Specialty (high-value) crops Crops with a limited number of producers and demand or those with high per acre production costs and value. Examples include most fruit and vegetable crops, ornamentals, greenhouse crops, spices, and low volume crops, such as artichokes.

Split application Breaking up the application of fertilizer into two or more applications throughout the growing season. Split applications are intended to supply nutrients more evenly and at times when the crop can most effectively use them.

Storage payments Annual payments per bushel or by weight made to individuals and corporations for the storage of commodities in the Farmer Held Reserve or placed under loan to the CCC.

Strip cropping A method of contour planting in conjunction with rotations that results in alternating strips of crops across the slopes of fields. When practiced with conservation tillage, strip cropping is an important and highly effective erosion control method.

System The overall approach used in crop or livestock production, often derived from a farmer's goals, values, knowledge, available technologies, and economic opportunities. A farming system influences the choice of methods and practices used to produce a crop or care for animals. Farming systems entail a combination of methods to accomplish farming operations. Conventional and alternative systems may use common practices or methods, but they usually differ in overall philosophy.

Systemic pesticide A pesticide that is absorbed within a plant system and distributed throughout the plant and fruit.

Systems research Interdisciplinary research that integrates knowledge from several fields of study into research projects designed to generate knowledge and understanding of farming systems.

Target price A commodity-specific price per unit of production (bushel or pound) for certain program commodities that is set by the Congress and administered by the USDA. Target prices are usually above market prices. They are used to determine deficiency payments.

Understory Vegetation growing in the shade of taller plants.

Universal Soil Loss Equation A = RKLSCP, where A is the computed soil loss per unit area over a specified time; it is usually expressed as tons/acre/year. The factors R, K, and S reflect characteristics of climate and land that generally cannot be modified by human activity designed to influence erosion rates: amount and intensity of rainfall (R), soil erodibility (K), and steepness of field slope (S). The factor representing length of slope (L) can be reduced by installing terraces, which effectively break the naturally occurring slope length into smaller segments. (The effective slope length can also be shortened by strip cropping and grassy waterways, but this is reflected in the P factor.) The remaining two factors reflect the effects of human activities on erosion rates: soil cover and management practices (C) and supporting conservation practices (P).

Use it or lose it   A characterization of water use by individuals or groups holding water rights contracts in western states. If the party with water rights uses less than its maximum allotment of water, subsequent rights to the unused portion of the full allotment can, under certain circumstances, be transferred to another party.

Variable costs   The portion of total cash production costs used for inputs needed to produce a specific yield of a specific crop. Variable costs typically include fertilizers, seed, pesticides, hired labor, fuel, repairs, and animal feed and drugs.

Vine dressing   The trimming of vines to maximize production.

Water depletion allowance   A provision of the tax law that allows for a tax deduction based on the depletion of certain aquifers used for agricultural irrigation.

Water-holding capacity   The ability of a soil and crop system to hold water in the root zone.

Index

R